Additional Praise for

DOWN TO THE SEA

"Bruce Henderson has produced a book that should sit proudly in the bookcase of any American. Get one for your favorite sailor."

—*Tin Can Sailors, National Association of Destroyer Veterans*

"Typhoon . . . It is nearly impossible for the landsman to conceive of the raw power and ferocity of these storms at sea. Henderson's graphic account, so skillfully written to appeal to seafarer and landlubber alike, is one of the best tutorials in my library. I relished every page."

—*Admiral James Holloway, former Chief of Naval Operations, and author of* Aircraft Carriers at War

"What stands out in *Down to the Sea* are those moments that transform the disaster into an epic of courage and sacrifice. This is a book for anyone who wonders what happens when a ferocious sea attacks a ship and her crew."

—*Naval History*

"War spawns both heroes and victims and *Down to the Sea* points to the fine line that sometimes ser͟ ͟ar II story that hasn't be͟

Journal Star

"My father, ͟ f more than 700 m͟ 944. In the war's af͟ d their sacrifice wa͟ k that will stand a͟ monu- mental stor͟

Kreidler

"Because of ͟ book I learned hov͟

Come

Smithsonian Books

COLLINS
An Imprint of HarperCollins*Publishers*

DOWN
TO THE
SEA

An Epic Story of
Naval Disaster and Heroism
in World War II

Bruce Henderson

HarperCollins books may be purchased for educational, business, or sales
promotional use. For information please write: Special Markets Department, HarperCollins Publishers, 10 East 53rd Street, New York, NY 10022.

First Smithsonian Books paperback edition published 2008

Designed by Jennifer Ann Daddio

The Library of Congress has catalogued the hardcover edition as follows:
  Henderson, Bruce B., 1946–
    Down to the sea : an epic story of naval disaster and heroism in
World War II / Bruce Henderson.—1st ed.
      p. cm.
    Includes bibliographical references and index.
    ISBN 978-0-06-117316-5
    1. World War, 1939–1945—Naval operations, American. 2. United
States. Navy—History—World War, 1939–1945. 3. United States—
History, Naval—20th century. 4. Destroyers (Warships)—United
States—History—20th century. 5. World War, 1939–1945—Pacific
Ocean. I. Title.

D.773.H45 2007
940.54'5973—dc22   2007031314

ISBN 978-0-06-117317-2 (pbk.)

08 09 10 11 12 ID/RRD 10 9 8 7 6 5 4 3 2 1

For the men of the sea who served and died,
for the loved ones they left behind,
and for the children they never knew.

They that go down to the sea in ships, that do business in great waters;
These see the works of the Lord, and his wonders in the deep.
For he commandeth, and raiseth the stormy wind,
which lifted up the waves thereof.
They mount up to the heaven,
they go down again to the depths . . .

—Psalm 107:23–26

Contents

CONTENTS

Prologue

ABOARD THE
DESTROYER *SPENCE*
IN THE
WESTERN PACIFIC

December 18, 1944

It was shortly after 11:00 A.M. when Water Tender 3rd Class Charles Wohlleb of West New York, New Jersey, left the after fire room and headed topside. He did so with two shipmates also not on watch that morning: Water Tender 3rd Class Cecil Miller and Boilermaker 1st Class Franklin Horkey. The three sailors had gone down to the fire room where there was always a pot of coffee brewing, but they found the crowded space too uncomfortable with the ship pitching and rolling in mountainous seas—the worst storm any of them had ever been through. Before they left, Horkey put on a sound-powered phone headset and told the men who remained at their duty stations that he would let them know what was happening topside. They climbed a ladder and went through a hatch to reach the main deck.

Wohlleb, a shy, soft-spoken twenty-year-old who after high school had worked in the Civilian Conservation Corps before receiving his draft notice and thereafter enlisting in the Navy in January 1943, would never forget the names and faces of the four men he last saw attending to their assigned duties that morning in the fire room. Standing atop the narrow steel platform in front of the control panel of the Babcock and Wilcox oil-fired boiler, Water Tender 2nd Class Frank Thompson operated the oil burners that fed the fire inside the firebox. Fireman 1st Class Norman Small, "a Nebraska farm kid who was not small," kept a close eye on the fuel-oil heater and pressure gauges. Fireman 1st Class Claude "Roy" Turner monitored the water levels and adjusted a valve whenever necessary to keep the right amount of water circulating through copper tubes above the boiler to ensure the generation of sufficient steam to drive the two General Electric geared turbines that powered the ship. Water Tender 1st Class Layton Slaughter, who was in charge, wore sound-powered phones in order to communicate with Horkey. Slaughter controlled the amount of air that went into the boiler and the color of smoke that came out the stack. In a war zone, releasing black smoke (not enough air in the boiler) was an unsafe proposition, as it could pinpoint the vessel's location to enemy warships and aircraft from miles away. As usual, no one on the "black gang"—so called from the days of coal-burning ships, when soot habitually stained the faces, hands, and clothing of men who worked in the fire room—was wearing a life preserver. The spaces where they worked were very hot, at times as much as 130 degrees, and they also wanted to be able to move around freely.

When Wohlleb and the other two sailors reached topside, they emerged under an alcove that blocked most of the howling wind and rushing seawater swamping the deck. Their protected location was fortuitous, as no one could have stayed in the open for long without being swept overboard—exactly why earlier that morning all hands not on duty had been ordered to remain in the berthing compartments two or more levels below the main deck. During heavy weather, men not on watch typically climbed into their racks—whether they could sleep or not—and braced themselves to keep from being thrown around.

There was no visible horizon; the driving rain and blowing spray obscured where the sea ended and the sky began. From this swirling, grayish spume a colossal wall of seawater, taller than the ship's 50-foot mast, emerged off the bow every twenty or thirty seconds. The destroyer was riding unusually high in the water due to being dangerously low on fuel, bobbing in the turbulent seas like a child's bath toy. Each time *Spence* ascended up another wall of water, she was inundated at the crest, where she teetered briefly before pitching forward. On the thunderous ride downhill, the ship rocked, rolled, and yawed precariously. Once at the bottom of the trough, *Spence* heeled steeply in the driving winds until the next onslaught.

Wohlleb and his companions watched in horror as a depth charge packed with 200 pounds of torpex—an explosive 50 percent more powerful than TNT by weight—broke loose from its rack nearby, skipped across the deck, and slammed into bulkheads before washing overboard. An acetylene tank broke loose and did the same precarious dance across the deck before taking flight in the wind.

Over the headsets, Horkey was receiving news from below. "Jesus!" he yelled, his voice sounding muted and far away. "After fire room— swamped!"

Wohlleb knew what that meant: seawater had gone down the stacks and probably also the fresh-air ventilators that went from the main deck to the fire room. Pumping would have to commence immediately to stop the rising water from shorting out the electric panels in the fire room and adjacent engine room, which could mean the loss of lights, power, steering—leaving *Spence*, a 2,100-ton *Fletcher*-class destroyer with a crew of 339 men, dead in the water.

Not more than a minute later, Horkey yelled: "Control boards—on fire!"

Right up until that moment, Wohlleb had given no thought to the ship sinking. He and the veteran crew had gone through too much together to worry about a storm. Over a fifteen-month period they had been in some of the toughest naval action yet seen in the Pacific, for which *Spence* had earned eight battle stars. They had come through their encounters with the Imperial Japanese Navy unscathed. Indeed, throughout the U.S. Third Fleet, *Spence* had a reputation for being not only a stalwart fighting ship but also a very lucky one.

"Lights out below! Losing power!"

Spence's run of good fortune was about to end.

One

PEARL HARBOR

December 7, 1941

The greatest generation's first day of war dawned bright over Oahu.

Although sunrise came officially to the Hawaiian Islands at 6:36 A.M. that morning, Pearl Harbor remained shaded to the east by the 2,000-foot volcanic twins, Tantalus and Olympus, for another half an hour. As the sun crested the low-slung mountaintops, its brilliance washed the sky with bold streaks of light and painted in emerald the endless sugarcane fields stretching up the lush slopes above the nearly landlocked home port of the U.S. Navy's Pacific Fleet.

The destroyer *Monaghan* (DD-354) was tied

up to a nest of three other destroyers: *Aylwin, Dale,* and *Farragut.* The four vessels, which made up Destroyer Division 2, were moored side by side in East Loch off the north end of Ford Island—less than one square mile of land situated in the middle of Pearl Harbor—home to a naval air station, warehouses, and oil storage tanks. Several dozen other ships, including three other destroyer divisions, were moored on that side of the island; however, most of the fleet's anchorages (including an impressive lineup of America's biggest warships on Battleship Row), dry docks, and repair facilities, along with a sprawling oil-tank farm, were located along the harbor's expansive southeastern shores.

Monaghan had been the ready-duty destroyer since 8:00 the previous morning, meaning that for twenty-four hours the ship was "in readiness to get under way on one hour's notice" should her presence be required outside the harbor. To ensure a quick getaway, *Monaghan* was moored in the outboard position of the nest and singled up (with only one mooring line rather than multiple tie-downs), with a fire under one boiler and the full crew aboard. In the event of hostilities, enemy submarines were believed to be the most serious threat to the flow of ships that came and went from the harbor, so there was always at least one destroyer patrolling outside the entrance. Another destroyer was on standby to assist with any emergency outside the harbor.

Monaghan belonged to the *Farragut* class (named for the first U.S. Navy admiral, David Glasgow Farragut, a Civil War hero credited with the legendary battle cry "Damn the torpedoes! Full speed ahead!"), which were the first modern destroyers built for the U.S. Navy since the end of World War I. A total of eight ships in this class were launched in 1934–35. Designed to carry a crew of 150 men (wartime complements exceeded 200), the vessels were dubbed by sailors as "gold platers" because they were so plush compared with their predecessors. Representing the peak of technology and naval design for their era, these 1,395-ton two-stackers with a flank speed of 37 knots*

---

* All ship speeds are given in knots, a nautical unit of speed; one knot equals approximately 1.16 miles per hour.

(43 miles per hour) were originally armed with five 5-inch deck guns (two forward, two aft, one amidships),* four .50-caliber mounted machine guns, eight torpedo tubes, and a pair of depth-charge tracks.

The last *Farragut*-class destroyer built, *Monaghan* was launched on January 9, 1935, in Boston and christened by Mary F. Monaghan, niece of its namesake. Like all destroyers, *Monaghan* was named for a hero; other ships were named for states (battleships), cities (cruisers), famous ships (aircraft carriers), and fish (submarines). Ensign John R. Monaghan had served aboard the cruiser *Philadelphia* during a native uprising in Samoa in 1899. Monaghan had joined a landing party assigned to restore order among the natives, and his small band was returning to the ship when they were ambushed, during which a lieutenant was badly wounded. Despite the lieutenant's order to leave him and save themselves, Monaghan and two sailors stood by their wounded officer, fighting until overpowered, killed, and beheaded by the natives.

Assigned to patrol duties outside Pearl Harbor that morning was the destroyer *Ward* (DD-139), "an old World War I vintage" vessel that could barely make 30 knots. The old ship had a new skipper, Lieutenant William W. Outerbridge, who had taken over this, his first sea command, two days earlier. Since the issuance of a war warning from Washington, D.C., in late November, the ships on offshore patrol were under orders to depth-charge any suspicious submarine contacts operating in the defensive sea area outside the harbor.

At 6:40 A.M., the crew of an auxiliary ship, *Antares* (AKS-3), towing a 500-ton barge toward the entrance to Pearl Harbor, spotted an object 1,500 yards off its starboard quarter. When the report reached *Ward*, the destroyer changed course to intercept the object, identifying it as a small

---

* Five inches refers to the width of the barrel; 5-inch guns were the largest weapons on World War II destroyers. U.S. cruisers were armed with 8-inch guns, older battleships with 14-inch guns, and newer ones with 16-inch guns. The largest ship guns on Japan's mighty *Yamato*-class battleships were 18-inch guns. The bigger the gun, the heavier the projectile it can fire, the longer the range, and the greater the destruction upon impact. A bigger ship, therefore, could disable or sink a smaller surface ship in battle before the latter came in range to fire its guns.

submarine attempting to enter the harbor behind the barge. Given his shoot-to-kill orders, Outerbridge did not hesitate to commence an attack. *Ward*'s forward deck gun fired a shell that struck the base of the sub's conning tower. The submarine submerged or sank, and as *Ward* passed close by, the destroyer's crew released a depth charge, rolling off a rack at the fantail a 600-pound cylindrically shaped "ashcan" packed with TNT and a fuse set to go off at a predetermined depth.

Outerbridge at that point radioed a report to Pearl Harbor communications: "We have attacked, fired upon, and dropped depth charges upon submarine operating in defensive area." The message from *Ward* filtered up the peacetime chain of command that Sabbath morning with glacial speed before orders went out to the ready-duty destroyer to assist *Ward*, which would be credited with sinking a Japanese midget submarine and firing the first shots of the war.* At 7:51 A.M. *Monaghan* received a dispatch from the Fourteenth Naval District Headquarters: "Proceed immediately and contact *Ward* in defensive sea area."

At 7:53 A.M., the first wave of 181 Japanese planes—launched in the predawn darkness from six aircraft carriers operating undetected 275 miles northwest of Pearl Harbor—began their coordinated attack on the ships in the harbor and surrounding military bases and airfields. To further confuse the situation and keep their carriers from being located, many of the attacking planes flew around Oahu and approached Pearl Harbor from the south.

With the sound of church bells ringing in nearby Honolulu for eight o'clock mass and sailors in dress whites coming on deck preparing to hoist the colors on many of the seventy combat ships and twenty-four auxiliaries in the harbor, little attention was paid to the circling and diving aircraft. Army and Navy pilots were often up at dawn buzzing

---

* On December 7, 1944, three years to the day after firing the opening salvo of World War II, *Ward* was taking part in the invasion of Leyte when attacked by several Japanese planes, one of which made a suicide dive into the old destroyer. When the resulting fires could not be controlled, *Ward*'s crew abandoned ship. To prevent *Ward* from falling into enemy hands, she was ordered sunk. Carrying out the task was the destroyer *O'Brien* (DD-725), whose skipper, William W. Outerbridge, watched from the bridge as *O'Brien*'s guns sank his first sea command.

the beaches and engaging in playful dogfights, and heavy bombers were regularly being flown over from the mainland. Then, across the island in paradise, ripped a cacophony of explosions.

Aboard *Monaghan* at 7:55 A.M., an excited crewman reported the air raid to the commanding officer, Lieutenant Commander William P. Burford, who before being notified about *Ward*'s message had been preparing to go ashore when *Monaghan* was relieved of duty at 8:00. In fact, the gig to take him ashore was already alongside. Stepping onto an outside catwalk, Burford saw a spiral of black smoke rising from the vicinity of the Army's Schofield Barracks. At the roar of a nearby plane, he turned to see a black plane with a bright red circle on its fuselage passing alongside *Monaghan* 50 feet above the water. The goggled pilot could be seen through the open canopy, and he lifted one hand in a wave as the plane swooped by. Burford realized the torpedo plane was taking aim on the battleship *Utah* (BB-31), moored several hundred yards to the south. Then, as if in slow motion, the plane dropped its torpedo.

"Sound general quarters!" the square-jawed Burford hollered.

A few seconds later, a loud explosion erupted from *Utah*.

Upon reaching the bridge, Burford's first order was over the junction box phone that served as the main shipboard communications system. Normally, a duty "phone talker" would relay the captain's orders to other sections of the ship, but this morning Burford took the phone himself. To the engineering officer, Ensign G. V. Rogers, already working in the engine room to get the ship under way to join *Ward*, the skipper ordered: "Get up steam on all boilers for emergency sortie!"

Picking up a radiophone receiver for the talk-between-ships (TBS) radio, Burford checked the status of the other three destroyers in the nest. Finding that most of the key officers and chief petty officers—the most experienced enlisted men and traditionally acknowledged as the backbone of any seagoing navy—were on weekend liberty, leaving junior officers in charge (a common situation aboard many vessels that morning), Burford directed the other ships to commence firing at the attackers as soon as they were capable of doing so.

At 8:27 A.M., *Monaghan* got under way—the first ship in the harbor to do so that morning—backing clear of the nest of destroyers, then turning to come about and head out the narrow north channel on a southwest course between Pearl City and Ford Island. The destroyer's guns were firing at last, the booming of her 5-inch deck guns interspersed with the throaty rat-tat-tat of mounted .50-caliber machine guns. Although the ship was in readiness to respond to an emergency outside the harbor, the live ammunition was stored in locked magazines several decks below. When no one could quickly find the keys, someone with bolt cutters had hurried to the magazines and snapped off the locks. A line of sailors quickly formed to pass the ammo topside.

Clearing a thick column of smoke, *Monaghan* slipped past *Utah*'s anchorage. Everyone topside was stunned by the sight before them: the 21,000-ton dreadnought had capsized, with only the bottom of her hull showing out of the water like the shell of a monstrous turtle. No one who saw the sight could help wondering how many men were still trapped inside the wrecked vessel. Adjacent to *Utah,* the light cruiser *Raleigh* (CL-7) was down by the bow after being hit by a torpedo, and several other nearby ships were listing badly. The tropical breeze, normally fresh and pristine, stank with smoke and oil—and death. Burned and broken bodies floated by like logs drifting to a downriver mill.

As *Monaghan* steamed down the channel with Burford at the conn "wanting to get out of that damn harbor as fast as possible," a signalman observed the seaplane tender *Curtiss* (AV-4) flying a pennant indicating the presence of an enemy submarine. *Curtiss* had fire under her boilers that morning, too, and had gotten under way soon after the attack began. (By then, many moored ships—desperate to get up steam so as not to remain stationary targets—were pouring out heavy black smoke from newly lighted boilers.)

"Well, *Curtiss* must be crazy," scoffed Burford, a former submarine commander, who knew well that standard submarines were too large to operate in Pearl Harbor's shallow waters.

Just then, *Curtiss*' deck gun boomed.

"Okay, Captain," said the signalman, "then what is that thing dead ahead of us?" He pointed off the starboard bow.

Burford was amazed to see through the smoke a small submarine moving toward them on the surface several hundred yards away. In the sub's bow were two torpedo tubes, not side by side as usual, but one directly above the other like "an over and under shotgun barrel."

Monaghan headed for the submarine. As the destroyer closed, Burford saw through his binoculars the submarine turn sharply toward them and fire a torpedo. The torpedo porpoised twice, then settled on a straight course. It passed just wide of the destroyer's starboard bow, ran up on a nearby beach, and exploded. Not intending to give the submarine a second chance, Burford ordered up flank speed. "Prepare to ram! Stand by the depth charges."

Passing over the midget submarine caused a slight vibration to be felt on the destroyer. Seconds later, the submarine's stubby bow was observed close astern in *Monaghan*'s roiling wake, canted up crazily out of the water. Two 600-pound depth charges were released in rapid succession off the destroyer's fantail. Burford knew that putting down depth charges in such shallow water risked blowing off his ship's stern, but he felt he "had to depth charge close to my own ship under the circumstances if I were going to destroy that sub." When the ashcans went off nearly simultaneously with violent concussive effect at a depth of 30 feet, the explosions lifted *Monaghan*'s stern clear out of the water and knocked down nearly everyone on deck. A cascade of blackish mud was thrown high into the air.

The sub popped to the surface, floating on its side like a dead animal.*

---

* Owing to "unexplained and almost incredible laxness," the gate to the antitorpedo net at the entrance to Pearl Harbor, which had been opened at 4:58 A.M. for the entry of two minesweepers, was not closed until 8:40 A.M.—in spite of *Ward*'s report two hours earlier of attacking an intruding submarine near the harbor entrance. None of the five two-man Japanese midget submarines—launched by full-size submarines a few miles off Pearl Harbor—assigned to sneak into the harbor and sink ships succeeded in their mission. All were lost or captured, three of them without firing a torpedo. One small sub beached on Oahu, and a surviving crewman was taken prisoner.

The destroyer sped on, going too fast to make the turn into the main channel leading to the sea. Burford at that instant realized they were about to collide with a derrick barge moored at the west side of the channel. "Full left rudder! All engines back emergency full!" Although his orders were carried out promptly, it was too late to check the ship's headway. *Monaghan* struck the derrick a glancing blow on her starboard hull, sustaining damage that was later found to have caused minor leakage below and salt water contamination in one fuel tank.

The destroyer came to a gradual halt as her bow struck bottom on the sandy shoal at Beckoning Point. Attempting to back clear, *Monaghan* became entangled in the derrick's mooring lines. Changing directions, the destroyer pulled ahead slowly and cleared the lines but was still aground. On the bridge, Burford ordered the ship backed slowly again to try to regain deeper water.

"Submarine!" hollered a lookout.

Even while stuck rather ignobly in the mud, *Monaghan* answered the call—a deck gun firing accurately on what turned out to be a harbor buoy.

Back in the channel a few minutes later, the destroyer turned her prow toward the entrance and stood out of the besieged harbor at 9:08 A.M.

Visible through the billowing smoke as the men of *Monaghan*—some with tears filling their eyes—peered back toward Pearl Harbor was the plight of two once-mighty battleships at the south end of Battleship Row: *California* (BB-44), after taking torpedoes amidships, was afire and listing badly, and *Oklahoma* (BB-37), after five torpedo hits, had rolled over and lay capsized. The scene "could scarcely be grasped" even by eyewitnesses, most of whom would be left with a "smoldering lust for revenge in their hearts." From such forceful feelings would soon emerge a wounded nation's new rallying cry: "Remember Pearl Harbor!"

As *Monaghan* proceeded on her way seaward, the torpedo bombers were finishing their work while dive-bombers continued to swoop down from the smoke-filled sky with their lethal loads, attacking any ships still afloat.

The long morning and the sudden dying were not yet over.

THE DESTROYER *Hull* (DD-350), undergoing an overhaul, was moored inboard in a nest of four other destroyers northeastward from Ford Island alongside the tender *Dobbin* (AD-3). There were no fires under *Hull*'s boilers. In fact, the interiors of the fireboxes were being rebricked with a 4-inch layer of fire-clay mortar that had to be hammered into place. All power for lights and equipment came via cables from the tender. Shipfitter 1st Class Robert Hill was suspended over *Hull*'s side on a scaffold, welding steel plates over a row of portholes whose glass panes had been removed. With war looming, the portholes had been ordered sealed on all ships to help maintain watertight integrity, even though it resulted in less fresh-air ventilation to spaces below deck. As the torpedoes and bombs fell, Hill, realizing the work had to be completed before *Hull* could go to sea, kept welding, only faster.

The third *Farragut*-class destroyer to be launched, *Hull* had slid down the ways at the New York Navy Yard in January 1934. The vessel was named after Commodore Isaac Hull (1773–1834), one of the most famous ship captains in Navy history. Although he had previously fought against Barbary pirates and the French, Hull distinguished himself in particular during the War of 1812. While in command of the frigate *Constitution*, he won one of the classic sea battles of all time, for which his ship earned the nickname "Old Ironsides" after British cannonballs bounced off her hull without causing severe damage.

Sister ships *Hull* and *Monaghan*, among the newer vessels in the Navy, had been sent to the 1939 San Francisco World's Fair to serve as public exhibits. While mooring at the historic Embarcadero, *Monaghan* hit the pier and wiped out about 50 feet of it. As a result, the two destroyers were kept anchored a hundred feet off the main exhibits at Treasure Island, and for three weeks motorboats ran visitors back and forth.

After that, *Hull* was ordered to perform the same ceremonial duties at the New York World's Fair, and had steamed two-thirds of the way to the Panama Canal before receiving a change of orders to Hawaii, which since then had served as the destroyer's home port.

About an hour before the Japanese struck on that December date "which will live in infamy," *Hull* received on board from the Dairymen's Association of Honolulu 7 gallons of ice cream and 15 gallons of milk. Officer of the Deck (OOD) Ensign Maury M. Strauss confirmed receipt as to proper quantity, and Pharmacist's Mate 3rd Class T. E. Decker, with trusty spoon in hand, "inspected as to quality." No sooner were the dairy products refrigerated below than the routine of the peacetime Navy, as all hands knew it, ended for good.

Asleep in his bunk after a rambunctious Saturday night ashore, Seaman 1st Class John R. "Ray" Schultz, a devil-may-care, twenty-one-year-old Kentuckian with a shock of wavy brown hair and a winning smile, had been trying his darnedest for some time to get kicked out of the Navy.

Schultz had joined up at seventeen. His coal miner father was killed in a mining accident when Schultz was two years old. After his mother suffered a nervous breakdown, the youngster was sent to a large orphanage with his two older brothers and sister. Following years of verbal and physical abuse from a sadistic headmaster, Schultz, black and blue from his neck to his knees as a result of the latest beating, ran away shortly before his thirteenth birthday. He did farm work for an uncle, then caught a bus to Arizona, where his mother had relocated. He enrolled in school but dropped out after the ninth grade and went to work as a carpenter's apprentice at Horse Mesa and later Bartlett dams. After being laid off, Schultz hitchhiked to California's Central Valley, where he found seasonal work in the canneries until enlisting in the Navy. Following boot camp in San Diego,* he was assigned to *Hull* in 1938. Although Schultz soon demonstrated a competency in completing tasks

---

* Recruit training in the Navy, Marines, and Coast Guard is called boot camp. In the Army and present-day Air Force, it is known as basic training.

assigned to him—and as a result earned regular promotions—he eventually grew disillusioned with the peacetime Navy's rules and regulations. Not that Schultz habitually followed them, of course, which earned him regular demotions, too. In fact, he had "quite a few times" made Seaman 1st Class (E-3), two ranks below what he had been four months earlier—before he and a shipmate took from its mooring the Pan American Clipper's high-speed crash boat so they could get back to *Hull* after an unauthorized absence and night on the town. Brought before their unamused commanding officer the next morning, the other sailor pleaded not guilty and was ordered to stand court-martial. Recognizing an opportunity, Schultz, who had been at the wheel of the crash boat as they were chased by harbor police, pleaded guilty. In the morning, the other sailor was kicked out of the Navy, but not Schultz, much to his dismay. Only later, Schultz figured out why. Captain of a 5-inch gun crew, he was the "only man aboard trained in night illumination" for nighttime fighting, which involved firing a pattern of flarelike star shells behind enemy ships so they could be seen.* That Sunday, Schultz was awakened by a "goofy guy" running through the compartment yelling, "The Japs are sinking the battleships!" It seemed like a bad joke, and besides, it was too early to be making such a racket. Schultz's shoes were hanging on the railing alongside his bunk. He grabbed them and hit the loudmouth on the back of the head as he passed by. Just then, Schultz heard a loud explosion, and knew the guy "wasn't off his rocker."

Dressing quickly, Schultz went topside to his deck gun station. *Dobbin*, in an effort to get under way, had disconnected the cables that gave power to the destroyers in the nest. That meant *Hull*'s 5-inch deck

---

* The U.S. Navy soon discovered it had devoted insufficient practice to nighttime tactics. In contrast, the Japanese practiced night actions "on a scale unheard of in other navies," most of which avoided night warfare at all costs. As a result, the initial nighttime engagements of the war favored the Japanese, as in the opening months of the Guadalcanal campaign in 1942. By day, the U.S. Navy controlled the seas around Guadalcanal. By night, the waters were controlled by the Imperial Japanese Navy, which when it came to night fighting was "still a couple of semesters ahead of the U.S. Navy." It took the U.S. Navy time to disseminate the necessary training and equipment and to learn how to fight effectively at night.

guns—loaded for each shot with a powder charge in a brass casing and a separate 55-pound projectile—had to be aimed and fired manually. Schultz's crew began doing so as ammo was hauled up by hand from a below-deck magazine, firing into a wave of torpedo planes that had made for their side of Ford Island.

As other ships joined in, the sky soon became pockmarked with bursting shells. The noise was deafening. For Schultz, the maelstrom of sound was like "one solid blast all the time." With torpedoes and bombs exploding, guns firing, ships blowing up, and planes crashing, "something was always going up." After expending about eighty rounds, Schultz realized the futility in trying to hit at close range fast-moving targets with a deck gun designed for longer ranges. Worse yet, he knew that some shells must be raining down on the streets of Honolulu.*

Following the first wave, which had been mostly "take and no give" for the U.S. fleet, there was a short lull in the attack around 8:30 A.M. Throughout the harbor, stunned sailors and their ships had time to replenish ammunition, organize defenses, and be ready to give something back in the next round. In the defense of Pearl Harbor, however, the battleships would play an almost insignificant role through no fault of their own. Most of them had been put "out of action or rendered incapable of retaliation" during the first fifteen minutes of the attack. Thereafter, all efforts aboard the largest ships in the harbor that morning were "directed toward saving lives, fighting the raging fires on board, and keeping them afloat." The destroyers unleashed most of the return fire.

Schultz secured the 5-inch gun and ran back to the .50-caliber machine gun mounted starboard amidships. Other men assisted by loading ammunition drums. Schultz snapped in place a full drum and cocked the weapon. He was a boatswain's mate—a jack-of-all-trades when it came to general seamanship—not a gunner's mate, but given

---

* Approximately forty antiaircraft shells from U.S. guns fell on Honolulu during the attack, including one that landed on the porch of the governor's mansion. Rather than explode in the air among enemy planes, the shells detonated on impact, with tragic consequences in some cases.

his gun-crew duties Schultz had done his share of target practice with weapons of varying sizes.

When a second wave of 170 attacking planes—launched from the Japanese carriers an hour after the first wave—appeared overhead at 8:45 A.M., Schultz opened up in short bursts, as he had been trained. To his surprise, the drum had been loaded with all tracer ammo. Normally, there was a tracer—a shell packed with white phosphorus, which burns brightly, making its flight visible to the naked eye—every fifth round to help the gunner's aim. Schultz found shooting all-tracer rounds to be like "squirting a garden hose."

Off *Hull*'s stern a motor launch with three men aboard was heading for the *Farragut*-class destroyer *Wordan* (DD-352) in the same nest. The next instant, the small boat and its occupants were gone—vaporized in a fiery explosion from a bomb that landed in the boat's engine compartment. As the low-flying aircraft that had released the bomb flew past, Schultz aimed the spray of his .50-caliber garden hose. Multiple hits tore into the bottom of the fuselage. The plane careened out of control and crashed in the harbor. A minute later, Schultz's machine gun tore off part of a wing of another attacking plane. A *Hull* gunner on the bow saw a plane he was firing at ignite and crash in a sugarcane field. When another aircraft—carrying a full load of bombs—crossed *Hull*'s bow, it was hit by fire from multiple guns before going down.

After that, the Japanese pilots seemed to avoid the hornet's nest of destroyers off the end of Ford Island that unleashed such withering fire.

AT FIRST word that the attack was "no drill" but "the real McCoy," civilian worker Thomas A. Stealey Jr., of Stockton, California, who had been waiting in Honolulu for a ship to take him to Wake Island for a construction job, climbed in the back of a truck rounding up Navy personnel to see if he could help out.

At the entrance to Pearl Harbor, they passed an empty guard gate. After they had gone as far as they could in the truck, Stealey, twenty-two, a muscular former high school football and baseball player, ran for the docks. The scene before him was "just a mess," with hordes of men—some burned or bleeding—running in all directions, and clouds of black smoke filling the sky.

Born in San Francisco and raised in northern California's Delta region, Stealey had hired on as a sheet-metal worker the day after he graduated from high school. In 1941, when the chance came to go to Wake Island for nine months and build airplane hangars and buildings while getting paid for a full year—with all his expenses paid for and his earnings held in a Honolulu bank until he returned—Stealey talked it over with his parents and his fiancée, Ida May Bryant. All agreed it was a good opportunity for him to save some money. On November 11, 1941, Stealey boarded a ship in San Francisco with other workers. After arriving in Hawaii, they settled into a barracks and waited. On December 5, they were told to pack their belongings and were taken to a docked transport ship. After loading guns, ammo, and other matériel, there was not enough room for all the construction workers. Those men whose last names started with *A* to *M* were boarded, and the rest were told to return to the barracks and await the next ship to Wake, which, given events, never did come.*

Stealey came upon a Marine urgently setting up a .50-caliber machine gun on a tripod. The leatherneck yelled that he needed ammo from a nearby warehouse. As attacking planes swooped overhead strafing anything in their gun sights, Stealey, his adrenaline pumping, ran back and forth several times carrying large canisters of ammunition— metal containers so heavy that when he tried to lift them later he could barely get them off the ground.

When Stealey reached the docks, he found more horror and confusion. Across the channel, the battleship *Oklahoma* (BB-37) had already

---

* Tom Stealey never learned the fate of his fellow civilian workers who boarded the ship on December 5, 1941, for Wake Island, one of the loneliest atolls in the Pacific, some 2,300 miles west of Honolulu. The Japanese bombed Wake on the same day as Pearl Harbor and, in spite of a heroic stand by Wake's defenders, captured the island two weeks later. In addition to 470 military personnel, 1,146 civilian workers at Wake became prisoners of the Japanese. The Wake POWs were so brutally treated—98 civilian workers were killed in one mass execution in 1943—that after the war the Japanese Army garrison commander was convicted of war crimes and executed.

rolled over at her mooring. The battleship *Pennsylvania* (BB-38), the flag-ship of the U.S. Pacific Fleet, in the dry dock next to Stealey, had been hit by at least one bomb and was afire. Two destroyers, *Downes* and *Cassin,* occupying the space at the head of the dock, were in worse shape. *Downes* had been literally blown in two by an explosion in an ammunition magazine, and *Cassin,* which lay alongside *Downes,* had also caught fire. Stealey joined a firefighting party working to bring under control the fires on the two destroyers, ablaze from stem to stern.

Shortly after 9:30 A.M., a thunderous explosion rocked the destroyer *Shaw,* about a quarter mile away. With the fires on *Downes* and *Cassin* mostly contained, Stealey and several others took off running to help with *Shaw.* When they got there, they realized that the destroyer was situated 100 feet offshore in a floating dry dock.

"I can swim," Stealey announced. "Anyone else?"

Another civilian worker came forward. They found a long line and Stealey tied one end around his waist, then dove into the water, which in spots was thickly coated with heavy oil from damaged ships. They swam out to *Shaw,* whose bow was engulfed in flames. Finding the dry dock submerged and *Shaw* afloat, they swam to the ship's stern and climbed up netting that had been thrown over the fantail.

Shaw seemed deserted. They inched their way closer to the fire, then waved to the men ashore, who had tied the opposite end of the line to a 3-inch fire hose. The two workers began pulling on the line. When they had enough hose hauled aboard, they signaled to shore for the water to be turned on. It took some time, and in the meanwhile Stealey dropped down a ladder to below deck. He was not prepared for what he found: bod-ies "all over the place." He checked pulses and tried shaking others that seemed lifelike, but the men were all dead, apparently killed by concus-sion. Some had lost blood from their ears, nose, and mouth. There were twenty-five to thirty bodies in all—some lying in their bunks where they had obviously been asleep when the morning was still peaceful, others in a compartment or passageway where they had fallen.

What Stealey did not know was that *Shaw* had taken three hits in the first wave: two bombs through the forward machine gun platform and another through the port wing of the bridge. As fire spread, the dry dock was flooded to try to contain the blaze. The order to abandon ship was given at 9:25 A.M. Five minutes later, the forward ammunition magazine had gone up in a thunderous explosion.

Stealey also did not know until later that after they got water flowing to the hose and began working the fire, he and the other civilian worker were standing amidships atop a full ammunition magazine, which could have blown sky-high had they had not succeeded in extinguishing the flames.*

By 10:00 A.M. the aerial attack was over, with the last of the surviving Japanese planes rendezvousing north of Oahu for the return flight to their carriers.

Never in the course of modern warfare had a war begun with "so smashing a victory for one side," and never had the victor paid for it so dearly in the end. That day, 21 ships of the U.S. Pacific Fleet were sunk or damaged and 188 military aircraft were destroyed. Personnel casualties were 2,403 killed and 1,178 wounded. Japanese losses in the Pearl Harbor attack were 29 aircraft and 55 airmen, along with 9 crewmen (and 1 taken prisoner) from the five midget submarines. In this one treacherous attack, the U.S. Navy lost 2,008 men, three times as many men as it had lost in enemy actions in the two previous wars: Spanish-American and World War I.†

When the first torpedoes and bombs fell, the aircraft carrier *Enterprise* (CV-6) was steaming for Pearl Harbor after delivering a squadron of Marine fighter aircraft to reinforce Wake Island. The carrier was still 200 miles out because of a delay fueling her escorting destroyers in heavy

---

* *Shaw* survived to fight another day. After temporary repairs were made at Pearl Harbor, the destroyer sailed to San Francisco, where the work—including installing a new bow—was completed. *Shaw* returned to service with the Pacific Fleet in fall 1942. The destroyer saw extensive action during the war, earning eleven battle stars.

† Nearly twice as many Americans were killed at Pearl Harbor as died in the first twenty-four hours of D-Day following the Normandy invasion, during which 1,465 U.S. servicemen were killed.

seas. The time lag proved to be fortuitous: *Enterprise* had originally been scheduled to enter the harbor entrance at 7:30 A.M. on December 7, 1941.

Flying his three-star flag on *Enterprise* was Vice Admiral William F. Halsey Jr., commander of a task force consisting of the 20,000-ton carrier, three heavy cruisers, and nine destroyers. Pugnacious, charismatic, and unpredictable, often all at once, Halsey was a 1904 graduate of the U.S. Naval Academy, where he had excelled in athletics—lettering in football as a hard-nosed fullback—but not in scholarship. At graduation, he ranked forty-third in a class of sixty-two. (A midshipman's class ranking is the product of a complex calculation that takes into account not only academic marks but also aptitude, leadership abilities, deportment, and adjustment to the Navy. This class number follows an officer throughout his naval career and determines his seniority in the service among other Academy graduates, including his own classmates.)* Halsey spent twenty-two of the next twenty-three years of sea duty in destroyers, known as "tin cans," he would explain, because the 3/8-inch steel-plated hulls were "too thin to turn even a rifle bullet." For sheer length of service in one type of ship, Halsey's time in destroyers was a record unmatched by most naval officers, given the prevailing notion that diversified seagoing duty was indispensable for career advancement. Recognizing the emerging importance of aviation, Halsey, already a captain, earned his naval aviator wings in 1935 at age fifty-two, although he did not find the year-long course easy. Given both poor eyesight and a demonstrated lack of hand-eye coordination, he "never mastered any [aerial] stunt that took delicacy." After flight school, Halsey was given command of the aircraft carrier *Saratoga*. Since 1940, he had commanded all the Pacific Fleet's carriers and their air groups.

Halsey proudly claimed a long lineage of "seafarers and adventurers,

* Ranking thirteenth in the class of 1904 was Husband E. Kimmel, the commander of the U.S. Pacific Fleet at the time of the Pearl Harbor attack. Thereafter, Kimmel was immediately relieved of his command (replaced by Chester Nimitz) and placed on the retired list in March 1942. Halsey later said although he knew the disaster would be formally investigated, he never would have "guessed that the blame would fall on Kimmel" because his Annapolis classmate did not deserve "any part of it."

big, violent men, impatient of the law, and prone to strong drink and strong language." No doubt he would have fit in at any ancestral reunion. "There are exceptions, but as a general rule I never trust a fighting man who doesn't smoke or drink," he stated with all sincerity. Now fifty-nine and graying, Halsey still displayed a "fighting-cock stance and barrel chest," along with "beetle brows [that] embodied the popular conception of an old sea dog."

At 6:00 on the morning of the sneak attack, Halsey had ordered eighteen of *Enterprise*'s planes to fly ahead to land at Ford Island. After the planes were launched, Halsey went below to his quarters, shaved, bathed, and put on a clean uniform. He then joined his flag secretary, Lieutenant Douglas Moulton, at breakfast. They were on their second cup of coffee when the phone in the flag wardroom rang. Moulton answered it. Seconds later, stricken, he turned to Halsey.

"Admiral, the staff duty officer says he has a message that there's a Japanese air raid on Pearl!"

Halsey leaped to his feet. "My God, they're shooting at our planes!"

Although based on incomplete information, Halsey's concern was not outlandish. *Enterprise*'s planes were fired on by anxious U.S. gunners at Pearl Harbor who mistook them for another wave of enemy aircraft. Five planes were shot down and three pilots were killed, including Ensign Eric Allen, who bailed out of his crippled fighter over the harbor channel near Pearl City and was shot as he drifted down in his parachute.

December 8, 1941

At 4:30 P.M., *Enterprise* entered the Pearl Harbor channel.

From the flag bridge in the island that towered above the flight deck, Halsey and his staff officers watched solemnly as they passed "scene after scene of destruction"—smoking hulks that had once been proud U.S. warships. It did not take long for Halsey to "see enough to make me grit my teeth."

As they proceeded up channel along the northwest side of Ford Island, "the worst sight" for Halsey was the capsized *Utah*, sunk at the berth *Enterprise* would have occupied that Sunday morning had weather conditions not delayed his task force. But there was more—the oily black vapors that hung over the harbor like a shroud, and the pungent stench of burned flesh.

Along with the commensurate loss of life, the major warships of the Pacific Fleet and its home port lay in ruins, and a dangerous and unpredictable enemy was somewhere over the horizon with little in the way of naval forces between Tokyo and the West Coast of the United States.

The admiral's staff officers heard the shaken Halsey mutter: "Before we're through with 'em, the Japanese language will be spoken only in hell."

Two

Bath, Maine

October 27, 1942

Near the mouth of the Kennebec River, 15 miles northeast of Portland, the yard whistle sounded at 3:40 P.M. at the Bath Iron Works, Maine's largest private employer. More than a hundred workers stopped what they were doing and assembled in front of a bunting-draped platform at the bow of a sleek gray warship aground in one of the yard's construction berths, over which towered a massive crane that could lift as much as 220 tons.

The purpose of the ceremony was three-fold: the first wartime observance of Navy Day—intended to acquaint the public with the Navy through parades and open houses—since its inception in 1922 on the birthday of the late president and ardent Navy supporter Theodore Roosevelt; the presentation of the Army-Navy "E" (for excellence) pennant to the shipbuilding plant; and the launching of the yard's newest destroyer, built in five months and one week.*

On the stand where the official party had gathered, Bath mayor Walker C. Rogers took the podium. "Our Navy has suffered some serious losses, and also has many outstanding victories to its credit." The men of the fleet, he went on, "are grimly determined to repay our enemies for their vindictive acts against us. It is the recognized duty of the men and women of Bath Iron Works to continue their fine labors in producing destroyers of the highest type as expeditiously as possible."

Bath had been a shipbuilding center since colonial times; the first English-built ship in North America was launched from this shore. Bath Iron Works (established 1823) had been building fighting ships of steel dating back to 1891, when the gunboat *Machias* (PG-5) slid down the ways to mark completion of the first steel vessel built in Maine. At the turn of the century the shipyard started building the U.S. Navy's new torpedo boats, and soon it gained a reputation for producing sturdy seagoing vessels. The complimentary term "Bath boats" stuck even in a navy that traditionally called vessels *ships*, not *boats*. From those small torpedo boats—designed to sink larger vessels with self-propelled torpedoes—evolved the bigger, well-armed, and speedy destroyers of World War II, which would become the fighting greyhounds of the sea. Destroyers were universally considered the most versatile vessel in any fleet—no admiral believed he ever had enough of them—serving in roles as diverse as fleet and convoy escort, screening and patrols, gunfire support, radar pickets, submarine hunting, and independent strike forces, often doing battle with larger vessels. And any new destroyer built here—a prized Bath boat—was considered "the Stradivarius of destroyers."

An account in the Bath *Daily Times* of the day's ceremony at the

---

* Bath's wartime record from keel laying to launching of a destroyer was 124 days.

shipyard noted that "throughout . . . could be felt the spirit of gravity and determination which seems to be settling over the people of this nation, a feeling which has not been lightened by the latest news from the Solomons."

Indeed, as new ships were being built, others were being sunk. Off the Solomon Islands in the southwestern Pacific the previous day, the aircraft carrier *Hornet* (CV-8), which four months earlier had participated in the U.S. victory at the Battle of Midway, during which the four Japanese carriers that had attacked Pearl Harbor were sent to the bottom along with most of their planes and experienced naval aviators, was so badly damaged by dive-bombing and torpedo-plane attacks that the ship had to be abandoned and sunk.

When the speeches ended, Mrs. Eben Learned stepped forward with a raffia-covered bottle of champagne in her kid-gloved hands. Wrapped against the autumn chill in a black cashmere coat with a Persian lamb collar, she wore a smart hat with a velvet bow and ostrich feathers placed forward on her forehead, the style of the season. A handsome and confident figure, she smashed the bottle with authority against the sculpted bow. The new ship's sponsor was the great-granddaughter of the ship's namesake, Captain Robert T. Spence (1785–1826), who during the War of 1812 successfully defended Baltimore harbor against a powerful British fleet.

Spence was the twelfth *Fletcher*-class destroyer launched at Bath since Pearl Harbor.* A third again as large as their immediate predecessors (the 1,400-ton *Farragut* class), the 2,100-ton *Fletcher*-class ships (fueled and fully loaded, they weighed 3,000 tons) were "fast, roomy, and capable of absorbing enormous punishment." At 376 feet in length, *Spence* was longer than a footfall field but with a narrow beam of only 39 feet. Two

---

* In all, Bath Iron Works built 31 *Fletcher*-class destroyers, more than any other shipyard. A total of 175 ships of the *Fletcher* class—"the heart and soul of the small-ship Navy"—were commissioned by the Navy between 1942 and 1945, forming the core of the World War II destroyer force. Twenty-five *Fletcher*-class destroyers were lost or damaged beyond repair in the war; 44 earned ten or more battle stars; 19 were awarded Navy Unit Commendations, and 16 received Presidential Unit Citations.

General Electric steam turbines on two shafts provided 60,000 horsepower and a rated top speed of 35 knots (in excess of 40 miles per hour). The *Fletcher*-class ships, with enclosed gun mounts and partial armor over vital spaces, were essentially seagoing gun platforms set atop thin-hulled power plants. In addition to five 5-inch deck guns and four mounted .50-caliber machine guns, *Spence* had four twin 40 mm and six 20 mm weapons, which could lay down a lethal blanket of anti-aircraft fire, and carried ten torpedoes and racks of 300-pound and 600-pound depth charges.

Launches at Bath were tightly run affairs, and this one was no different. They were carefully timed during slack water at high tide, when there was no strong current to carry the new vessel—as yet without machinery, controls, or crew—into the nearby twin-spanned Route 1 bridge over the Kennebec.

Released from her restraints, the ship slid down the ways, slowly at first, then building up speed. Plunging into the water for the first time, the destroyer sent up an enormous splash, then rose and shook as the spectators ashore cheered and whistled. Her momentum carried *Spence* to midriver, where two tugboats waited adjacent to the bridge to collar the drifting vessel. Such was the frenetic schedule of wartime ship-building that a huge crane held the keel of the next ship to be built, ready to swing it into place on the now-empty ways.

"We will fight them as well as you build them," Chief Quartermaster R. E. Furry, representing "the enlisted men of the United States Navy," had promised the onlookers from the platform shortly before the launching.

A new warship had been born, and she was soon to join the fight.

The Battle of Komandorski Islands
March 26, 1943

On March 26, 1943, six U.S. Navy warships—the heavy cruiser *Salt Lake City,* the light cruiser *Richmond,* and four destroyers—were strung out in a scouting line 180 miles west of Attu, at the farthest end of the Aleutian Islands, and 100 miles south of the Komandorski Islands, a group of treeless Russian islands set in the Bering Sea east of Kamchatka Peninsula. In the vanguard of the 6-mile-long column was the destroyer *Coghlan*, with *Monaghan* in the rear. They were making 15 knots and zigzagging. The air temperature hovered around freezing, and the water temperature was 28 degrees.

Nine months earlier, the Japanese had invaded the Aleutians—a chain of more than 300 small volcanic islands extending for 1,200 miles from Alaska to Kamchatka—capturing Attu and Kiska Islands. While the Japanese had no plans to invade Alaska, and U.S. strategists had ruled out attacking Japan via this isolated northern route, this region of "perpetual mist and snow" would become a costly sideshow in the war in the Pacific, with neither side willing to relinquish it to the other.

The six American warships formed a task force under the command of Rear Admiral Charles H. McMorris, who flew his flag in *Richmond*. For several days they had been patrolling on a north-south line to prevent the Japanese from reinforcing and supplying their garrisons on Attu and Kiska.

As the crystalline Arctic morning broke, light winds blew from the southeast and a gently heaving sea sent an occasional wave swishing over *Monaghan*'s bow. Visibility was excellent; lookouts on the bridge and forecastle could observe fish broaching miles away. Just as the crew finished morning chow and was preparing to set the morning watch (8:00 to noon), the lead ships in the column made radar contact on five enemy vessels bearing due north, ranging between 8 and 12 miles.

Lookouts and radar operators determined the vessels to be two transports or cargo carriers screened by two destroyers and a light cruiser. On the flag bridge of *Richmond*, McMorris anticipated a "Roman holiday"—sinking soft-target auxiliaries while outgunning a few escorts.

In *Monaghan*'s fire room, Fireman 2nd Class Joseph J. Candelaria, nineteen, of Bakersfield, California, who had been aboard ship fourteen months, still found himself "just thrilled by the power surge" of a destroyer whenever the throttle man in the engine room opened it up when the bridge rang up flank speed. It wasn't just the increased speed but the sound that came with it. On a *Farragut*-class destroyer, the four fire rooms, which each housed a boiler, were pressurized to prevent blowback from the fireboxes; the forced-air blowers providing the air to the fire rooms were mounted in pairs on the main deck, port and starboard. When the speed increased, the boilers gulped air in huge quantities. To meet the demand, the blowers rapidly "revved up to a high-pitched whine that could be heard throughout the ship."

Of Italian and Mexican heritage, the handsome, dark-haired Candelaria, with a likeness to screen star Ramón Novarro, had dropped out of high school and enlisted ten days after Pearl Harbor. Only days later he was saying farewell to his mother and father, three brothers, a sister, an uncle, and assorted cousins on the platform of the Bakersfield depot as he boarded a train for San Diego. Following boot camp, Candelaria caught a Navy freighter bound for Hawaii to meet *Monaghan* at Pearl Harbor in January 1942.

As *Monaghan* sped up to join the task force, Candelaria manned the oil burners that fed the boiler. Supervising the fire room operation was Chief Water Tender Martin Busch, an old sea dog who mesmerized young sailors with his story of escaping from the battleship *Oklahoma*, which in the first twenty minutes at Pearl Harbor was hit by three torpedoes and rolled over until her masts touched bottom, taking more than 400 sailors to their deaths.

"Cut 'em in!" Busch yelled to Candelaria, who opened more burners to spray oil into the fire so added steam could be generated and superheated to 850 degrees Fahrenheit before being directed into the geared turbines.

Word passed among *Monaghan*'s crew that the task force was closing on some enemy freighters and an attack was imminent. From the sound of it, Candelaria thought the action would "all be over in a few minutes. Duck soup."

As the range between the two forces decreased, new reports reached McMorris. Any hope of pouncing on an overmatched enemy soon vanished as one Japanese heavy cruiser was sighted, then a second— followed by two additional light cruisers. When the enemy ships were all accounted for, the U.S. ships were outnumbered two to one. At that point, the Japanese warships turned to engage, while their two transports headed in another direction.

At 8:40 A.M., enemy cruisers opened fire from 12 miles away, straddling *Richmond* with salvos that caused 60-foot-high waterspouts.

Realizing there was now slim hope of catching the freighters, and fearing unacceptable losses if he engaged the superior enemy force, McMorris ordered his ships to retire at high speed. With the Japanese in pursuit of the fleeing U.S. ships, the two forces continued to slug it out through several course changes, which found the Americans heading away from their nearest Aleutians base at Adak (600 miles away), from which they could have expected air support by land-based aircraft, and toward a Japanese base at Paramushiro at the southern tip of Kamchatka (400 miles away). If that wasn't bad enough, the Japanese were making a couple of knots more speed than the U.S. ships and were gaining.

At 9:10 A.M., an enemy shell landed on *Salt Lake City* amidships, followed ten minutes later by another hit on her quarterdeck. It was clear the Japanese were aiming to take out the largest U.S. ship. The cruiser fought on valiantly with some fancy shooting, landing 8-inch shells squarely on the decks of two Japanese warships.

At 10:02 A.M., *Salt Lake City*'s steering gear failed, restricting course changes to only 10 degrees and limiting her ability to zigzag to thwart enemy gunners. During the next forty minutes, the cruiser took two more direct hits, which left her after engine room flooded. As engineers counterflooded to correct the ship's list, they accidentally let water into the fuel oil, which extinguished the burners and stopped the production of steam. A few minutes before noontime, *Salt Lake City* went dead in the water, leaving the cruiser with her crew of 612 men a "sitting duck" with "little chance" against the closing enemy. A smoke screen was ordered, and the destroyers released from their stacks thick columns of black smoke to obscure the big warship—now a stationary target—from enemy gunners.

To buy his fire room crews time to restart the burners, *Salt Lake City's* commanding officer, Captain Bertram J. Rodgers, asked McMorris to order the destroyers to unleash a torpedo attack on the advancing enemy, and the admiral complied. Any admiral would trade a destroyer for a cruiser, which came as no secret to fleet destroyermen, who knew their smallish ships were "expendable" in such situations. The destroyers *Monaghan, Bailey,* and *Coghlan*—after hours of steaming with their sterns to the enemy force—now reversed course. In a "magnificent and inspiring spectacle," they charged at flank speed toward two Japanese heavy cruisers 17,000 yards distant.

Monaghan had come close to making a similar sacrifice during the Battle of Midway in June 1942. After being struck by several enemy torpedoes, the aircraft carrier *Yorktown* lost power and went dead in the water. The crew abandoned the badly listing ship, but the carrier somehow stayed afloat through the night. The next morning a 170-man salvage party went aboard in a last-ditch effort to save the vessel. Brought alongside was the destroyer *Hammann,* which tied up to the crippled carrier's starboard side to furnish pumps and electrical power. Shortly after 3:00 P.M., a Japanese submarine lurking nearby released a salvo of torpedoes that churned toward *Yorktown's* starboard beam. Standing lookout watch on *Monaghan's* bow, Candelaria spotted the white wakes in the water. He looked up toward the bridge, where he saw standing on an outside platform Lieutenant Commander William Burford, who had earned the nickname "Wild Bill" as a result of his aggressive attack on the submarine at Pearl Harbor. Before Candelaria could sound the alarm, another lookout yelled, "Torpedoes!" Almost immediately, *Monaghan* sped up. To his disbelief, Candelaria realized Burford was trying to maneuver *Monaghan* into the path of the torpedoes to keep the aircraft carrier from being hit. Had he succeeded, Candelaria knew, the bow would have been blown off and "I wouldn't be here," but the spread of torpedoes missed *Monaghan*. Two impacted *Yorktown*, sealing her fate, and a third struck *Hammann* amidships, breaking the destroyer's back in a fiery explosion and causing her to sink in less than a minute, killing many of her crew.

Now, eight months after that near miss and with a new skipper in command, Lieutenant Commander Peter H. Horn, all hands aboard *Monaghan* knew the torpedo run on the enemy cruisers was a suicide mission. In all likelihood, they would be blown out of the water before reaching torpedo range. At best, they would provide a diversion, their deaths serving as a delaying action.

In the engine room, the chief engineer went down the line of men who worked under him, shaking their hands. "I should have sent you to diesel school," the chief told Machinist's Mate 1st Class Ernest Stahlberg. Had he done so, they both knew, Stahlberg would have been "in the States right now."

The chief and others aboard ship—from mess cooks to gun crews—were saying their goodbyes. They all figured they were about to die one way or the other, either after *Monaghan* received a hit from an enemy cruiser or when they abandoned ship in freezing seas, where they would last only minutes before hypothermia set in. "We were goners," Candelaria remembered.

As the destroyers bore in, the Japanese ships fired nonstop at them, "smothering them with splashes" from near misses. To observers on the other U.S. ships, it "seemed impossible" that the three destroyers would survive.

Topside on *Monaghan*, Seaman 1st Class Joseph Guio Jr., of Holliday's Cove, West Virginia—25 miles west of Pittsburgh—would not have disagreed with the prevailing pessimism. Having turned twenty-five a week earlier, Guio was older than many of his shipmates. A husky six-footer, Guio had worked in the steel industry since high school. Although defense industry deferments were available to steelworkers, Guio had walked into a Navy recruiting office three months after Pearl Harbor and signed up. After boot camp, he went to gunnery school in

Great Lakes Naval Training Center in Illinois, graduating with a final mark of 93.2 (out of 100). His battle station was a forward 5-inch deck gun. It was from there Guio and the other dozen men in his gun crew were slinging everything they had at the charging enemy ships—fifteen 55-pound rounds per minute—and "boy, what a sight" Guio had as the opposing forces took dead aim on each other like Dodge City gunslingers.

The Japanese flagship, the heavy cruiser *Nachi,* received several direct hits from the "outstandingly valiant" American destroyers, one of which, *Bailey,* received two hits that killed several men and cut all electrical power. The destroyers still had some distance to cover under the blistering enemy fire before moving into optimal range to release their torpedoes. Then there occurred "what seemed almost a miracle." In the face of the onrushing destroyers, the Japanese force turned to a new course, breaking off the action.* Soon after, *Salt Lake City* had her boilers back on line, and the U.S. ships headed for the safety of Dutch Harbor, midway up the Aleutians.

That morning's four-hour engagement would be the last surface battle of the war between naval forces without the use of air power or submarines.

More noteworthy to the crew of *Monaghan* was the beer party at Dutch Harbor hosted by *Salt Lake City* for the men of the three destroyers who had bravely put everything on the line to protect the damaged cruiser. It lasted all day and well into the night, providing a break in the war they would all remember. There were "no fights, nothing" like the brawls that the sailors often engaged in when they drank. "Just having a good time," Candelaria recalled, "because we'd been goners but didn't die."

---

* Thwarted in their supply mission, the Japanese returned to Paramushiro. Their commander—unaware of *Salt Lake City*'s predicament—broke off the attack because his ships were running low on ammunition and fuel, and also for fear of being attacked by U.S. land-based aircraft. The Americans fought a "brilliant retiring action against heavy odds," according to naval historian Samuel Eliot Morison, and Komandorski "should make a proud name in American naval history." The Japanese made no further attempts to supply their Aleutian bases by surface ships.

Three

On August 7, 1942, the First Marine Division landed at Guadalcanal, a little-known island in the Solomons that would be in the news for the next six months as the scene of a fierce fight to keep the Japanese from building an air base that could threaten shipping between the United States and Australia.

The early morning landings went surprisingly smoothly given that they were "the first amphibious operation undertaken by the United States since 1898." The only opposition came from labor troops and engineers engaged in airfield construction. Although initially "surprised and overwhelmed" by the invasion, the Japanese would not take long to send in reinforcements and counterattack, beginning the long campaign for Guadalcanal that would eventually be considered one of the turning points of the war in the Pacific.*

On D-Day of the first U.S. offensive against the Japanese, *Hull* was part of a protective screen of warships covering fifteen transports as they disgorged troops into landing craft. An hour into the operation, a large formation of enemy planes appeared at 10,000 feet and began bombing the invasion force. As U.S. carrier-based fighters ferociously engaged the Japanese aircraft overhead, *Hull* and other warships joined the fray with guns blazing. An hour and a half later, another swarm of enemy bombers attacked, hitting the destroyer *Mugford* (DD-389), killing eighteen men and wounding seventeen. Numerous enemy planes were shot down by fierce antiaircraft fire. At 7:15 P.M., *Hull*'s crew finally was allowed to secure from battle stations, which they had been at for a numbing eighteen hours.

Offshore, the night passed quietly, but everyone expected an all-out enemy air strike the next day. Shortly after sunrise, *Hull*'s sailors heard the amplified *oooga-oooga-oooga* of the general alarm reverberate throughout the ship, calling them to battle, followed by "All hands man your battle stations!" Men ran from wherever they were—in their bunks, taking showers, eating breakfast—some dressing along the way. What could have been chaotic with 200 men scrambling to different parts of the ship was made manageable by a rule to ensure they did not all run into each other: those going forward and up took the starboard side, and those heading aft and down went on the port side.

At 10:40 A.M., a radio dispatch warned that many aircraft were en route from Rabaul in New Guinea, the major Japanese stronghold in the region. As *Hull*'s helmeted gun crews waited for targets to appear overhead, the destroyer crisscrossed the harbor at varying speeds. Shortly before noon, a swarm of enemy aircraft swooped down in a

---

* For six months Guadalcanal was bitterly contested by naval, air, and ground forces of the United States and Japan. There were numerous pitched battles in the jungle interior of the 2,510-square-mile island, as well as six major naval engagements fought in the surrounding waters, which became the ocean graveyard for so many ships that American sailors dubbed the region Ironbottom Sound.

coordinated assault on the invasion force, utilizing bombs, torpedoes, and cannon fire. One low-flying plane lined up on *Hull* and released its torpedo, which became hung up at one end. The torpedo dangled precariously as the plane zoomed past. *Hull*'s gunners pumped a salvo of shells into the plane as it went by, and the torpedo fell away. Both plane and torpedo careened into the transport *George F. Elliott* (AP-13), exploding in a fireball.

In quick order *Hull*'s well-trained gunners splashed two more aircraft. A strafing run by an enemy plane landed hits on the destroyer's bridge and upper decks, and falling shrapnel made "various holes topside." Six men were wounded during the air strike, two seriously; one sailor lost an arm at the elbow in a "traumatic amputation," and another lost part of one hand.

A piece of shrapnel found deck gun captain Ray Schultz, who since Pearl Harbor had ceased his one-man campaign to get kicked out of the Navy, accepting that he was "in for the duration." He squeezed a shard of red-hot metal from the lump it had raised in his arm, and kept his crew firing away. (Later, he would drill a hole in the shrapnel and put it on a key chain as a souvenir.)

When the attack ended, *Hull* went alongside *George F. Elliott* to assist. Schultz was among the damage-control party that boarded the transport and helped put out a fire in the engine room. Topside, Schultz told *Elliott*'s commanding officer that there appeared to be a fire in the number two hole.

"What makes you think so?" asked the officer.

"There's smoke coming out of your ventilators," Schultz said.

The CO decided that smoke from the old fire must still be circulating through the ship's ventilation system.

Hull's men departed. Thirty minutes after the destroyer pulled away, a distress call came from *Elliott* asking for assistance in extinguishing a fire in the number two hole. *Hull* quickly returned and provided extra hoses, water, and men to the effort, but it was too late. The transport had to be abandoned, and her crew was received aboard *Hull*.

As midnight approached, *Hull* was ordered to sink the gutted transport lying close in to shore. Over the span of an hour the destroyer fired four torpedoes, but all missed their target. On *Hull*'s bridge there had been an ongoing spat between the torpedo officer and his chief about how to operate the new torpedo directional system. The officer kept winning the argument, but the torpedoes kept missing. Shortly after one torpedo passed under the transport, a truck parked on the beach behind *Elliott* blew sky-high. Word spread among *Hull*'s crew that their torpedomen had "sunk a truck." (So unsatisfactory was their performance against the stationary target, the torpedo officer and chief were soon transferred off *Hull*.) At sunrise, deck gunners had their turn and punched several holes in *Elliott* at the waterline. The gunners had another opportunity to prove their prowess that afternoon when they plummeted a schooner believed to be a liaison vessel for Japanese ground forces, although much to their surprise they were soon to discover that the small craft was already sitting on the shallow bottom of a lagoon, having been sunk earlier.

Thereafter forming up with a group of empty transports, *Hull* departed Guadalcanal waters early evening, although her crew had not seen the last of the Solomons. In September *Hull* returned thrice to the waters off Guadalcanal, escorting supply ships and conducting patrols to interdict the flow of reinforcements and supplies to the enemy garrison. Once, after other U.S. ships departed, *Hull* was ordered to stay behind and provide daytime bombardment for the Marines ashore, and several times "ran Japs off the hills" overlooking contested Henderson Field. For a week, *Hull* sat alone in the harbor—keeping a "short chain" on the anchor for a quick getaway if necessary. When a major Japanese naval force—consisting of two battleships, several cruisers, and a dozen destroyers—approached offshore to shell Henderson Field, *Hull*, "no match" for the enemy fleet, found deep water only 50 feet from shore and allowed "the trees to hide her mast." After a week, with "nothing to eat but rice and beans," *Hull* was relieved of her "artillery duty."

During a layover in Pearl Harbor, *Hull* picked up a handful of new crew members, including Seaman 1st Class Michael Franchak, twenty-three, of Jermyn, Pennsylvania, a small ethnic community where daily life centered around the local coal mines and the Russian Orthodox Church. Just shy of six feet and solidly built, Franchak was one of nine children born to immigrants from Galicia, Spain. He had quit high school after his second year and worked locally to help support his family. He eventually moved to New York City, where several of his siblings and friends had relocated, and found work behind a soda fountain and as a dishwasher. When war broke out, he followed two brothers into the Navy instead of waiting to be drafted into the Army.

Before being assigned to *Hull*—where his shipmates soon shortened his name to "Frenchy"—Franchak attended fire-control school upon graduation from boot camp, both at Great Lakes Naval Training Station. Fire-control men operated the primitive computer that targeted a ship's guns, and it was from this group of technically trained personnel that the Navy helped fill the ranks of a new technical rating to operate a new piece of equipment being installed on more ships in the fleet: radar.* Franchak eventually became a radarman after the high school dropout worked hard to pass the training course alongside classmates with more education, including some with college degrees.

The day after Franchak reported aboard, *Hull* went to sea. Several days later—after day and night maneuvers and gunnery practice—they put in at picturesque Lautoka in the Fiji chain. Most of the excitement surrounding Franchak's first liberty in the tropics involved watching gaming cocks fighting to the death in the street while locals cheered. That, and natives scaling trees and tossing down fruit to the sailors. Much to the disappointment of Franchak and his shipmates, the only available beer was "British and warm."

Hull joined the battleship *Colorado* (BB-45) in the vicinity of Fiji and the New Hebrides, helping protect the sea routes to Australia and New

---

* Radar, an acronym for "radio detection and ranging," was developed independently in the United States, England, France, and Germany during the 1930s. The foundation for this discovery was half a century of radio development, plus early suggestions that because radio waves are known to be reflected, they could be used to detect obstacles in fog or darkness. World War II led to fast-track research to find better resolution and more portability.

Zealand used by Allied shipping. In January 1943, *Hull* escorted a convoy to San Francisco and then entered Mare Island Navy Yard for repairs and alterations, including removing the plate-glass bridge windows and replacing them with portholes (reduced visibility in exchange for increased protection) as well as the installation of radar and sonar gear.

Although *Hull*'s crew had anticipated a return to the South Pacific, it wasn't long before they got the impression they were headed elsewhere. One piece at a time, they were issued foul-weather gear: heavy peacoats, wool-lined gloves, fur-lined boots, and watch caps. Destroyer duty in the balmy Pacific was considered part of the "dungaree Navy"—smaller ships that did away with much of the ceremony of larger ships, such as wearing white (summer) or blue (winter) uniforms, inspections, and constant saluting.* On and off duty, *Hull*'s men were accustomed to spending their days dressed casually in denim trousers and white T-shirts or blue chambray shirts.

Sonarman 2nd Class Pat Douhan, twenty, of Fresno, California, a recent addition to *Hull*'s crew, conjectured with his shipmates that the new issues meant they were "headed up north somewhere." To their amazement, they were even given permission to grow beards as further protection from the cold. Guys who joked about going to the North Pole were closer to the truth than they knew.

Most members of Douhan's October 1942 boot camp company in San Diego went directly to the fleet to help ease the Navy's shipboard manpower shortage. Tall and thin, Douhan, who showed the world the twinkling eyes of an Irishman with the gift of good humor, was one of only a few in his company to be sent to two advanced schools before receiving a fleet assignment. A review of his records showed that Douhan, who had attended Salinas Junior College for two years, was working on a seismographic crew for Shell Oil when he enlisted. "You're going to sound

---

* In the U.S. Navy, Marine Corps, and Air Force, hand salutes are traditionally given only when a cover (hat) is worn, while the U.S. Army gives salutes both covered and uncovered. In wartime on many smaller ships where officers and enlisted men were constantly passing each other in tight spaces, saluting was not required.

school," he was told. During fifteen weeks of training, he had learned how to operate and maintain shipboard sonar.*

The retaking of Attu and Kiska (less than 200 miles apart)—anticipated since the day in June 1942 when the Japanese invaded the sparsely populated Aleutians, where there had been no U.S. military presence— was finally at hand after giving way to higher-priority operations elsewhere in the sprawling Pacific. Attu would come first, it was decided, because it lay on the westward side of the Aleutians and would put U.S. forces in position to block any reinforcements from Japan headed for Kiska. As March "thawed into April" (1943), the attack fleet assembled in aptly named Cold Bay at the tip of the Alaskan Peninsula—three battleships, an escort carrier, five minesweepers, four transports carrying 3,000 troops of the Army's Seventh Infantry Division (which, incomprehensibly, had trained in the Nevada desert and on sandy stretches of California coastline under conditions "as unlike those at Attu as could well be"), and twelve destroyers, including *Hull* and *Monaghan*. There the task force waited out inclement subarctic weather; mountainous seas, and high surf that made amphibious landings impossible. By nightfall on May 10, the ships had groped their way into position a short distance off Attu "in a fog as opaque as cotton batting."

Hull was assigned to antisubmarine screening duty for the battleship *Nevada* (BB-36), which would be providing bombardment support with her ten 14-inch deck guns, as directed by Army Rangers ashore.

As the landings got under way—much like Guadalcanal, there was little opposition at the beachhead, with Japanese defenders dug in farther inland, where they waited to make a defensive stand—*Hull* received an urgent "man overboard" flashing-light (blinker) signal from *Nevada*. Lookouts on *Hull*, which had been following *Nevada*, soon spotted a man bobbing in the water.

The destroyer came around to make the rescue but in the choppy

---

* Sonar, an acronym for "sound navigation and ranging," uses sound propagation under water to detect submerged submarines as well as to navigate around sunken obstructions. Although developed by the British in 1912 and improved during World War I, sonar was not put into fleet-wide service by the U.S. Navy until World War II.

seas missed on the first pass. By the time *Hull* was in position for another try, the man in the water was unable to grab hold of the lifeline thrown to him. A boatswain's mate tied a line around his waist and jumped into the freezing sea. Both men were hauled aboard, but for *Nevada's* crewman it was too late. He was already "frozen stiff" and not breathing. Extensive artificial respiration elicited no response. The body of the sailor, "about twenty-one and redheaded," was stored in *Hull's* meat locker until given an at-sea burial the next day.

For a week after the invasion, *Hull* stood by *Nevada* as the battleship provided devastating fire support. At one point, when U.S. troops were stalled by stiff opposition from an entrenched enemy on higher ground, the Rangers ashore radioed quadrants to *Nevada,* whose big guns soon found the range and "wiped out an entire mountainside" filled with Japanese troops.

After intense fighting during which the Japanese attacked in one of the largest banzai charges of the Pacific campaign—ultimately it became necessary to land 12,000 U.S. troops—Attu was secured by the first of June. With the exception of twenty-eight prisoners, the entire Japanese garrison of some 2,600 troops was wiped out (hundreds committed suicide with hand grenades rather than surrender). American casualties were 600 dead and 1,200 wounded.

Following the fall of Attu, *Hull* joined a destroyer blockade around Kiska, a narrow, 22-mile-long island with an east-facing harbor midway down its length. *Hull* steamed several miles off the harbor, where most of the Japanese installations were located. Each night around midnight, the destroyer moved closer to shore and fired at least a hundred rounds of 5-inch shells, outbursts which were often met by return artillery fire from ashore. With all deck-gun batteries firing simultaneously, the noise was deafening and "the whole ship vibrated, with dust and debris flying everywhere." Then *Hull* turned and steamed out of range, resuming her patrol. This went on until they ran out of ammunition and stores and were relieved by another destroyer, at which point they headed for Dutch Harbor to take on more ammo and supplies before going back out. For weeks, as "fog hung over the island like a shroud of mystery," Kiska was shelled by the patrolling destroyers, which regularly faced blinding snowstorms and pea-soup fog that obscured rocks, barrier reefs, and treacherous shallows and made maneuvering hazardous even for ships equipped with radar and sonar.*

On patrol off Kiska shortly after midnight on June 22, *Monaghan's* radar picked up a contact at a range of 14,000 yards. Steaming to investigate, the destroyer closed to within 2,300 yards. With the night "thicker than coagulated ink," the target still could not be seen visually. The destroyer's guns, directed by radar control, opened fire. There was an immediate response from the unseen foe: machine-gun fire raked across *Monaghan's* decks.

Several slugs missed Seaman 1st Class Joseph Guio's head "by three feet" as he and his deck-gun crew fired away. "But I guess that's as good as a mile because I came out of it safe," Guio later wrote to his younger brother, Bill, back home in West Virginia. "I trembled like a leaf on a frosty morning. Will tell you all about it some day."

The blind duel continued for twenty minutes, the destroyer's salvos lighting the night with blinding flashes and sending booming echoes across the harbor, until a *Monaghan* lookout reported a "glowing mushroom" on the horizon. Radar reported the target blip was still on the screen but moving toward the enemy-held harbor. The destroyer broke off contact, ending a surface action in which her gunners never once glimpsed their target. Aerial photographs taken later revealed a large enemy transport submarine piled up on the jagged rocks outside the harbor "like a dead whale."† *Monaghan's* reputation as a sub killer— first earned at Pearl Harbor and enhanced weeks later when the

---

* The *Farragut*-class destroyer *Worden* (DD-352), *Hull* and *Monaghan's* sister ship, had been lost in January 1943 due to the same impediments. Leaving Dutch Harbor, *Worden* struck a submerged rock, tearing a huge gash in her hull. Floundering in heavy surf, the vessel was pounded against the rocks until her seams split open. When the flooding could not be stopped, the crew abandoned her; all but fourteen were rescued by nearby ships.

† The Japanese submarine engaged by *Monaghan* was the *I-7*, a 2,500-ton long-range fleet submarine with a crew of 100 that could transport another 100 combat troops and a Yokosuka E14Y seaplane for reconnaissance flights. Eighteen months earlier, the *I-7* had participated in the attack on Pearl Harbor.

destroyer attacked and damaged an enemy submarine—went up another notch.

Based on the stiff enemy resistance at Attu, there was reason to believe that the Kiska invasion would be a costly action. But when D-Day arrived in early August and 35,000 troops landed on Kiska, they found the island abandoned. Under the cover of impenetrable fog, thousands of enemy troops had been evacuated over a period of weeks by surface ships and transport submarines—a "remarkable exploit" that succeeded in spite of the watchful patrols by radar-equipped U.S. warships. For the lack of a fierce fight for Kiska, however, U.S. military strategists were unapologetic. The Japanese were out of the Aleutians for good without the loss of additional American lives.

Hull and *Monaghan, Farragut*-class sister ships launched two months apart whose hull designations were separated by only four numbers (350 and 354, respectively), had shared similar experiences since the opening day of war, when they had been moored not more than 200 yards apart at Pearl Harbor. For a few weeks more they continued escort and antisubmarine patrol duties along the cold, foggy Aleutian chain. Then, with their crews worn down from the "long, tedious grind" of conducting operations in "possibly the world's worst weather," *Hull* and *Monaghan* received new orders.

The two destroyers turned their prows toward warmer climes.

Four

Spence went to war in the South Pacific in fall 1943.

Three months after her launching that brisk autumn day in Bath, Maine, the new *Fletcher*-class destroyer was commissioned in January 1943.* A month-long West Indies shakedown cruise followed, with a complement of 350 officers and enlisted men, most of whom had never before been to sea and were "green as grass." It did not help that *Spence* got

---

* A ship's commissioning is a ceremony where the officers and crew formally take charge of a vessel on behalf of the Navy. It differs from a launching and christening ceremony in that the new warship is deemed fully operational and ready for service. By long tradition, members of a new ship's first crew are called plank owners. In the days of wooden ships, plank owners upon transfer or retirement were awarded a piece of wood from the ship.

caught in a winter storm off the coast of the Carolinas; so many in the crew became seasick that it was difficult finding enough men to stand four-hour watch sections. In the opinion of 2nd Class Yeoman Alexander "Al" Bunin of Roselle Park, New Jersey, a veteran of three years' sea duty when he became a *Spence* plank owner, "5 percent of us put that ship into commission." Other than the chiefs and a handful of other senior personnel, the enlisted men with few exceptions were right out of recruit training depots, while most of the officers came from civilian colleges with only ninety days of training at a reserve officer midshipman school before being sent to the fleet (hence the origin of the wartime term "ninety-day wonders"). For weeks, special signs had to be posted showing the way to various compartments and spaces below deck. So "screwed up" was the crew that the first time a 5-inch deck gun—with a range of nearly 10 miles—was fired, the projectile went only a few hundred yards. The novice crew was eventually "whipped into shape" by *Spence*'s first commanding officer, Lieutenant Commander Henry J. Armstrong Jr., an experienced destroyer skipper known as "tough but fair." Armstrong, forty, a native of Salt Lake City, Utah, told the crew in no uncertain terms that he was training them for war and "some of us won't be coming back." Adequately preparing his men was a task that Armstrong took seriously. "You have over three hundred mothers' sons aboard this beautiful ship," he wrote in his journal. "What are you going to do about that responsibility?" The transformation of those mothers' sons from seasick polliwogs to a capable crew took place only after months of hard work. *Spence*'s first assignment was escort duty in Caribbean waters, followed by several convoys across the Atlantic to Casablanca. Then, in July 1943, they were ordered to the Pacific via the Panama Canal. En route, the determined Armstrong commenced a "refresher course in battle readiness," ordering drills and gunnery practice at all hours.

Soon after arriving in the Pacific, *Spence* was assigned to Destroyer Squadron 23, commanded by Captain Arleigh A. Burke, who was destined to become the war's most famous combat commander of destroyers. Before long, it became apparent that the hard-charging Armstrong had turned *Spence* into one of the fastest-shooting and smoothest operating ships in the squadron. Burke, who possessed his own "hard-driving, meticulous style," heartily approved.

Born "a thousand miles from the sea on a hardscrabble farm at the foot of the Rocky Mountains" in Colorado, Arleigh "Ollie" Burke gave every indication of belonging to his paternal line of hearty Swedes (his immigrant grandfather had changed the family name from Bjorkegren). This forty-two-year-old American Viking was tall, husky, blue-eyed, and blond. At Annapolis, where he admittedly had to "work like hell" to earn good grades, he competed in boxing, fencing, and wrestling—not surprisingly, all martial arts—and finished in the top 20 percent of his graduating class (1923).

Believing that his squadron needed a "trademark that all the ships can be proud of," Burke decided that the youthful Navajo character Little Beaver, a courageous, pint-size sidekick to fighting cowboy Red Ryder of a popular comic series of the same name, would make a suitable moniker. His destroyers would be "the little beavers for [the] cruisers" and other large warships. Burke, who following his first destroyer command in 1939 had spent the succeeding years at the Bureau of Ordnance in Washington, D.C., where he built a reputation as one of the Navy's best at ordnance and gunnery, had insignias painted on the superstructure of each of his destroyers. Sporting a cartoonish Indian attired in moccasins and headband and toting a cocked bow and arrow—"how he loved to hunt Japs, day and night"—the eight ships of Destroyer Squadron 23 soon became popularly known as the Little Beavers.*

On November 1, 1943, the Third Marine Division landed on the enemy-held island of Bougainville in Papua New Guinea—the opening gambit of an operation designed to put pressure on the Japanese stronghold at Rabaul by building an airfield on Bougainville (200 miles away) from which air strikes could be launched against Rabaul. The landings were supported by four light cruisers and Burke's destroyers. The Japa-

---

* In addition to *Spence*, the other seven ships—all *Fletcher*-class destroyers—of Arleigh Burke's renowned Little Beavers squadron were *Stanly* (DD-478), *Converse* (DD-509), *Foote* (DD-511), *Thatcher* (DD-514), *Charles F. Ausburne* (DD-570), *Claxton* (DD-571), and *Dyson* (DD-572).

nese answered with air attacks from Rabaul, and a naval unit composed of two heavy cruisers, two light cruisers, and six destroyers raced toward Bougainville to break up the nascent landings. In the early hours of a moonless November 2, the ships of the two opposing forces sliced through the pitch-dark sea toward a showdown in Empress Augusta Bay, west of Bougainville. The lead U.S. destroyer made radar contact with the Japanese force at 2:29 A.M. from a range of 20 miles. While the four U.S. cruisers—*Montpelier, Cleveland, Columbia,* and *Denver*—took positions to block the enemy ships from attacking the beachhead, Burke declared on the short-range TBS radio to anyone listening: "I'm heading in!" Rear Admiral A. Stanton Merrill, in command of the twelve-ship force and flying his flag in *Montpelier,* was not surprised by Burke's pronouncement.

At the beginning of a surface battle, U.S. destroyers traditionally had been kept close to the main battle line of bigger ships (cruisers and battleships). Such a defensive strategy had come at a high cost in the Battle of Tassafaronga, fought a year earlier in the channel between Guadalcanal and Savo Island when an American force of five cruisers and four destroyers was badly mauled by a squadron of eight Japanese destroyers. The commander of the U.S. task force withheld permission for his destroyers to launch their torpedoes and instead kept them closely tied to his column of cruisers. As a consequence, the Japanese destroyers "administered to the U.S. Navy the most humiliating defeat in its history." One U.S. cruiser was sunk and three others were so badly damaged they were out of the war for a year, against the loss of a single Japanese destroyer. (Casualties totaled 395 Americans and 197 Japanese.) U.S. Navy officers from the highest echelons in Washington, D.C., to the smallest ships of the most backwater fleets had analyzed the tactics and tragic results of Tassafaronga, in which a "superior cruiser force was defeated by an inferior destroyer force," with the hope of preventing further lopsided naval losses.

Burke had studied Tassafaronga, too, and as a result had developed strong notions about how destroyers should be used more independently and aggressively in battle. He had discussed his ideas at length with Merrill, who agreed that when the opportunity presented itself the Little Beavers could go in first, firing torpedoes—a destroyer's most lethal weapon against other ships. Before departing for their first major battle, Burke had gathered his skippers and issued a standing order: once they were detached from the task force, their destroyers were "to attack upon enemy contact" without awaiting further orders. Burke wanted his destroyer commanders to know they had freedom of action at such times. He did not want them steaming around aimlessly awaiting orders. If they erred on the side of being overly aggressive, he would stand by them. Burke made it clear he would brook no excuses for inaction by anyone in his command.

As more images came into focus on the radar screens, Burke, flying his command flag aboard *Charles Ausburne,* realized they were about to face down three enemy columns. He picked the nearest one, consisting of three ships—he did not yet know their type—and ordered his destroyers to prepare to fire half their torpedoes. Minutes later, when the three enemy ships came within torpedo range at about 4,000 yards off the port bow, some two dozen 25-foot-long torpedoes—each packed with 500 pounds of explosives—jumped from their port launch tubes amid hisses of highly compressed air and dove into the sea with their foot-long propellers whirring. When an enemy ship fired star shells skyward, the U.S. destroyers and the telltale torpedo wakes were illuminated in bursts of white light. The Japanese column wheeled to a new course, causing the torpedoes to miss.

Opening up with 5-inch guns, *Spence* was "the first to put a Japanese cruiser on fire." The cruiser *Sendai* also came under devastatingly accurate fire from other U.S. ships, including the cruisers. Still in midturn, *Sendai* was "wracked by a murderous salvo . . . that virtually disemboweled her." As other Japanese ships maneuvered desperately to avoid the same fate, the destroyers *Samidare* and *Shiratsuyu* collided with each other. At almost the same instant, the heavy cruiser *Myoko* rammed her escorting destroyer, *Hatsukaze,* slicing off her bow.

Although ablaze and turning in circles with a jammed rudder, *Sendai* was still dangerous. The enemy cruiser released a spread of deadly 30-foot-long Model 93 "Long Lance" torpedoes—the most advanced torpedo in the world—each packing more than 1,000 pounds of explosives, double the charge of U.S. torpedoes, and with nearly quadruple the range (11 miles at 49 knots for the Model 93 compared with 3 miles at 46 knots for U.S. torpedoes). One Long Lance hit home, blowing the stern off *Foote*, killing nineteen, wounding seventeen, and leaving the destroyer with no power or steering and her main deck awash aft. With *Foote* out of action, her crew would have a long struggle on their hands to keep her afloat until she could be taken under tow later.*

At 3:10 A.M., Burke radioed his destroyers to execute a turn to starboard on his count. Upon Burke's command—"Execute turn!"— Armstrong ordered "right standard rudder," then checked to make sure that the rudder-angle indicator confirmed the rudder had gone over to the desired position. *Spence* began a smooth swing to starboard, which, if the other destroyers made the same maneuver simultaneously, would keep adequate spacing between them.

Lieutenant Jared W. Mills, standing at the open port hatch of the bridge, suddenly cried out: "Ship approaching sharply on port side, close aboard!"

"Full right rudder!" Armstrong yelled.

As added rudder was applied, *Spence*'s turn tightened. But it was too late—*Spence* and the destroyer *Thatcher* were on "roughly parallel courses headed in opposite directions" in frightful bow-on positions. The two ships struck bow to bow, and "sparks flew wildly into the night" as they raked each other's hull from stem to stern at a combined speed of 60 knots.

The impact sent men and loose gear flying, and for some terrible moments there was the screeching sound of steel against steel. *Spence* carried a "handsome silver St. Christopher's medal affixed to her foremast," and Armstrong surmised later that "the good saint must have been working overtime" to prevent the ships from slicing the bow off

---

* After major repairs in San Pedro, California, *Foote* returned to the war in time for the invasion of Leyte in late 1944, and saw action at Okinawa six months later. The destroyer received four battle stars for World War II service.

each other. Neither vessel was put out of action by the mishap, although
the list of damage to *Spence* would fill two typed pages. At-sea collisions
had long been considered career-ending occurrences for commanding
officers, but Burke would subsequently attach no blame to either skipper due to the incident taking place "in night actions while operating at
high speeds under enemy gunfire."

Slugging it out at close range with *Sendai*, *Spence* was hit by an 8-inch
round on the starboard side. The projectile bounced off and fell into the
water without exploding. However, an 18-by-6-inch hole was punched
into the ship at the waterline, causing a crew compartment below to
flood and salt water to leak into two fuel oil tanks. A mattress and blankets were stuffed into the hole as a makeshift measure, but the fuel
contamination was a larger problem. The fires in two of the four boilers
went out, reducing the destroyer's speed accordingly—potentially a
death sentence in a sea battle. Hustling boiler room personnel shifted
fuel oil suction for the two boilers to a standby tank. Then, with their
ship's fate and their own hanging in the balance, they rapidly restored
the fires to the boilers.

In the aft engine room, Machinist's Mate 2nd Class Robert Strand,
twenty-two, of Ridgway, Pennsylvania, was experiencing his first major
battle and learning how loud the sounds of combat could be even so far
below and while working next to one of the ship's two powerful General Electric steam turbines that generated *Spence*'s combined 60,000
horsepower. Strand, fine-boned but well-coordinated and athletic—
back home he played on a traveling all-star baseball team and was competitive in tennis, basketball, and especially bowling (and hoped one
day to own his hometown bowling alley)—knew his way around motors of varying sizes. After graduating from high school in 1938 and
working as a butcher's apprentice, he had been hired by one of the
town's biggest employers, the Elliott Company, an electric motor man-
ufacturer that had a contract to provide motor drives for submarines.
When the war began, Strand was given a draft deferment because of his
job in the defense industry. One day he found that someone had slapped
a fresh coat of yellow paint on his lunch box, no doubt implying cow-

ardice on his part for not going into the military. Strand immediately went to the Navy recruiting office and enlisted, and left a few weeks later, in August 1942. With the skills he had learned working with motors, he was sent to a Navy school following boot camp and graduated with high enough marks to earn his machinist's mate 2nd class rating. His assignment to *Spence* had followed in March 1943, and he boarded the new destroyer at Boston's Charlestown Naval Base shortly after *Spence* returned from her shakedown cruise.

Worse than the noise of gunfire in battle, Strand was learning, was no sound at all in the cavernous engine room—as when the huge turbine whined to a halt when the boiler fires went out. But now, with superheated steam again coursing through its feeder tubes, he and the other machinist's mates had gotten the turbine back on line—and *Spence* was back in the fight at full speed.

The "feisty *Spence* and her fighting crew" again went after *Sendai,* which was still turning in circles with guns blazing. *Spence* released four torpedoes and was rewarded with four waterline hits resulting in "columns of fire, water and debris fountaining skyward." The battered and blazing *Sendai* soon went down before their eyes. Spotting the two damaged Japanese destroyers that had collided earlier now attempting to flee, Armstrong set off after them but had to give up the chase, as *Spence* was running low on fuel.

The destroyer was heading back through columns of smoke to rejoin the squadron when Burke came on the radio: "We have a target smoking badly at 7,000 yards. We're going to open up."

Suddenly, tall waterspouts erupted all around *Spence*—the destroyer was being straddled by incoming shells. "Cease firing!" came the urgent call from *Spence* over the TBS. "Cease firing! Goddammit, that's us!"

"Are you hit?" an anxious Burke radioed.

Told by *Spence* they had been spared by near misses, Burke came back with a classic naval repartee: "Sorry, but you'll have to excuse the next four salvos. They're already on their way."

The destroyer zigzagged to avoid the shells raining down.

When the Battle of Empress Augusta Bay ended shortly before sunrise, the bloodied Japanese force was limping its way back to Rabaul in disarray without having fired a shot at the troop transports or the Marines ashore.*

At 8:00 A.M., two formations of Japanese bombers and torpedo planes appeared overhead. A total of sixty enemy planes from Rabaul carried out a coordinated attack, some bombing from high altitude while others dropped low to release torpedoes and strafe the decks of the American ships, which put up a thick blanket of antiaircraft fire and downed twenty planes while taking no serious hits. *Spence*—nearly out of ammunition after expending some 2,000 rounds of 5-inch shells in the battle—was credited with three shoot-downs.

Strand and his shipmates had been "scared plenty for a while" that morning, but they "overcame fear, and the encouraging results gave strength, courage and high spirits to the crew. A good deep breath was drawn by all."

A week later, a *Spence* lookout sighted a life raft with men aboard. As they approached, seven Japanese who appeared to be downed aviators could be seen lying "sprawled over . . . in grotesque positions." As *Spence* closed on the raft with the intention of snagging it with a grapnel and bringing the bodies aboard, the Japanese "suddenly came to life." One man who seemed to be the officer in charge stood up and yelled something. Making it clear they would not be taken prisoner, he pulled out a machine gun and pointed it at *Spence*. All hands on deck scurried for cover as a 5-inch deck gun swung into position and was trained on the raft. Before the disbelieving eyes of *Spence* crewmen, the officer turned toward a raft companion, who put the muzzle of the weapon in his mouth. The officer pulled the trigger, blowing off the back of the man's skull. Two others followed suit until one man balked. He was held down while the officer did the brutish deed. When the officer was the last one

---

* Rear Admiral Sentaro Omori, in charge of the Japanese forces in the Battle of Empress Augusta Bay, was relieved of his command upon his return to Rabaul, so displeased were his superiors with his failure to get at the American transports or otherwise disrupt the landings. U.S. Navy Rear Admiral A. Stanton Merrill's victory was credited to his releasing Burke's destroyers to fulfill a primary offensive function as well as the task force's "swift continuous turns to avoid enemy torpedoes" while "pouring out continuous rapid fire"—tactics characterized by naval historian Samuel Eliot Morison as "masterpieces of maneuver."

left, he made a brief "fanatical speech" in Japanese directed at the officers on *Spence*'s bridge, then shot himself, toppling into the bloodstained
sea, which by then was "swarming with sharks." Those who witnessed
the bizarre incident came away realizing they faced an enemy capable of
"the most weird things." A search of the raft revealed that the airmen
had left classified publications and maps. While *Spence* had no prisoners
to turn over for interrogation, a "very good intelligence haul [was] made
just the same."

On the afternoon of November 24, the Little Beavers got under way
from New Georgia Island for a high-speed sortie after receiving urgent
orders from the headquarters of Admiral Halsey, who in October 1942
had been placed in charge of the South Pacific command based at New
Caledonia.* Naval Intelligence, which had broken the Japanese code
and was reading intercepted radio dispatches, had learned of a planned
evacuation of 1,500 military personnel off Buka, an island north of Bougainville. Burke's mission was to get his destroyers quickly into position
off Buka to intercept the enemy naval force of unknown size and composition.

Only five ships in Burke's squadron were operational; *Foote* and
Thatcher were on their way stateside for major repairs, and *Stanley* was
alongside a tender undergoing needed work. That left *Charles Ausburne,
Dyson, Claxton, Converse,* and *Spence*—and Burke hadn't been sure about
Spence due to problems with one of her four boilers, which was unable to
generate steam. Burke's point-blank question to Armstrong before departing was whether *Spence*'s skipper wanted to come along on the mission or stay behind and have the boiler fixed. "Please, Arleigh, we want
to go!" Armstrong had pleaded, promising that by cross-connecting his
propulsion plant *Spence* would lose no more than 3 or 4 knots of speed.

Burke well knew that Armstrong was suggesting a violation of regulations, which specified that a ship in combat should have her propulsion

---

* Admiral Halsey was not present at his New Caledonia headquarters on November 24, 1943, as
he was in Brisbane, Australia, for one of his periodic conferences with General Douglas Mac-
Arthur, the supreme commander of Allied forces in the Southwest Pacific Area. Standing in for
Halsey at New Caledonia was his operations officer, Captain Harry R. Thurber.

system "split" so that if one half was knocked out, the other half could still power the ship. But with the squadron already shorthanded, Burke agreed.

En route to Bougainville, Burke was asked by Halsey's headquarters for his location and speed. In his answering radio message, Burke reported a speed of 31 knots, which seemed odd to Halsey's staff since *Fletcher*-class destroyers were capable of reaching 35 knots. The next message from Halsey's command—written by Halsey's operations officer, Captain Harry R. Thurber—was addressed to "31-Knot Burke," which was a "gentle reproach" to Burke from his old friend (and one-time squadron mate) for proceeding on an urgent mission at something less than top speed. For better or worse, "31-Knot Burke" was a sobriquet that would famously stick with Arleigh Burke all his life, and one that the press and public would long and erroneously assume was a tribute to his blazing speed rather than a "sardonic rib" for his slowness.

Arriving 55 miles off the west coast of Bougainville two hours before midnight, Burke, "with true instinct for the chase," decided there was a better chance of intercepting the Japanese ships if he patrolled farther to the northwest. Under overcast skies and frequent showers, Burke proceeded with his destroyers into hostile waters and airspace nearer Rabaul than "any Allied surface craft had penetrated since the Japanese had seized that port."

At 1:42 A.M., approximately 50 miles southeast of New Ireland's Cape St. George, *Spence* reported a surface radar contact ahead at 22,000 yards. It was the initial contact with what would turn out to be a column of five enemy destroyers en route from Buka to Rabaul on the evacuation mission. Betting that the enemy was not yet aware of his presence—most Japanese warships did not yet have radar, and although their crews were experts in nighttime gunnery, they faced "a foe with long-range vision"— Burke allowed the distance to close over the next fifteen minutes. He then ordered his ships into position for a torpedo attack and launched fifteen deadly "fish." The torpedoes had an estimated run time of nearly five minutes to target—"a wait that stretched suspense to the limit of endurance." Right on cue, "detonations boomed" and "orange flame spouted against the sky." The destroyers *Onami* and *Makinami,* screening for a column of three other destroyers astern, were caught completely by surprise and set afire. Burke sent *Spence* and *Converse* to finish off *Onami* and *Makinami* while he went "in hot pursuit" with the rest of the Little Beavers for the other column, now running desperately for the sanctuary of Rabaul.

In a stern chase, the pursuers normally can bring only their forward guns into play, as the rear guns cannot fire past a ship's superstructure at a target off the bow. Burke, however, fishtailed his ships several times—sacrificing forward speed for firepower—in order to open up from one side or the other with both forward and aft guns. In this manner, hits were scored on *Yugiri* that caused violent, fiery explosions. The blazing ship began circling out of control in a kind of slow death dance before sinking.

Meanwhile, *Spence* and *Converse* had carried out their mission. At 2:54 A.M., they radioed Burke: "One more rising sun has set." Gunners from *Spence* and *Converse* had committed to the deep both *Onami* and *Makinami.**

Shortly past 4:00 A.M., Burke reluctantly ended the chase, knowing he could not continue after the remaining two enemy ships unless the Japanese would be willing to refuel his ships at Rabaul—where, he deadpanned to another officer, he doubted the "fuel line connections" would fit the U.S. destroyers. There was also the real threat of a vengeful air strike launched from Rabaul—now not much more than a hundred miles away—after sunrise.

Battle-fatigued crews prepared for the inevitable onslaught from the sky. First light brought the ominous drone of aircraft overhead. However, they turned out to be a swarm of P-38s from the U.S. airfield at Munda, sent to fly cover for the returning destroyers. "Never had the white star on a wing meant so much to tired sailors," observed Burke,

---

* *Onami* and *Makinami* went down with all hands and a large complement of military personnel from Buka, other than a few who managed to reach shore on rafts. After the battle, Japanese submarines rescued from the water 289 survivors from *Yugiri,* which carried a crew of 197 as well as 300 soldiers being transported on her deck.

who within months, as word spread in the press of the exploits of the Little Beavers, would be labeled "King of the Cans" by *Time* magazine.

Three Japanese ships were destroyed and another damaged with "not even a hit" on a U.S. ship in the Battle of Cape St. George, which would be deemed by the Naval War College—where the tactics of the battle would be taught for years—"an almost perfect action that may be considered a classic."

Burke, however, knew how favored they had been by the "fortune of war." A fifteen-minute delay "would have prevented the battle from being fought." Even such a short interval would have allowed the enemy ships to be out of range. Because of their reduced speed due to *Spence*'s boiler problem, "we reached the enemy by the narrowest of margins."

Pulling into Purvis Bay in the Florida Islands north of Guadalcanal at 10:00 P.M. on Thursday, November 25, the Little Beavers found themselves "hoisted into the celebrity class." The ships in the harbor had stayed abreast of the battle with piped-in radio reports their crews eagerly followed with something akin to the drama of a World Series game. The illuminated decks and rails of those same ships were manned by waving and cheering sailors. Ship whistles tooted and horns blared in salute to the little band of destroyers.

They had missed Thanksgiving dinner, but no one complained. As Strand, aboard *Spence*—mindful of the censorship rules against being specific regarding locations and activities since one of his letters had been returned by the ship's censor—wrote (in part) to his parents the next day:

> *Yesterday was Thanksgiving, but we didn't celebrate . . . We were not in a position to eat a big chow and everyone was thankful just to be O.K. Today is the day!!! You can see by the menu I am sending just how much we had to eat.*
>
> *I'm feeling swell, and hope every one is the same at home.*
>
> > *As always,*
> > *Your Loving Son,*
> > *Bob*

Five

Built by Brown Shipbuilding of Houston, Texas, the destroyer escort *Tabberer* (DE-418)—a new breed of ship designed primarily to protect convoys—was named for one of the Navy's first heroes of the war in the Pacific.

Shortly before the U.S. Marines hit the beach at Guadalcanal on August 7, 1942, the aircraft carrier *Saratoga* (CV-3), steaming offshore, turned into the wind to launch a combat air patrol. Among the planes airborne that morning was an F4F Wildcat—the Navy's best carrier-based fighter at the time but inferior to Japan's top fighter, the Zero, in speed, maneuverability, and range—piloted by Lieutenant (j.g.) Charles Arthur Tabberer, twenty-six, of Kansas City, Kansas. As Tabberer and his VF-5 squadron mates climbed for altitude to be in position to dive on enemy aircraft, a flotilla of invasion ships took up positions below.* When the aerial attack came an hour after the landings began, the initial wave consisted of twenty-seven twin-engine bombers escorted by seventeen Zero fighters, both types of aircraft built by Mitsubishi. They were pounced on by eighteen Wildcats swooping through the clouds like shrieking birds of prey. Leading a two-plane section, Tabberer, although "viciously intercepted" by Zeros, "gallantly pressed home his attacks" against the bombers. Last seen "dogfighting a Zero," Tabberer was reported missing in action; neither his body nor his aircraft was found. He was awarded the Distinguish Flying Cross posthumously for his "courageous fighting spirit and resolute devotion to duty," which resulted in the destruction of at least five enemy bombers and "played a major role in disrupting the Japanese attack."†

During christening ceremonies on February 18, 1944, an overcast day on the Houston Ship Channel, 50 miles up from Galveston Bay, a bottle of champagne was broken against *Tabberer*'s bow by Mary M. Tabberer, the brunette widow of the war hero, with his mother, Mrs. S. W. Tabberer, serving as matron of honor and cradling a dozen red roses. Their husband and son had been gone for a year and a half, but for the two women the pain of their loss, not camouflaged by the ceremony, lingered on unsmiling faces.

Early in the war, the Allies were in short supply of armed escort vessels, and enemy submarines in both oceans exacted a heavy toll on merchant shipping. With full-size destroyers desperately needed to op-

---

* Among the Guadalcanal invasion flotilla that morning was the destroyer *Hull*, her crew at their battle stations ready to help protect the fifteen transports off-loading troops into landing craft.

† Participating in this first air battle between land-based Zeros and U.S. carrier fighters was Japan's ace of aces, Hiroyoshi Nishizawa, known as the "Devil of Rabaul." Credited with destroying six F4Fs in the fight—including four from VF-5—Nishizawa in all likelihood shot down Tabberer. Tall, lanky, and an expert in judo and sumo, Nishizawa, whose flying skills would become legendary, believed that he could never be shot down in aerial combat. En route with other pilots to pick up replacement aircraft at an airfield on Luzon, he died in October 1944 at age twenty-five as a passenger on a twin-engine Nakajima Ki-49 bomber downed by two Navy fighters. Nishizawa personally claimed eighty-seven aerial victories, although some sources credit him with more than one hundred aerial kills.

erate with the fast-attack fleets, not enough could be allocated for convoy duty. The solution to the escort shortage came from a new design: a ship that was easier and less costly to construct, and while smaller and slower than a destroyer, it was an equal in antisubmarine warfare capabilities. A destroyer escort needed to be able to maneuver only relative to a slow convoy—merchant marine or Navy supply vessels—which generally traveled at 10 to 12 knots, and defend itself against enemy aircraft while detecting and destroying submarines, which averaged less than 10 knots submerged and about 20 knots on the surface. The first destroyer escort was commissioned in April 1943; nearly 500 were built during the war, with 78 of them going to the British Royal Navy, which designated them captain-class frigates. There were six destroyer-escort classes in all, with the main differences being the power plants (diesel or steam) and armament. The Navy's smallest major combat vessel, destroyer escorts ranged in size from 1,140 to 1,450 tons displacement. The destroyer escorts—with a much tighter turning radius—were more maneuverable than conventional destroyers and carried all the latest antisubmarine warfare equipment. In heavy weather, however, it was always a rough-and-tumble ride aboard a destroyer escort, rolling one minute and plunging the next, with often either the bow or the fantail rising out of the water and the foredeck swamped. Destroyer escort sailors joked that they should be receiving both flight and submarine pay since they were in the air or underwater much of the time.

It had taken less than five weeks to build the 1,350-ton *Tabberer*, which at 306 feet in length was markedly shorter than a destroyer but similar in beam with a width of 38 feet.* *Tabberer*'s two steam turbines produced 12,000 horsepower for a flank speed of 24 knots, slightly faster than the submarines she was designed to hunt and kill but at least 10 knots slower than full-size destroyers. *Tabberer* was one of the new so-called 5-inch

---

* *Tabberer* was 70 feet shorter but only a foot narrower than the *Fletcher*-class *Spence*, and although 35 feet shorter than the *Farragut*-class destroyers *Hull* and *Monaghan*, *Tabberer* was 3 feet wider. The length and narrowness of conventional destroyers—built for speed, not stability—could make them top-heavy in heavy seas, while destroyer escorts, with their shorter and wider hull design, exhibited no such troubling tendencies.

destroyer escorts (*John C. Butler* class), so named because she was fitted with the bigger gun mounts (forward and aft) instead of the 3-inch guns of earlier classes. Additional armament included four 40 mm and ten 20 mm antiaircraft "flak guns," and three torpedo-launching tubes amidships. *Tabberer* carried a full complement of new Mark XI depth charges, equipped with electric-powered fins to give the "ashcan" rotation in the water, allowing it to drop quicker and straighter with less chance of drifting off target. Located near the bow were new British-developed Hedgehog launchers capable of throwing a pattern of 35-pound torpex-filled charges about 250 yards directly ahead of the ship in order to bracket a submerged target.

In May 1944, three months after her christening, *Tabberer* was commissioned with Lieutenant Commander Henry L. Plage, of Atlanta, Georgia, commanding. Plage, twenty-nine, had an athletic frame that stretched to six feet one inch. Oklahoma-born and Georgia-raised, he had the southern drawl to prove it. Easygoing, with a prominent jaw and high cheekbones, Plage had received his commission through the Naval Reserve Officer Training Corps (NROTC) at the Georgia School of Technology, where he was captain of the swimming team for two years and a member of Omicron Delta Kappa, a national honor society. Upon graduating in 1937 with a B.S. in industrial management (his highest NROTC marks were in seamanship and navigation), he stayed in the naval reserve on inactive status and went to work in sales and later accounting for a national retail credit company with headquarters in Atlanta. In January 1941, believing war was inevitable and desiring to get a head start on his naval career before a flood of new officers inundated the service, Plage volunteered for "immediate active duty," asking to be an "assistant athletic officer" at one of three naval air stations in Florida or Texas. Ordered to report in March 1941, he was instead anchored to a desk job at the Navy Yard in Charleston, South Carolina, for a year before finagling sea duty three months after Pearl Harbor. When it came time to list three types of ships he wished to serve on, his order of preference was "cruiser, battleship, destroyer." After two months of Sub Chaser Training School in Miami, he received orders to a substantially smaller vessel: the diesel-powered 280-ton, 175-foot-long patrol craft *PC-464*, which carried a crew of five officers and fifty enlisted men. There were, however, consolations for Plage: the submarine chaser, in navy lingo, was "new construction," and he would be the commanding officer, an unusual billet for a young officer's first sea duty. After escorting convoys along the East Coast and in the Caribbean, Plage was transferred in May 1943 to the destroyer escort *LeHardy* (DE-20), also new construction, as second in command. A short five months later, he was ordered to put into commission the destroyer escort *Donaldson* (DE-44) as commanding officer. During these early and varied assignments, Plage's learning curve was a steep one, and he came to rely on chief petty officers—the seniormost enlisted rank—almost all of whom had extensive sea duty, finding truth in the axiom that chiefs run the Navy.* He also, in the seclusion of his stateroom, "engulfed himself in books," learning more about navigation, seamanship, and gunnery in an effort to stay "one step ahead of the crew." *Donaldson* arrived at Pearl Harbor in February 1944 and soon took part in the invasion of the Marshall Islands. Detached from *Donaldson* in March after earning an "above average" evaluation on his fitness report, Plage spent a month at Fleet Sound School in Key West, Florida, taking an advanced course in sonar before traveling to Houston to meet *Tabberer*.

By now accustomed to commissioning new ships and training novice crews, Plage, who had shown not only a natural aptitude for command but also inspiring leadership qualities, was nonetheless surprised by what he found when he took command of *Tabberer*. While all the spaces and equipment were new and sparkling clean from the bridge down to the engine room, the crew was younger and less experienced than any he had seen. Most of the more than 200 enlisted men were "boots," just out of recruit training; only some fifty were rated petty officers (3rd, 2nd, or 1st class), and of those a scant twenty had served on

---

* This is as true today as yesteryear. On August 1, 2005, Chief of Naval Operations (CNO) Admiral Mike Mullen told a gathering of chief petty officers at the Senior Enlisted Academy in Newport, Rhode Island: "I believe chiefs run the Navy. You may think that I run the Navy, but I assure you that the Navy runs because of what you do."

ships. The manpower requirements for the large number of new ships being launched had stretched thin the Navy's ranks of experienced officers and enlisted men. Of *Tabberer*'s fifteen officers (all were naval reservists), only four had been to sea. Plage found the others to be "90 Day Wonders at sea for the first time," including one nineteen-year-old ensign who "couldn't get a drink in a civilian bar." As a result, Plage again relied on his chiefs—usually eight or nine were assigned to *Tabberer*—to "run the ship because they had the knowhow." He also counted on the chiefs to help turn into seagoing sailors the preponderance of young landlubbers in the crew—so many were teenagers, some of whom did not even have to shave on a daily basis yet.

Lieutenant Howard J. Korth, twenty-four, of Saginaw, Michigan, was one of *Tabberer*'s few experienced officers. A 1941 graduate of Notre Dame University, where he majored in engineering and played guard on the Irish's nationally ranked football team for head coach Elmer Layden (the all-American fullback in Notre Dame's famed backfield known as the "Four Horsemen"), Korth had been offered a football coaching position at a small Catholic college in Kansas City but instead signed up for a Navy commission a month after Pearl Harbor. Called to active duty in March 1942, he was sent to California—Treasure Island and San Diego—for training. In line for an engineering position on an aircraft carrier due to his academic background and expressed interest in aviation, Korth, upon being informed that gunnery officers were sorely needed by the fleet, volunteered to switch to gunnery. The handsome, square-jawed, solid six-footer, whose four brothers had nicknamed him "Hutch" after a favorite action-movie hero, Hurricane Hutch, soon found himself on stormy Atlantic crossings as the officer in charge of the gun crew aboard the troop transport *John Lykes*. Before being assigned to *Tabberer* as gunnery officer, Korth completed special gunnery work on the old battleship *Wyoming* (BB-32), converted to a training ship operating out of Chesapeake Bay, and packed with the latest fire-control radars and mounted guns from 5-inch to .50-caliber. (*Wyoming*, which would never fire a shot in anger, expended more ammunition than any Navy ship in the entire war, and in the process trained some 35,000 fleet gunners and gunnery officers.)

Plage wasted no time in fitting out *Tabberer* for sea. The day after he took command, the final loading of equipment, spare parts, and assorted stores was completed under his watchful eye. The last of these included 204 pounds of bread, 300 dozen eggs, 20 gallons of milk, 90 pounds of butter, 40 pounds of raisins, 100 pounds of tomatoes, 60 pounds of cantaloupes, 5 cases of Camel cigarettes, 5 cases of Chesterfields, 5 cases of Lucky Strikes, 3 cases of Philip Morris cigarettes, 25 boxes of Life Savers, and 10 boxes of spearmint gum. Two days later, *Tabberer* was under way in the Houston Ship Channel, headed for the San Jacinto Ordnance Depot dock, where the crew loaded ammunition. The following day, *Tabberer* pulled into Galveston and tied up to Pier 37, with new members of the deck crew shown how to secure the six wire-rope lines—three forward and three aft—each doubled to prevent drifting. The next day, *Tabberer* nudged away from the pier and headed for open waters.

Although Plage was on the bridge, a Coast Guard officer charged with piloting the ship out of the harbor was giving the orders. Twenty minutes after leaving the pier, *Tabberer*, with "engines going full astern and rudder amidships," struck the outboard wall of a floating dry dock. Before the harbor pilot was finished, he steered *Tabberer* into a motor whaleboat moored alongside another ship, capsizing the whaleboat. Returning to the pier, *Tabberer* remained dockside for the next three days as damage was assessed and collision reports written. When *Tabberer* got under way again, Plage was at the conn.* There would be no more collisions, and it would not take long for his crew to realize that in Plage they had a "very capable, very good captain" who also happened to be a "great ship handler."

The new destroyer escort cleared the harbor after midnight on June 11. The log entry signed by the officer of the deck (OOD)—an officer

---

* Contrary to what many people think, a ship's captain never actually operates the wheel or helm but leaves that job to an enlisted sailor, most often a quartermaster-rated petty officer. Whichever officer "has the conn"—usually either the captain, executive officer, officer of the deck (OOD), or junior officer of the deck (JOOD)—gives orders to the helmsman and others to direct the engines and rudder, thus controlling the ship's speed and direction.

standing watch on the bridge who is in charge of the navigation and
safety of the ship unless relieved by the captain—for the midnight-to-
4:00 watch read: "Cruising in the Gulf of Mexico for training purposes.
War cruising condition of readiness, split engineering plant, ship darkened topside." *Tabberer* and many of her crew were at sea for the first
time.

When they pulled into Bermuda three weeks later in the middle of
an exhausting shakedown cruise to test men and equipment alike, more
of *Tabberer*'s seamen were sick than able-bodied. The North Atlantic near
Bermuda—a British territory located 650 miles due east of Cape Hat-
teras, North Carolina—was famously rough, which was why Navy ships
often trained crews hereabouts. Accepted theory held that new sailors
had to get seasick before they could find their sea legs, and thereafter
they would not get seasick again. With so many men too ill to carry out
their duties—others shakily stood their watches with a pail close at
hand—the ship at times could "hardly stay up with the maneuvers" being conducted. For a month the ship was in and out of Bermuda. At sea,
the crew ran through constant drills such as torpedo runs and gunnery
practice, as well as learning how to tow another ship and be towed.
When in port, the men relaxed ashore, enjoying this exotic island of
pink sand, turquoise seas, and bronzed women; for most of them it was
their first trip to a foreign port.

Plage quickly set out to show his young sailors what was expected of
them at sea and on the beach, as well as the penalties for errant behavior, much of which he privately considered "youthful exuberance aboard
ship and ashore," while maintaining the image of a firm but fair discipli-
narian. In his first captain's masts, he meted out attention-getting punishments: the loss of three liberties for being absent without leave
(AWOL) less than two hours, five liberties lost for being nine hours
AWOL, two days' solitary confinement on bread and water for being
absent from duty station and insubordination, twenty hours' extra duty
for the men of a gun crew who were caught gambling while on duty,
and four days' confinement for direct disobedience of orders.

The period of extensive training and maneuvers began to mold the young bluejackets of *Tabberer* into a functioning crew. Sonarman 3rd Class Frank Burbage, eighteen, of Newark, New Jersey, was impressed by the "very high morale" aboard the new ship, which the crew soon fondly nicknamed *Tabby*. Burbage, who had left high school before graduating to enlist in the Navy rather than be drafted into the Army when he turned eighteen, had attended Fleet Sonar School after boot camp. Aboard *Tabberer*, the sonar station was on the bridge, so he stood watch in close proximity to Plage and the other officers, all of whom "treated the enlisted men very well" and seemed to genuinely care about them, an attitude that permeated the ship from the top down.

Gunner's Mate 3rd Class Tom Bellino, eighteen, raised on a dairy farm near Boise, Idaho, had been sent to two gunnery schools after boot camp, emerging with his petty officer's "crow," although it seemed for a time to be "attached to a zipper" because he kept getting written up for minor infractions and busted in rank. Still, Bellino soon joined his shipmates in their appraisal of the man who would take them to war: "everybody loved the skipper."

A boon to shipboard morale was the quality of food served on *Tabberer*, which according to one old salt was the "best chow in the Navy"— surely an exaggeration but a fine compliment nonetheless for the man who did most of the cooking, Ship's Cook 2nd Class Paul "Cookie" Phillips, nineteen, of Texarkana, Texas. A butcher in civilian life, Phillips had enlisted a week before Christmas 1942 and after boot camp went to the Navy's cooking and baking school in Alameda, California. Phillips then spent ten months on the destroyer escort *Stanton* (DE-247), which was assigned to convoy-escort duty in the Atlantic and Mediterranean. There he learned how to cook and serve meals in varied sea conditions.

A high-energy, wiry type, Phillips had been the flyweight amateur boxing state champion back home before his seventeenth birthday and was still close to his fighting weight of 120 pounds. He just "couldn't get fat on my own food," although it was his observation that a lot of Navy cooks did exactly that. Like every ship's cook, Phillips had been trained to follow the *Cook Book of the United States Navy* (1932 edition), which listed 170 pages of recipes—each designed to make 100 servings—for such

naval classics as Bean Soup No. 1, Chopped Beef en Casserole, and Fried
Chicken. The latter called for 80 pounds of chicken, but Phillips soon
learned that to feed *Tabberer's* enlisted complement, which averaged
about 210 men (the officers had a separate mess with their own cooks),
he needed 180 pounds of chicken and proportional increases to the
other ingredients: flour, eggs, cracker meal, pork drippings, and fat for
deep frying. Phillips soon found that the crew liked beef dinners
best—Minced Beef, Pot Roast, and Roast Beef were favorites—and for
breakfast, fresh eggs when available, coffee cake, and hot cinnamon
buns, the latter made by Chief Commissary Steward/Baker Alan Lumb
and acclaimed as the "best cinnamon buns in this man's Navy." As was
the case on every Navy ship, *Tabberer* was allocated $1.09 per man to feed
each enlisted man three hot meals a day. Phillips found it "not too
tough" to stay on his budget and was held accountable if he exceeded
the weekly allotment, as it meant "cutting back next week," which seldom happened on *Tabberer*.

Tabberer carried a full-fledged mascot: a rat terrier mix that one of the
sailors had brought aboard as a puppy and who promptly won the hearts
of all hands, including the captain, who overlooked regulations against
having a pet aboard ship. Named Tabby, the little white dog with a blackish snout and straight-up ears was becoming a sailor in his own right.
Already he was able to scamper on his short legs up and down the steep
ladders between decks, something "not many dogs could do." He knew,
too, the best times to stand outside the galley hatch and yip persistently
until Cookie Phillips arrived with leftovers, usually slices of bologna or
meat scraps. Tabby had also needed to find his sea legs, once losing his
balance and falling overboard as the ship pulled up to a pier. With no
thought to his own safety, a sailor dove into the murky water and
grabbed the frightened mutt. A line was dropped and they were hauled
aboard, so covered in oil that "they didn't look like themselves."

Tabberer pulled into Boston harbor on August 4. The crew went on
alternating duty sections, and those who did not have the watch could
go ashore at the end of each workday. One night, a handful of *Tabberer*
sailors, as Plage would later recall, were visiting an "establishment that

sold liquid refreshment" when some sailors from a destroyer "made a few remarks about destroyer escorts" and those who served on the smaller ships. Five *Tabberer* sailors "took on about twenty" destroyer sailors and came back to the ship "bloody, with a black eye or two, but with big grins and slapping each other on the back." In Plage's view, his sailors had become "a real crew of a fighting ship." While he was still having to remind them that they were "not supposed to beat-up on the Shore Patrol," Plage let the incident pass without dispensing any punishment. Plage for "the first time fully realized" that his men had "developed real pride in themselves," which made him "so very proud." He knew they were ready at last for whatever awaited them at sea.

After two weeks at Charlestown Navy Yard, which included a few days atop keel blocks in dry dock to inspect her underwater exterior— such as valves, propellers, rudder, and other fittings—*Tabberer* was deemed ready for fleet assignment. It did not take long. Orders were received to get under way on August 16 to escort the oiler *Severn* (AO-61) to Hawaii.

Tabberer arrived at Pearl Harbor on September 7 and for the next five weeks conducted underway training, including antisubmarine and gunnery exercises, in Hawaiian waters. The ship also screened and served as plane guard for the aircraft carriers *Ranger* (CV-4) and *Coral Sea* (CVE-57) during night-flying qualifications involving extensive launching (takeoff) and recovery (landing) operations. *Tabberer* would be stationed astern of the carrier, ready to pick up any pilot who went in the water—and pick up numerous drenched pilots they did, in the process working out what Plage considered a "pretty good system." When a pilot went down, as did Ensign James Brenner of *Coral Sea* shortly before 9:00 P.M. on October 9, *Tabberer* hurried to the crash site. After Plage maneuvered the ship into position, the motor whaleboat was lowered. Brenner was picked up and aboard *Tabberer* in under twenty minutes. After being checked out by the pharmacist's mate—who served as *Tabberer*'s medical staff in lieu of a doctor (which smaller ships did not normally carry)— the pilot's clothes were rushed to the laundry as he took a hot shower, then he was served coffee and a sandwich in the officer's wardroom.

When flight operations ended two hours later, *Tabberer* approached the starboard side of *Coral Sea* and returned via high-wire chair their orphaned pilot dressed in freshly ironed clothes suitable for liberty ashore and clutching a gallon of his favorite flavor of ice cream.

The latter touch was undeniable grandstanding; traditionally, a carrier would gratefully send over ice cream to any smaller ship that rescued a pilot. Ships the size of destroyer escorts were not allocated ice cream makers. The machine that found its way to *Tabberer*—meant for another ship but not yet picked up by its crew—was liberated one night from a Pearl Harbor pier by Cookie Phillips, who also served as the ship's unofficial scrounger, and his galley gang. They also grabbed several cartons of the dry mix, which came in three flavors: vanilla, chocolate, and strawberry. When a little water was added to the mix, the machine produced in about forty minutes a gallon of soft ice cream that could be eaten immediately or frozen. Plage at the time did not ask Phillips how they suddenly had an ice cream machine aboard, but he did drop by the galley most nights after the crew had eaten to see if there was any left over. Not surprisingly, "there always was some ice cream for the skipper." And every Friday night, the skipper was invited to the galley to enjoy broiled filet mignons expertly sliced off quarters of beef by Phillips, the former butcher who kept his own set of sharpened knives. The menu never varied: steak, ice cream, cinnamon buns. When it was time and the smells were wafting to the bridge directly above the galley, Plage would turn to the officer of the deck and say, "Sir, you have the conn. I'm off duty." The skipper would soon be knocking at the locked galley hatch. Also invited on Friday nights was the ship's mail clerk, Seaman 1st Class William A. McClain, nineteen, of Knoxville, Tennessee, in exchange for giving the galley crew their mail right after the officers and before everyone else.

As life aboard *Tabberer* settled into a routine in the Pacific, one officer on the ship had added reason to be grateful for having traded the often inhospitable Atlantic for another, balmier ocean. Executive officer (and, as such, Plage's second in command) Lieutenant Robert M. "Dusty" Surdam, twenty-seven, of Hoosic Falls, New York, sporting crew-cut blond hair, was six foot two with the lean, sinewy build of a runner. The son of a local bank president, he had prepped at Deerfield Academy, where he competed in track and soccer. He turned down an appointment to West Point in favor of Williams College, a small liberal arts school in western Massachusetts, where he majored in economics and graduated cum laude in 1939. After a vacation that summer to England, Germany, and France as the clouds of war in Europe became darker by the day, Surdam went to work as a bank clerk in Albany, New York. He enlisted in the Naval Reserve in October 1940 and five months later was sent to Midshipman School aboard the old battleship *Illinois* (BB-7)—renamed *Prairie State* (IX-15) and serving as a floating armory and naval school in New York harbor. After three months of training, Surdam was commissioned an ensign and received orders to the destroyer *Warrington* (DD-383), a *Somers*-class destroyer of 1,850 tons that had just finished an overhaul at Charlestown Navy Yard when Pearl Harbor was attacked and was ordered the next day into the South Atlantic to patrol "under war conditions during the national emergency." In Surdam's first fitness report submitted that same month, *Warrington*'s commanding officer noted: "Ensign Surdam is exceptionally quick witted and has unusually sound judgment for one having so little experience. He is willing to assume responsibility and has the intelligence and ability to do so with excellent results. An officer of distinct value to the Navy." Two weeks into the new year, *Warrington* was sent to the Pacific, where the destroyer remained until mid-1944, when she returned to the East Coast to undergo routine repairs.

Surdam had several times requested other assignments—to the Naval War College, where many of the brightest officers ended up in preparation for command and top staff positions,* and also to flight training—but had been turned down each time because he was too valuable aboard ship. Comments on fitness reports such as "As officer of

---

* A notable alumnus of the U.S. Naval War College, established in 1887 and traditionally open to naval officers from other countries, was Isoroku Yamamoto, fleet admiral and commander in chief of the Imperial Japanese Navy, architect of the December 7, 1941, surprise attack on Pearl Harbor.

the deck during action against Japanese aircraft, Lieutenant Surdam proved himself practically indispensable to the Commanding Officer" did not help him get his ticket punched off the ship. After more than two years on *Warrington,* Surdam finally received new orders: to Submarine Chaser Training in Miami. Thereafter, he was ordered to Brown Shipbuilders in Houston for "the fitting out of USS *Tabberer* and for duty as executive officer of that vessel when placed in commission."

On September 10—three days after *Tabberer* arrived in Hawaii— *Warrington* departed Norfolk for Trinidad, in the southern Caribbean. Two days out to sea, the destroyer encountered a violent storm along the Florida coast. After receiving word that they were steaming directly into a hurricane, *Warrington* tried to change course, but it was too late. In the early morning hours of September 13, the destroyer lost part of her bow and water flooded the forward engine room, knocking out the ship's electric power. The main engines shut down and the steering mechanism failed. By noontime the following day, it was apparent that *Warrington,* foundering badly in mountainous seas, would not survive. Almost immediately after the abandon-ship order was given, the destroyer went down. A prolonged search over the next two days found only seventy-three survivors—more than 250 men perished.

When word circulated of *Warrington's* loss, Surdam, like other mariners who left a ship shortly before it went down, would long recall the names and picture the faces of lost shipmates. At the same time he mourned their deaths, Surdam counted his own but-for-the-grace-of-God blessings, grateful not to have been on his old ship in the Atlantic, with its unpredictable weather and dreadful storms.

Six

When their Alaskan duty ended in fall 1943, *Monaghan*'s crew returned to Pearl Harbor for a month of liberties on the beaches and in the bars of Honolulu while their ship underwent routine maintenance. The Royal Hawaiian Hotel had been taken over by the Navy, and some of the crew were able to spend a day or two at a time ashore. However, a strict curfew was in place in Hawaii—including a blackout—restricting night travel and activities.

Even better news soon came for the home-sick sailors: assigned to West Coast escort duty, *Monaghan* returned to the shores of the main-land United States for the first time since the war began. The visit stateside turned out to be all too quick, however, and there was not enough time for men to take leave for visits home.

After brief port calls in California, *Monaghan* joined a convoy steaming to Espiritu Santo in the South Pacific. En route with three escort carriers conducting training, *Monaghan* was called upon to rescue five pilots who ditched at sea during flight operations. They reached their destination on November 5 and waited a week as the largest U.S. invasion force yet assembled for a single operation gathered: 35,000 soldiers and Marines in thirty-six transports, seventeen aircraft carriers, twelve battleships, twelve cruisers, and sixty-six destroyers. Their objective: a little-known atoll named Tarawa in the Gilbert Islands.

Joseph Candelaria, the Ramón Novarro look-alike who had thought they were "goners" as *Monaghan* carried out the brazen torpedo attack in the Battle of the Komandorski Islands in March 1943, had been promoted to water tender 3rd class. Now, as they waited for the invasion of Tarawa, Candelaria received word that his mother had died unexpectedly. He immediately requested emergency leave to go home. The captain declined, telling Candelaria that he was "more valuable down in the fire room for the upcoming invasion" and that he was sure his "mother would understand." Candelaria went away without another word but headed to the fire room and asked Water Tender 3rd Class Louie W. Childers for torpedo juice—denatured alcohol used as fuel for torpedoes mixed with fruit juice to make it palatable.* Childers, a Tennessean who had run moonshine into Georgia until caught and given the choice between jail or the military, usually had a batch ready to go. Not one to let a man drink alone, Childers joined Candelaria, who had taken his first sip of liquor on a troop train with other recruits headed to boot camp and was still a lightweight drinker. He got very drunk "thinking of my mother."

Early on the morning of November 20, 1943, *Monaghan* and other ships fired on Tarawa for nearly three hours before the assault troops started ashore. Although the bombardment and air attacks turned

---

* Denatured alcohol is 95 percent ethyl alcohol (used in beer), 4 percent methyl alcohol (highly toxic), and 1 percent pyridine. The last is added to give the mixture a vile taste so as to discourage its consumption, as it can be fatal.

much of the one-square-mile island—at most places no more than a few hundred yards wide—into something resembling moonscape, the troops found stubborn enemy resistance and suffered high casualties, so many that the costly battle would be remembered in military annals as "Bloody Tarawa." Although costly in lives lost, it was here that the Marines learned how to press the attack and beat the Japanese in their preferred style of warfare: hand-to-hand jungle fighting.*

Stunned *Monaghan* sailors stood at the fantail and watched as landing craft, hopelessly hung up on coral reefs and fighting fluctuating tides, dumped out their troops in shallows some 500 yards from shore, requiring them to struggle toward the beach under "murderous fire." Anyone who saw "the Marines in those shallow waters at Tarawa" would never forget.

Not far away, the aircraft carrier *Independence* (CVL-22), one of the new smaller carriers converted from the hulls of other ships (usually cruisers and tankers) to meet the demand for greater naval air power in the Pacific, was attacked by a group of enemy planes zooming low over the water.† Six attackers were splashed, but not before several torpedoes were released, one of which struck the carrier, causing severe damage. *Monaghan* escorted the crippled carrier to Pearl Harbor, and once at the command center of the U.S. Pacific Fleet, the destroyer received surprising orders.

As deck gunner Joseph Guio, who had narrowly missed being shot in the head during the running gunfight with an enemy submarine off Kiska, wrote to his younger brother, Bill, in West Virginia, "the President of the United States is sending us to San Francisco to have a good

---

* The high casualty rate suffered by the numerically superior U.S. force in taking this small island—half the size of New York's Central Park—ignited a storm of protests in the press and Congress. Tarawa was one of the two most heavily defended atolls (the other was Iwo Jima) to be invaded by American forces in the Pacific, and it turned into one of the bloodiest battles per square foot in the Pacific. More than 3,000 U.S. Marines were killed at Tarawa; only 17 enemy soldiers from a garrison of nearly 5,000 were still alive at the end of the three-day battle.

† Light CVL or escort CVE carriers were capable of carrying thirty to forty planes, compared with ninety planes on full-size carriers.

time on Xmas and maybe New Year's." Guio, who had advanced to gunner's mate 3rd class and become one of the most popular members of the crew as a result of his "never-ending cheerfulness and sense of good humor," knew that they would not be out of action for long. After the holidays, Guio told his brother in the same letter, "we are going to the Solomons to do more fighting." Fighting was something the veteran crew knew about, as the ship had already earned a string of battle stars. "The *Monaghan* is known for her fighting crew," Guio proudly wrote, admitting that "lots and lots and lots of action and some very narrow escapes" had taken their toll.

> *Bill, I've changed a lot since I left home. I say my prayers every night before going to bed. A night never passes that I don't pray for myself and all you people back home. There are many nights I lay in bed thinking of you, Mom, Dad, Luck, and [all] my sisters and brothers with big tears running down my weather beaten cheeks. This is a very tough life but I'm willing to fight so this Damn War will end so I can return home to Luck and settle down.*
>
> *Don't ever tell mother and Dad about this letter. It's best they never know about it. I wouldn't have them worrying about me for the world . . . Bill, I want you to call Luck on the telephone then go up to see her and show her this letter. I also want you to take Luck to a movie once a week for me. She will be more than glad to go and see a movie with my kid brother. I'll provide you with the money. Here is ten dollars.*
>
> *Love and kisses to all.*

The letter, which Guio posted on the day *Monaghan* arrived in San Francisco, was so detailed in listing specific battles fought by the destroyer—in violation of military censorship regulations—that he explained to his brother he was "putting a fake address on the envelope so it can't be tracked."

The day before steaming under the Golden Gate Bridge, *Monaghan* held a change-of-command ceremony at sea, with Lieutenant Commander Waldemar F. Wendt, thirty-one, a 1933 graduate of the United States Naval Academy and a veteran destroyerman, relieving Lieutenant Commander Peter H. Horn, whom the crew had come to like as a "real laid-back guy" known for talking to his young sailors "like a father." In the months ahead, Wendt would earn a reputation among his fellow destroyer skippers for handling *Monaghan* "exceptionally well." As for his crew, they quickly nicknamed him "GQ Wendt" because "every five minutes, every time you turned around," according to Water Tender 3rd Class Joseph C. McCrane, "he'd sound general quarters, sending all hands to their battle stations."

At twenty-seven, McCrane, of Clementon, New Jersey, was older than many of his shipmates. When he enlisted in 1941 shortly after Pearl Harbor, McCrane was informed by the Navy recruiter that "everyone wanted to be in aviation." Told that if he picked aviation and washed out he could likely end up in the medical corps, which did not interest him, McCrane took the required exams without stating a preference. That found him after boot camp being sent to the Navy's boiler school, from which he emerged a rated petty officer. His first day on *Monaghan* in October 1942, McCrane had been assigned one of the dirtiest jobs in the fire room: scraping out the tubes in the big boilers. There were about five or six sailors inside the boilers, "all wrapped up in cloth because of the dirt flying all over the place." When they emerged for lunch, the guys got to talking, and someone asked McCrane his rate. When he said 3rd class, they looked at him as if he were crazy, explaining that scraping was routinely assigned to nonrated firemen. McCrane shrugged and went back into the boilers to finish the grimy job. Before long, the short, stocky McCrane, "a very religious man," was being called "Mother McCrane" because of his propensity for doing the right thing and helping others.

As Joe Guio hoped, he and his shipmates spent New Year's as well as Christmas seeing the sights in San Francisco, a Navy-friendly town that looked like all the postcards, especially to wide-eyed bluejackets trying to forget a far-off war as they took their first rides on a Fisherman's Wharf cable car.

Monaghan took part in fleet exercises off San Diego the opening weeks of 1944. No one was surprised when the ship was sent westward with a task force for the invasion of the Marshall Islands, the next step in the island-hopping campaign to establish forward bases to support wider operations in the mid-Pacific. After a quick stop at Pearl Harbor for replenishment, the invasion force pushed onward, its objective 2,400 miles away.

On January 25, during routine flight operations aboard the escort carrier *Sangamon* (CVE-26)—recently converted from an oil tanker—a returning plane's tail hook failed to catch any of the recovery wires stretched across the flight deck. Crashing through reinforced barriers, the plane careened into aircraft parked on the forward flight deck. Upon impact, its belly tank tore loose, spewing aviation fuel, which ignited. Although the blaze was quickly brought under control, eight crewmen died and seven were seriously injured. To escape the flames, fifteen other sailors leaped off the flight deck into the sea below.

Rapidly responding to the red and yellow diagonal signal flag Oscar for "man overboard," *Monaghan* carefully searched in the carrier's wake, rescuing eleven of the thirteen men who were picked up (two were never found), and eventually transferring the survivors to *Sangamon*.

Four days later, the invasion of Kwajalein in the heart of the Marshalls began. For a week *Monaghan* protected a group of carriers launching daily strikes against enemy positions. Offshore, things usually quieted down at night, and when they did someone in *Monaghan's* radio shack would often tune in Tokyo Rose and pipe the sentimental music into compartments for the crew to hear. Of course, those torch songs came with propaganda—such as how the Japanese fleet "knows where you boys are and we are going to annihilate you." More difficult for the men to hear was her chatter about what the women at home might be doing. "Now I'm going to play you something on the lonely side. Think about where you girlfriend is tonight." And the men, lonely and homesick, *did* wonder: "Is she dancing the night away while I'm out here?"

The same day Kwajalein was stormed by Army and Marine units, Majuro Atoll, an undefended island group 300 miles southeast of Kwajalein, was also seized. It was soon turned into a major resupply and repair anchorage for U.S. ships, and *Monaghan* would visit it often over the next several months while participating in a series of invasions and strikes at places that most Americans had not heard of but which soon became familiar names—Eniwetok, Palau, Yap, Hollandia, Truk. On May 4, *Monaghan* again put in at Majuro and this time stayed several weeks for routine repairs and maintenance to ready the ship for the upcoming invasion of Guam in the Mariana Islands.

On May 31, Louie Childers was at the fantail sleeping off another bout with torpedo juice when it started to rain. He was seen to rise and stagger across the wet deck. When Childers didn't show up for duty in the fire room at 4:00 A.M., Candelaria, now water tender 2nd class and in charge of the morning watch section, didn't think much of it and assigned another man. The next morning, Childers was still missing and Candelaria had to explain his failure to report Childers' absence. "I thought he was sleeping it off somewhere," he said.

Childers was never seen again. An official investigation would declare him dead a year later, the most likely cause "falling over the side" and drowning, as at the time he went missing *Monaghan* was moored in a berth in the middle of Majuro harbor with land "over a mile away" and no other ships "closer than 600 yards." Candelaria was never completely convinced, however, and it "wouldn't have surprised" him at any time to see "Louie walk in right now because he was streetwise and no one ever got the best of him." He even pictured Louie having made it back home to the Smoky Mountains, running moonshine.

Another *Monaghan* crewman also got a one-way ticket off the ship shortly before the invasion of Guam. As far as Joe McCrane was concerned, it was high time for the young fellow called Sailor. It wasn't until his older brother, a Marine Corps officer, came aboard and took the youth away that the bureaucratic process was begun to boot the fourteen-year-old out of the Navy. How "they took the kid in the Navy" to begin with McCrane could not figure, since Sailor, who worked in the fire room, "*looked* like he was fourteen."

Arriving off Guam on July 17, *Monaghan* took up position screening a force of cruisers as they "pounded the enemy's beach defenses at Agat Bay on the island's west side." That night, *Monaghan* moved close to shore to protect underwater demolition teams as they blasted passages through dangerous reefs for the landing boats. *Monaghan* then went back to bombarding "all night and all day and all night the next night." The bone-jarring salvos went on for such a long time they became "monotonous" for the shell-shocked crew; "back and forth, guns flashing." It would have been worse had they known at the time that the naval bombardment was "not doing much good because the shells would ricochet and bounce off" without killing many of the dug-in defenders, as the soldiers and Marines who hit the Guam beaches on July 21 found out for themselves.

By the time Guam was finally secured two weeks later at a cost of 3,000 American lives, *Monaghan* was steaming for Puget Sound Navy Yard at Bremerton, Washington, to undergo a scheduled major overhaul.

Monaghan floated into dry dock midafternoon on August 14, 1944, and shipyard workers began pumping out the seawater. Within two hours the destroyer was riding high and dry on keel blocks, receiving all shipboard power and fresh water from cables and hoses attached to the dock. As the crew hoped, twenty days' leave was granted to nearly half of the enlisted crew, and when they returned the others would be allowed to go home on leave as well.

Among the first group to leave was Fireman 1st Class Evan Fenn, twenty, of Pomerene, Arizona. Fenn, a sturdy cowpoke type raised around livestock on his family's 40-acre grain and alfalfa farm, was working in construction "when the Japs bombed Pearl Harbor." He and a buddy had hightailed it 50 miles to Tucson to join the Navy but found the service had "filled their quota." Not long after, Fenn followed his father to Utah. After working in a Provo steel mill, Fenn joined a crew building several 200-foot smokestacks, and while he was not keen on heights, he found it easy to get used to because they "started at the ground and worked up." When he received his draft notice, Fenn again tried the Navy and was accepted. He went to boot camp in May 1943 and boarded *Monaghan* at Pearl Harbor three months later. His first impression was of a "good ship" that had been kept "pretty busy" in the war. Nothing had happened in the past year to change his mind on either count.

Their second day in dry dock, 114 enlisted men in dress blues stepped one at a time or in small groups onto *Monaghan*'s quarterdeck, saluted the junior officer of the deck, then the U.S. flag on the fantail, and departed.

For many of them, it would be their last visit home.

Seven

Since coming aboard *Hull* shortly after Guadalcanal, Michael "Frenchy" Franchak, now a
radarman 3rd class, who had been struck on
his first liberty in the tropics at the sight of
gaming cocks fighting in the street, had in the
past year seen men, ships, and planes engage
in similar struggles to the death.

One incident that proved difficult for
Franchak to shake was a mercy mission to Espiritu Santo to pick up 200 wounded Marines
and transport them to a naval hospital in
Suva. Each crew member was ordered to clear
his rack and go topside in order to make room
below for the wounded, which included "quite
a few amputees," along with doctors and
nurses. The ship was so overcrowded there
was hardly room to get through the narrow

passageways. For the wounded, the 800-mile trip turned into a "murderous journey in sweltering heat" from which there was no relief, and some died en route.

For *Hull's* crew, more nightmarish memories ensued.

On the day *Monaghan* was covering the Tarawa invasion in November 1943, *Hull* had the same assignment at Makin Island, also in the Gilbert Islands. Unlike Tarawa, Makin was lightly defended—the 800 enemy defenders were overrun by 6,500 U.S. troops. The snappy radio message sent by Major General Ralph C. Smith, commander of the U.S. Army's 27th Infantry Division, on November 23—"Makin taken!"—was quoted in newspaper headlines across America announcing the first capture of a Japanese-held island. Although fewer than eighty American troops were killed at Makin, many times more that number of sailors went to their fiery deaths in full view of *Hull* the morning after Makin's capture.

Hull and three other destroyers had formed an antisubmarine screen around three escort carriers. Shortly before sunrise, a "dim, flashing light"—which turned out to be "a float light" dropped by an enemy plane—was reported on the water's surface not far away. One destroyer went to investigate, leaving "a hole in the already thin screen" and "presenting a perfect target" for the submerged Japanese submarine *I-175*.

That morning, *Hull* sonarman Pat Douhan, the California Irishman who had worked on a seismographic crew before enlisting, awakened early with "an uneasy feeling." Looking at his watch, he knew the crew would soon be called to general quarters prior to the carriers beginning flight operations. He decided to dress and head for the bridge, where he served during battle stations as the captain's talker—conveying the skipper's orders to various sections of the ship over a sound-powered telephone, which meant he had to be calm, reliable, and clear-spoken even in the midst of battle. Coming from below, Douhan had just emerged from the hatchway to the main deck when the escort carrier *Liscome Bay* (CVE-56) "blew up in front of me." By now Douhan had seen and heard his share of blasts and explosions, "but never one that big," which looked more like an ammunition dump going up than a ship.

When *Liscome Bay* was hit at 5:13 A.M. by a single torpedo amidships, a "column of bright orange flame" shot a thousand feet in the air. Within seconds, bombs lined up on the flight deck detonated, and with a "mighty roar the carrier burst apart as though she were one great bomb," tossing high in the air "men, planes, deck frames and molten fragments." Ships in the vicinity were showered with "fragments of steel, clothing, and human flesh." Commissioned only a few months earlier, *Liscome Bay*'s first battle proved to be her last. Twenty-three minutes after being hit, the doomed ship "flared up for the last time and sank hissing." Those who made it off the sinking carrier into the water found themselves immersed in a "spreading pool of burning oil."

Hull, which had closed on the burning carrier before she sank, now "entered the oil slick" and lowered her whaleboat to search for survivors, many of whom were in "frightful condition, with shattered limbs, internal hemorrhages, head concussions and horribly disfiguring burns." Within minutes, *Hull*'s whaleboat was filled with survivors, most of them "thickly covered . . . by viscous, stinging fuel oil." Boats from other ships joined *Hull* in the rescue operation; in all, 272 survivors were pulled from the water, while 644 of their shipmates—including task group commander Rear Admiral Henry M. Mullinnix (number one in his Annapolis class of 1916) and *Liscome Bay* commanding officer Captain Irving D. Wiltsie—lost their lives that morning.*

Back at Pearl Harbor two weeks later, *Hull*'s crew was called to quarters on December 8, 1943, for a change-of-command ceremony, with Lieutenant Commander Charles W. Consolvo, thirty-two, a dark-haired, blue-eyed Virginian and graduate of the U.S. Naval Academy, taking over. Although he had finished in the bottom 15 percent of his Annapolis class (1935), Consolvo was already carving out a reputation as a highly effective and charismatic officer. In his previous assignment—as executive officer of the *Fletcher*-class destroyer *Ammen*

---

* With *Hull* and other destroyers preoccupied with rescuing survivors, the enemy submarine escaped undetected. However, *I-175* and her crew did not have long to live. Three months later, the submarine was destroyed in a depth-charge attack by the destroyer *Charrette* (DD-581) and destroyer escort *Fair* (DE-35) near Kwajalein.

(DD-527)—Consolvo had received outstanding marks in his latest fitness report for his "ability to command" and "ship handling." Asked in the same report for his preference of next assignment, Consolvo wrote: "Remain in destroyers—Pacific Fleet." During Consolvo's tenure as *Ammen*'s second in command, the destroyer participated in the capture of Attu, for which the commander of amphibious operations in the Pacific had commended the ship's officers and crew for their "splendid contribution . . . to the accomplishment of the mission," as well as the occupation of Kiska and "task force sweeps and ocean escort operations in North Pacific, Bering Sea and Aleutian Areas." Consolvo had previously served as gunnery officer on the destroyer *Anderson* (DD-411), during which he had the unusual assignment of firing on another U.S. warship. Operating with the aircraft carrier *Hornet* (CV-8), notable for launching the Doolittle Raid from her flight deck (sixteen U.S. Army B-25 bombers led by Lieutenant Colonel James Doolittle in their raid on Tokyo) and participating in the victorious Battle of Midway, *Anderson* was ordered to sink the abandoned *Hornet* when the carrier was badly damaged during the Battle of the Santa Cruz Islands in October 1942.

Aboard *Hull,* it soon became apparent that the seasoned Consolvo had a talent that did not come naturally to every commanding officer— in addition to his athletic parlor trick of "holding a broomstick in his hands and jumping over it forwards and backwards without letting go." Consolvo was skilled at the training of junior officers to become competent ship handlers and qualified "officers of the deck under way," capable of operating a vessel at sea in the absence of the captain from the bridge. Consolvo was a bit "didactic" by nature, which aided his tutorial role. To teach others, it also helped that he was a good ship handler, and in that arena Consolvo soon demonstrated his impressive abilities.

Consolvo showed his touch at the conn the first time he brought *Hull* alongside another ship at sea, which Navy vessels often did to take on fuel and supplies. Many commanding officers exhibited overriding caution when approaching a tanker or supply ship for an underway replenishment, a "high-stress situation" for everyone from the bridge down to the engine room given the potential for a collision. Consolvo was bolder than most skippers, but he also knew what he was doing. Consolvo would overtake the supply vessel from the rear with unusual rapidity, approaching at two-thirds or full speed.* Then he threw the engines into reverse. At the right moment—this was a combination of experience and feel—he ordered ahead one-third, bringing *Hull* along-side the other ship on a parallel course. Consolvo's string of orders to the helmsman and engine room was clear, timely, and precise. Soon, the two ships—with lines tethering them together—were steaming along in unison. Done right, this high-speed maneuver was like "a car traveling at sixty miles an hour as it approaches a parallel parking space," and the driver slamming on the brakes and pulling into the spot "cleanly, without an inch to spare." Consolvo's deft handling of the old, prewar destroyer was an amazing sight to behold—awing even other ship captains—and something *Hull*'s crew soon took pride in.

Ensign Lloyd G. Rust Jr., twenty-four, of Wharton, Texas (50 miles southwest of Houston), came aboard *Hull* on December 9, 1943—the day after Consolvo assumed command. Signing up right after Pearl Harbor for the Navy's V-7 program, in which college students were al-lowed to continue their education and receive commissions upon grad-uation, Rust, a pre-law major, graduated from the University of Texas in August 1942 and received a deferment to study for the state bar exam to be given in three months. Upon taking the exam, he was ordered to active duty in December and sent to U.S. Naval Reserve Midshipman School on the shore of Lake Michigan at Chicago's Northwestern University, which during the war would train nearly 26,000 naval officers—including a future PT boat commander named John F. Kennedy—thereby earning the nickname "Annapolis by the Lake."

For Rust, a solid six-footer with brown hair and eyes to match, and a

---

* Standard speeds on naval ships are ordered by the bridge on an engine order telegraph that rings in the engine room (thus the term "rang up" knots). Speeds are shown in increments of 5 knots; forward one-third (5 knots), two-thirds (10 knots), standard (15 knots), full (20 knots), and flank (25 knots), and reverse one-third (5 knots), two-thirds (10 knots) and full (15 knots). Other speeds—17 knots, for example—were achieved by the bridge ordering on the revolutions indicator a specific number of engine revolutions per minute (17 knots equated to 165 rpm on *Hull*, although computing exact revolutions to specific speeds varied from vessel to vessel).

gregarious nature that included a Texas-size laugh that shook "his whole body and could be heard across a room," it was not only his first Christmas in uniform and away from home but also his first trip outside his "beloved" Lone Star state. Any pangs of homesickness were eased by his meeting and beginning a serious romance with Dee Dee Wrigley, the attractive scion of America's leading chewing-gum family, with whom he spent the holidays.*

Although possessing excellent study habits forged by the rigors of law school, Rust found at Midshipman School a curriculum crammed with complex subjects such as celestial navigation, ordnance, and seamanship. For example, one had to solve problems involving math and geometry just to grasp the principles upon which celestial navigation is based. Yet in a scant few months, the Midshipman School turned out newly minted ensigns ready for the fleet. The same process took four years at Annapolis—the main reason for the resentment that graduates of the Naval Academy harbored toward the "ninety-day wonders." After graduation, Rust went to antisubmarine warfare school in San Diego, then to Pearl Harbor for a brief stint as a decoder—until the Navy finally agreed with him that he was a terrible typist—before being assigned as assistant to a commodore embarked on the destroyer *Phelps* (DD-360) in Alaskan waters. Not long after, Rust received a card in the mail notifying him that he had passed the state bar examination. After years of hard work to become a lawyer, which had always been "so important" to him, he had made it—yet Rust did not feel like celebrating. After all, he was in the middle of a war with no end in sight, and had already seen enough to know there was no guarantee he would "make it home alive," let alone ever step into a courtroom as a barrister.

It did not take long for Rust to form an opinion of his new commanding officer on *Hull*. Rust found Consolvo to be a "tremendous seaman" and "100 percent competent." Consolvo let it be known he believed

---

* During the war, Wrigley's best-selling brands—spearmint, Doublemint, and Juicy Fruit—were removed from the civilian market by company president Philip Wrigley to dedicate their entire production to the U.S. armed forces. After the end of the war, the brands were again available and quickly exceeded their prewar popularity.

it "his duty to teach every man on the ship as much as he could," and that if he was successful it would not only "help the Navy" but also assist "the war effort and the whole country." Consequently, if they were not in a "serious situation," Consolvo often stepped aside at the conn and let his junior officers "handle the ship" so they could gain experience. Rust and the other officers knew they had "a whole lot to learn," but they did not find it a chore to be taught by Consolvo. Quite the contrary, they found it "a privilege to talk to him for five minutes about anything." As for the enlisted men, Consolvo "never treated them differently than the officers" and showed that every man regardless of rank was important to him. This went far toward building good morale on *Hull,* where some of the officers and enlisted men became friends, talking about wives or girlfriends, families, home, hopes, and dreams as they stood their round-the-clock watches side by side at various stations on the ship.

Another of the young officers Consolvo set out to mentor was Lieutenant (j.g.) Greil I. Gerstley, twenty-three, a five-foot-eleven, slender Philadelphian with a charming smile, dark curly hair, and "Jewish movie-star good looks." Prepping at William Penn Charter School and graduating in 1941 from Cornell University, where he was president of Zeta Beta Tau—and known with other "natty ZBTs" as models of the "urbane, suave, genteel" man—Gerstley had grown up in a "privileged environment in terms of money and social status in Philadelphia's Jewish community." His father had made a fortune in liquor distribution and securities, and his mother came from a prominent Montgomery, Alabama, merchant family named Greil—hence his rather unusual first name. Gerstley, who had taken two years of Army ROTC in a field artillery unit at Cornell, applied for a commission in the Naval Reserve shortly after Pearl Harbor. In the months since his college graduation, Gerstley had been working as a clerk in his father's securities firm, which held a seat on the New York Stock Exchange. It was no secret that his father was grooming his son to take over the family's businesses; however, any such talk would have to wait, for young Gerstley was eager to offer his "services to the country in this time of need." Writing

Surprise attack on Pearl Harbor: "A day that will live in infamy." U.S. NAVY

In front of battleship *Pennsylvania*, burned hulks of destroyers *Downes* and *Cassin*, fires which civilian worker and future *Hull* sailor Thomas Stealey fought. U.S. NAVY

USS *Hull*

1. The destroyer *Hull* (DD-350), commissioned in 1935, winner of 10 battle stars. U.S. NAVY 2. James A. Marks, Annapolis Class of 1938, the last commanding officer of *Hull*. U.S. NAVY 3. Lloyd Rust, Jr., reserve officer and Texas lawyer, *Hull* CIC officer. FAMILY PHOTOGRAPH

2

5

4. Greil Gerstley, of Philadelphia, second in command of *Hull*, on the day of his September 1944 wedding in San Francisco to Eleanore Hyman. FAMILY PHOTOGRAPH
5. C. Donald Watkins, reserve officer from Columbus, Ohio, *Hull* torpedo officer. FAMILY PHOTOGRAPH
6. Edwin B. Brooks, Jr., *Hull* sonar officer and "southern gentleman." U.S. NAVY 7. Archie DeRyckere, "never been whipped" after four years in the fleet, *Hull* chief quartermaster. FAMILY PHOTOGRAPH 8. John "Ray" Schultz, boatswain's mate and member of *Hull* crew since 1938. FAMILY PHOTOGRAPH

Kenneth Drummond, *Hull* storekeeper who stood watches on the bridge. FAMILY PHOTOGRAPH

Thomas Stealey, of Stockton, California, *Hull* fireman who as a civilian worker helped fight fires during the attack on Pearl Harbor. FAMILY PHOTOGRAPH

Patrick Douhan, of Fresno, California, *Hull* sonarman. FAMILY PHOTOGRAPH

Michael "Frenchy" Franchak, of Jermyn, Pennsylvania, *Hull* radarman. FAMILY PHOTOGRAPH

Portia and John Kreidler, *Hull* chief sonorman, one month after their August 1944 wedding. FAMILY PHOTOGRAPH

Carl T. Webb, an Oklahoma cowboy turned *Hull* seaman. FAMILY PHOTOGRAPH

USS *Monaghan*

Joseph "Mother" McCrane, of Clementon, New Jersey, *Monaghan* water tender. FAMILY PHOTOGRAPH

L. Bruce Garrett, graduation photo, Annapolis class of 1938, the last commanding officer of *Monaghan*. U.S. NAVY

354

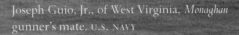
Joseph Guio, Jr., of West Virginia, *Monaghan* gunner's mate. U.S. NAVY

Joseph Candelaria, Jr., of Bakersfield, California, *Monaghan* water tender.
FAMILY PHOTOGRAPH

The destroyer *Monaghan* (DD-354), commissioned in 1935, winner of 12 battle stars. U.S. NAVY

Four of the six survivors of *Monaghan*; from left, Evan Fenn, William F. Kramer, Doil Carpenter and James T. Story, being handed their leave papers by Ensign Maragaret Harrison upon their arrival at Los Angeles harbor. LOS ANGELES EXAMINER

Last living survivor of *Monaghan* today: Evan Fenn, 84, of Saint David, Arizona. CHRIS DAVOBICH, SAN PEDRO VALLEY NEWS-SUN

recommendation letters to the Navy on behalf of Gerstley were a Cornell English professor ("a young man of lively personality, good intellectual ability, and dependable moral character"), Cornell's director of public information ("member of an old American family, steeped in the finest traditions of our land . . . his loyalty and dependability can be counted on 100 percent"), and a justice of the Supreme Court of Pennsylvania ("unimpeachable integrity . . . has physical courage, is ambitious to duty . . . a fine specimen of young American manhood"). Commissioned in March 1942, Gerstley wound up at Northwestern's Midshipman School, too. Upon graduation, he drove a new Chrysler sedan to San Francisco on the first leg of a journey that took him nearly halfway around the world to meet his ship. The greenhorn ensign caught up with *Hull* in September 1942, losing his cover on deck to a stiff breeze and watching forlornly as his cap drifted in the water between the ship and pier. Notwithstanding that rookie move, Gerstley was soon writing glowing reports to friends and family about the Navy and his ship: "I am getting an enormous kick out of the Navy with plenty of thrilling experiences and fascinating times daily" and "I wouldn't swap destroyer duty for anything else, except maybe for subs, but the folks wouldn't like the latter." At various times, Gerstley handled myriad duties on *Hull*— including gunnery, communications, and navigation—and his superiors found him to possess "good judgment" and a "cheerful disposition."

As Christmas (1943) approached, *Hull*'s crew received word that they were heading for California, where for a few weeks they would take part in amphibious exercises off the coast. *Hull* pulled into San Francisco on December 21. There was insufficient time for leaves home or for a shipyard stay even though everyone knew the well-traveled ship "had to get some work done before long." Nonetheless, the crew enjoyed a week's worth of holiday liberties.

For some, the brief stateside visit was time enough to start a serious romance, as it was for Greil Gerstley. Asked by a local friend if he would agree to a blind date, Gerstley demurred, explaining that he wanted to "get a look at her first." He went to the department store where his would-be date worked and "looked her over," which Eleanore Hyman "greatly resented," although not enough to turn down the handsome naval officer when he asked her out. The beautiful twenty-year-old brunette, who had attended Stanford University, and the debonair Philadelphian "fell in love very quickly."

On January 13, 1944, *Hull* sailed with Task Force 53 (which included *Monaghan*) for the invasion of the Marshall Islands. Two weeks later, *Hull* was screening transports off Kwajalein, and more screening and patrol duties continued throughout February off Eniwetok and Majuro. Antisubmarine patrols could be "dull business," but aboard *Hull* the routine would never again seem rote or mundane to men who had witnessed the sudden sinking by a single torpedo of *Liscome Bay* with its horrendous loss of life.

The next action for *Hull* came in March, when she joined the aircraft carrier *Lexington* (CV-16), two battleships, and several destroyers in the bombardment of enemy-held Mille Atoll in the Marshall Islands. *Lexington,* in commission for only a year, had already survived one near-fatal attack three months earlier off Kwajalein, when an enemy torpedo knocked out her steering gear and ruptured fuel tanks. Listing and ablaze, the crippled carrier made it to Pearl Harbor for repairs, while Tokyo Rose jubilantly reported her sunk with all hands. Painted dark blue and the only carrier not to wear camouflage colors, *Lexington* was soon dubbed the "Blue Ghost," as Tokyo Rose would report the carrier sunk a second time in April during the devastating raid on the major Japanese base at Truk, where *Hull* once again stayed protectively close to *Lexington,* untouched even as her aircraft downed seventeen enemy planes. *Hull* was also in *Lexington*'s company for the June 15 invasion of Saipan in the Marianas, during which the U.S. task force was attacked by waves of Japanese planes. Close on *Lexington*'s port quarter, *Hull* helped bring down an enemy dive bomber heading for the carrier "either to land or crash on it." *Lexington*'s air group played a major role in the resultant Battle of the Philippine Sea (also known as the "Marianas Turkey Shoot") on June 19–20, during which the Japanese lost three carriers and 600 planes, some when they ditched at sea with nowhere to land after their carriers were sunk. While again emerging unscathed, the

"Blue Ghost" for a third time was reported by Tokyo Rose to have been sunk during battle.

Hull had not seen the last of the Marianas, serving in the naval force for the invasion of Guam in July and afterward patrolling off the island for nearly two weeks. Then came the best news possible for the crew, who reacted with unbounded delight: they were scheduled for a major repair and refit at Puget Sound Navy Yard in Seattle, Washington. The long months of operations in the Pacific had worn out the ship, equipment and men alike.

Ray Schultz, the seaman who before the war had tried his best to get booted out of the Navy but succeeded only in being regularly demoted, was now on an opposite promotional path. He had made 1st class boatswain's mate, considered the most versatile shipboard rate in the Navy. Masters of seamanship, boatswain's mates perform almost any task in connection with deck maintenance: loading and securing cargo, setting gangplanks, standing security watches, small-boat operations, handling ropes, lines, and cables, and operating hoists, cranes, and winches. No captain of any Navy ship, large or small, would want to go to sea without qualified boatswain's mates. Since Pearl Harbor, Schultz had gone from incurable cutup to indispensable sailor.

Schultz had now been on *Hull* nearly seven years—an uncommonly long stretch on one ship and testament to previous commanding officers' desire to keep him aboard. Schultz had seen all types of skippers come and go and had learned to assess their strengths and weaknesses. In Consolvo, he believed they were being led by not only a "regular guy" who didn't put on airs like many commanding officers but also the "best ship handler" Schultz had known. At the conn, this skipper could "make the ship do tricks." Schultz particularly liked the way Consolvo brought *Hull* into a buoy—always on his first try, which was appreciated by Schultz's deck force, standing by with lines to secure the ship. Consolvo let out either the port or starboard anchor chain until the anchor was swinging just above the height of the buoy, then aimed that side of the bow for the buoy. At the right moment, he'd drop anchor and hook the buoy like a marlin, reverse engines, "take the rear buoy, and that's it."

When it came to handling characteristics, *Hull* tended to be top-heavy, as were all the six surviving *Farragut*-class destroyers (put into service in 1934–35). First and foremost, it had to do with their narrowness of beam: 34 feet 3 inches. In comparison, the newer *Fletcher*-class ships (commissioned 1942–44) were 39 feet 6 inches abeam and the *Sumner*-class destroyers (1941–44) had 40-foot beams. Nearly 6 feet less in the beam made a distinct difference in handling, seaworthiness, and ride.

Soon after coming aboard in 1938, Schultz found that "40-degree rolls" were common, and *Hull*'s side-to-side swaying even in moderate seas could be "terrific." He was told by some of the original crew who had put *Hull* into commission back in 1935 that upon returning from their shakedown cruise—during which *Hull* visited ports in the British Isles, Portugal, and the Azores—all the chief petty officers except one put in for a transfer, even though in those prewar days such voyages tended to be "rather relaxed affairs" designed to give the new crew time to "get acquainted" and start "developing teamwork." The chiefs were unanimous that *Hull*, although at the time one of the Navy's most modern destroyers, was "too top-heavy" and would "never make it in the China Sea," known by mariners for high, turbulent seas.

Farragut-class ships had other basic design flaws, particularly the "fashionable broken-deck design of the hull," which gave the vessels a low, sleek look in the water, but also less freeboard—the distance from the waterline to the main deck level—than the older four-stack destroyers that sat higher in the water. This meant the low-slung main deck amidships was often swamped in heavy seas, and it was at this critical point where the access hatches to engineering spaces were located. At such times, fire room watches could not be relieved without taking water down opened hatches. Another problem was the absence of a safe, covered way to move between the forward and after sections. With the cavernous spaces for the boilers and engines taking up the middle of the ship, the only way to go from one end to the other was by coming topside and crossing an open deck—a potentially dangerous passage in stormy seas.

Now a decade old, the *Farragut*-class ships "were not up to the newer and more glamorous *Fletchers* and *Sumners* in their armament, communications or appearance," although the aging destroyers were still "dependable performers" and their "services were needed" by the Navy in the two-ocean war. Every time a *Farragut*-class vessel went into the shipyard for a scheduled overhaul the ship was fitted with newer and additional equipment—such as the latest in radar, radio, antiaircraft weapons, and automated fire-control systems—much of it topside on the superstructure and 50-foot mast. While it was anticipated that such additions and modifications would be offset by weight reductions in other areas, such alterations in busy shipyards crowded with battle-damaged vessels were considered a low priority and often did not get done. This extra weight only exacerbated the built-in topheaviness of the *Farragut*-class ships. Designated as 1,500-tonners at the time they were built, by 1944 the ships of this class were "well over [their] designed tonnage" and being operated at weights "greatly in excess of those for which the ships were built." In fact, in May 1944, the Navy's Bureau of Ships set a new maximum loading limit of 2,255 tons for the *Farraguts*.

Earlier in the war—during *Hull*'s first trip to the South Pacific after the Aleutians campaign—the destroyer was bombarding enemy fortifications on Wake Island, and the Japanese were firing back. Enemy shells came so close that Sonarman Pat Douhan swore they were dropping "between our stacks." Zigzagging to avoid incoming fire, *Hull* crossed the wake of a cruiser and rolled over so far—"probably 65 degrees"—the crew was "climbing the bulkheads," certain the ship was going to capsize in "perfectly calm" seas. Since then, even more topside equipment and weight had been added.

En route to Seattle, Schultz found out he had made chief petty officer, which meant he would be trading his blue dungarees for the khakis worn by chiefs and officers. It also meant *Hull* had one too many leading boatswain's mates given the size of the ship. One of them had to go.

"I'm leaving the ship when we get to Seattle," longtime Chief Boatswain's Mate E. M. "Doc" Toland told Schultz.

Schultz saw an opportunity. "You've been on here nine years, Doc," he said. "It's your home. Why don't you let me go and you stay?"

Toland wouldn't bite. "No, I'll be glad to get off. You're taking over. What's the first order you want to give?"

Schultz shook his head. "I don't know. Let me think about that." He admitted there was one thing Toland had always taken care of that he wasn't sure he could handle. "Ordering supplies for the ship. Never done that."

"You won't have no trouble with that," Toland promised. "It's just like ordering out of the Sears catalog. I'll go over it with you before I leave."

Not long after, Toland did as promised, showing Schultz the book and forms for ordering winches, lines, parts, tools, and other supplies for the deck force specifically and for the crew at large. Schultz now told Toland that he had thought it over and wanted his first order to be getting everyone in the crew kapok-filled canvas life jackets to replace the old inflatable rubber belts that were worn around the waist whenever the crew went to general quarters. The old belts were creased and cracked from being folded and were badly worn out, and Schultz suspected that many had developed leaks.

"Half the belts leak," Schultz said. "That means half the guys think they have a life jacket that will keep them up in the water but they don't."

Toland doubted so many of the rubber belts leaked, and said he was concerned the full-chest kapok life jackets were "too bulky" and would "get in the way when passing ammo."

Schultz, captain of the number four gun crew, disagreed about the life jackets interfering with ammo passing. "When the Japs are coming at us," he said, "nothing slows down my guys." He did grant that men who worked in some jobs—like in the engine and fire rooms—wouldn't be able to wear the jackets. But he added that "they can keep them close by" and have them when needed.

Toland proposed a test. They would select a number of rubber belts at random, inflate them, and check for leaks.

"Let's do it right now," Schultz said.

Nearly half of the twenty-five belts they tested had leaks. Schultz was soon filling out his first requisition—promptly approved by Consolvo—for more than 200 kapok life jackets, along with whistles and small one-cell battery-powered lights to be affixed to each one.*

Hull arrived in Seattle on August 25, 1944, and moored portside to the destroyer *Wilson* (DD-408) in berth number one at Todd Pacific Shipyard. A storage barge moved into place alongside *Hull*'s starboard side, and the ship began receiving electricity, water, steam, and telephone service from the dock. The extensive overhaul, which would cost $10 million, commenced immediately. In the eyes of Schultz, who knew the ship as well as anyone, *Hull* was "pretty well beat up and run down." They had even "wore out the guns," and all four of the 5-inch deck guns were scheduled to be replaced in the yard.

Over the next three days, more than a hundred officers and enlisted men departed the ship to begin twenty days' leave. Most would be heading home by bus or train to visit parents, grandparents, wives, girlfriends, and hometown friends. For others, such as Greil Gerstley, promoted to full lieutenant the previous month, his entire family would be coming from Philadelphia for his San Francisco wedding—an event he and his fiancée, Eleanore Hyman, had planned through letters over the past eight months since having to part soon after their not-so-blind first date. Gerstley invited several *Hull* officers to the wedding, and also asked Signalman 2nd Class Robert Coyne, whom he had come to know during long watches together on the bridge. Coyne, who had been on *Hull* since before Pearl Harbor and would be transferred off in a month's time to attend a Navy school, explained that he would rather not attend

---

* Kapok is a light, resilient, very buoyant fiber from a tropical tree cultivated in Asia, the Philippines, South America, and other humid climes. The kapok is the official national tree of Puerto Rico. Although previously used in life jackets and similar flotation devices, the fiber has been largely replaced today by man-made materials.

because he would be "the only enlisted man there with all that brass." Gerstley, nicknamed "Gabby" by the crew and well-liked by officers and enlisted men alike, said disappointedly, "I understand."

As half the crew left, hordes of civilian workers—welders, electricians, pipe fitters, and a host of other trades—crowded aboard with boxes of tools, welding torches, metal saws, and drills. Tangles of rubber hoses and electrical cords soon stretched across the decks and passageways.

Not far from where *Hull* was moored, Tom Stealey, the civilian worker who had been at Pearl Harbor awaiting transport to Wake Island at the time of the Japanese attack and ended up swimming out to the burning destroyer *Shaw* to help extinguish the raging fire, was searching the waterfront for his own ship. Stealey was finally in the Navy, although it had been a long and circuitous route. He had attempted to enlist in Honolulu within days of the attack but was told he would have to find his way back to the States in order to do so. He was still trying to get back in February when he hired on as a gun guard manning a .50-caliber machine gun on a luxury liner (soon to be converted to a troop ship) heading for San Francisco. He had "never fired a machine gun in my life," but of course he didn't tell anyone. Once home, he went to his draft board and again tried to enlist in the Navy but was told the quota was filled, "without enough training barracks built for those we've got." When they found out he was a sheet-metal worker, he was advised to get a defense job. He and his fiancée, Ida May Bryant, soon married, and a year later they had a baby boy. Then in June 1944, when Stealey "didn't want to leave" his young family, his defense-industry deferment was cancelled and he was told to report to his draft board. After the group physical examination, he did not step forward when they asked for volunteers for the Marines, but when a Navy guy asked for volunteers Stealey "stepped right out." After boot camp in Idaho, he was assigned to a destroyer—by luck or happenstance, the ship type he had requested. He was soon on his way to meet the ship in Seattle.

By now Fireman 2nd Class Stealey had several times sought directions to find the ship. He had taken the ferry across Puget Sound to

Bremerton only to be directed back to the Seattle side. Carrying his loaded seabag balanced on one shoulder, he spotted a destroyer moored at a long pier. *Oh, what a beauty,* he thought. Looking at the hull number, however, he saw this was not his ship. When he reached the next ship he was pleased to see another "beautiful destroyer"—but also not his ship. Third in line was "an old, beat-up, battle-scarred, dirty thing." Checking his orders, he confirmed the hull number: 350. Sure enough, this "messed-up ship," the destroyer *Hull,* was his new home.

When Stealey crossed the gangplank, he found the ship dotted with alternating patches of old blue-gray paint, rust, sanded spots, fresh green primer, and splotches of newly applied multicolored camouflage paint. The place was so cluttered with lines and hoses that he walked along certain the soles of his shoes "never touched the deck." The noises, filth, and acrid smells were overpowering. It took Stealey some time to find anyone official with whom to check in. He was then dispatched ashore to a nearby barracks, where crew members were being housed during the yard overhaul.

The next day, Stealy reported back to the ship and went to work on rusty decks and railings with a metal scraper and wire brush. Before long, he was asked by a petty officer what kind of work he'd done before. He proudly answered, "Sheet metal," thinking there might be similar work aboard ship.

"You're an oil tender now," the petty officer said.

"What will I be doing?"

"Feeding the boilers down in the fire room."

Stealey didn't like the sound of that. "Is it hot down there?"

"Yeah, but you'll survive."

Stealey could not possibly know all that would entail.

Eight

After fighting "the nearest thing to a perfect naval battle produced by World War II" in the "utterly one-sided" Battle of Cape St. George in November 1943—in which three Japanese vessels were destroyed and upward of 1,500 enemy died, with no U.S. losses—*Spence* and the Little Beavers under Captain Arleigh "31-Knot" Burke had stayed exceedingly busy.

To Torpedoman 3rd Class Albert "Al" Rosley, nineteen, of Frostburg, Maryland, who had worked in a sawmill before joining the Navy six days before his eighteenth birthday in November 1942 and who had been aboard *Spence* since her commissioning, it seemed as if "we were into something every night."

The turn of the calendar year to 1944 marked *Spence*'s first anniversary. In the year since her commissioning, the destroyer had steamed 67,050 miles. Although still a "young ship and a young crew," they were now wily veterans of the Pacific campaign, with battle stars to show for it. "We have proved our worth more than once and will do it again when necessary," wrote Chief Yeoman Harold L. Bryant in the ship's monthly newsletter, *Ye Olde Dis-Spence-er,* which included features such as "Man of the Week" and "Captain's Corner." "We have a name now, let's enlarge it until it fairly rings on everyone's lips."

For the Little Beavers, the new year began with escort duties and antisubmarine patrols, followed by operations in the Bismarck Sea, ringed with formidable enemy bases such as Rabaul and Kavieng. On February 18 the destroyer squadron undertook the first bombardment of the Kavieng airfield on the northern tip of New Ireland, "plastering" a supply dump and a fuel dump and hitting aircraft and runways. Enemy shore batteries opened up, and salvos landed "uncomfortably close on both sides and astern" of *Spence.* "By radical maneuvering the ship escaped damage," reported *Spence* commanding officer Henry Armstrong, "though she used up a few of her lives."

Four days later, the Little Beavers came upon *Nagaura,* a Japanese merchant vessel of 5,000 tons "engaged in evacuating aviation personnel" from the Bismarcks. As they approached bow on bow—making visual identification difficult—the enemy captain evidently assumed the destroyers were friendly because Allied shipping had "not dared penetrate these waters." Burke fanned out his five destroyers in echelon formation and closed before hoisting the international signal flag demanding surrender. The merchant ship's reply was the chattering of deck-mounted machine guns. On Burke's command, the destroyers turned "their full broadsides simultaneously" on the ship. *Spence* shot eight booming salvos from her 5-inch guns in a minute, then ceased firing as the "target listed to starboard smoking heavily from many hits." Four minutes later, the ship sank.

Assigning two ships to screen against submarines, Burke directed the other destroyers to search for survivors, many of whom swam away rather than be picked up by U.S. ships. Nevertheless, seventy-three survivors were boarded. Burke then gave a curious command for warships operating in enemy waters: in an "act of respect the rescued Japanese appreciated," the U.S. sailors held a one-minute prayer service for the "gallant enemy captain" who had "opened ineffective fire against overwhelming odds."

The day's fighting was not over. Burke's destroyers moved in for another strike on Kavieng, again finding the return fire withering. To avoid the incoming shells, *Spence* started "salvo chasing"—steering for where the last splash hit the water on the theory that the next salvo would land elsewhere—and "thanks to judicious fishtailing and changes of speed" was able to avoid several "straddling salvos." Without continuous efforts to "keep clear," reported Armstrong, *Spence* would have "quite possibly been hit repeatedly." At dusk, the marauding destroyers overtook and sent to the bottom a minelayer, and after nightfall they sank a small freighter and several barges—all filled with Japanese reinforcements and supplies.

By the end of their Bismarcks operations, the Little Beavers had added to their "remarkable record" as a group of destroyers that often worked independently. By March 1944, however, opportunities for such small-scale naval operations by destroyers and cruisers would be limited, with the war spearheaded by the mighty force of U.S. aircraft carriers as it "conquered its way across the Pacific." Through no choice of his own, Burke would be part of it—unexpectedly transferred (even though he was not an aviator) from destroyers to chief of staff for one of the Navy's top carrier commanders, Admiral Marc A. Mitscher, in charge of a fast-attack carrier division that was to take part in nearly every major battle remaining to be fought in the Pacific. Upon first learning the news, Burke was "devastated" and said angrily: "Somebody's trying to railroad me out of these lovely destroyers." But he had no choice in the matter, and with the transfer would soon come a promotion to rear admiral. On March 27 Burke climbed into a high-line chair to be transferred at sea to the carrier *Lexington*. Trying to "conceal the emotion that his moist eyes betrayed," Burke told those around him he would "always keep track" of the squadron. "Tell the boys if any of them ever is in Washington where I live to look me up. They'll be welcome. Goodbye now—and for God's sake don't drop me in the drink."

In April 1944, *Spence* conducted antisubmarine screening for aircraft carriers as they struck targets on New Guinea in support of landings at several locations. At the end of the month, the carriers struck enemy shipping and installations at Truk, the Japanese bastion in the Caroline Islands. *Spence* then returned to the naval base at Majuro for a period of routine upkeep. The first week of June found *Spence* again in action with a carrier group for the assault on the Marianas. As aircraft struck various islands, *Spence* and other ships bombarded enemy positions on Saipan and Guam in preparation for landings. At the "Marianas Turkey Shoot" in June, *Spence* was a plane guard for carriers during flight operations, rescuing several pilots down at sea.

Lieutenant (j.g.) Alphonso S. "Al" Krauchunas, twenty-four, of Kalamazoo, Michigan—he had been born in a Wisconsin farmhouse to Lithuanian parents—came aboard *Spence* in April 1944. A supply and disbursement officer, he was in charge of "S Division," which had become the "pride and joy of the ship's crew because of the excellent food provided them daily by a dedicated group of cooks and bakers," and he was also the ship's paymaster. Those duties made Krauchunas popular; in fact, his nickname among other officers soon became "Pay." Beyond his dispensing food and money, all hands learned that the husky officer—five foot ten, 200 pounds—was a stalwart shipmate.

A graduate of Western Michigan College (later Western Michigan University) in Kalamazoo, where he starred as a hard-hitting, smooth-fielding shortstop, Krauchunas was drafted by the Chicago White Sox and played second base for their farm team (batting .284) until "the war interrupted his dream to play professional baseball." He enlisted in the Navy in February 1942 and went to boot camp as an enlisted man before receiving an ensign commission in the supply corps. His "athletic training"—a physical education major, he was "a strong swimmer" and competed in basketball, too—played a role during the

Marianas campaign when Krauchunas twice "dove off the ship and swam 75 to 80 yards to assist floundering Navy pilots." The second time he did so he "narrowly missed being attacked by a shark."

At 8:00 A.M. on July 8, while moored at Eniwetok Atoll in the Marshall Islands, *Spence* "executed colors." Following the raising of Old Glory at the fantail, the crew was called to general quarters for drills. They went through their paces under the watchful gaze of a special guest who had been aboard for two days to learn the ship and observe the crew: the next skipper. At 9:15 A.M., the crew was "mustered at quarters and marched to the forecastle," where "pursuant to Bureau of Personnel dispatch orders" Commander Henry J. Armstrong was relieved of command. The "impressive ceremony" was short—less than fifteen minutes. After a brief speech to the crew praising their accomplishments, Armstrong, who had shown an inclination for being "taut in all things pertaining to duty," stepped back. The new commanding officer stepped forward and read aloud his official orders to command the vessel. Afterward, he made an about-face, saluted Armstrong, and said, "I relieve you, sir." Armstrong returned the salute, and with that the deed was done. Armstrong, forty-one, was widely respected by the crew, who considered him a "very good captain," "all business," and "not afraid of anything." After a round of goodbyes and a final meal with officers in the wardroom, Armstrong headed down *Spence*'s gangway for the last time. "I trained the crew to go to war," the seasoned skipper said before departing, "and we went to war." Gone was the man who had turned a new ship and her neophyte crew into decorated combat veterans, keeping them alive and well in the process.

Spence's future now rested with Lieutenant Commander James P. Andrea, thirty-one, of West New York, New Jersey, for whom taking command of his own ship represented the pinnacle of his naval career. Five foot nine, 160 pounds, with thick dark hair, brown eyes, and a ruddy complexion, Andrea was one of eight children born to Teresa (Favoino) and Michele Andrea, a cement mason; both came from the same region on the east coast of Italy, although they had not met until migrating to America. Growing up on the Jersey side of the Hudson River, Jim

Andrea "wasn't a natural student but he studied hard and got good grades" and was his class salutatorian in high school, where he also "distinguished himself as a basketball player." After graduating at age sixteen and unable to afford college, Andrea went to work for Macy's department store for two years until receiving an appointment to the Naval Academy. Congenial and with a "keen but gentle sense of humor," Andrea made friends easily at Annapolis. "An actor on the side," he spoke several languages, loved Italian opera, and was known to sing impromptu arias. He was playfully dubbed "the Dorothy Dix of the Academy"—after the popular author of the syndicated "Dear Dorothy" advice column—due to his willingness to "listen to everyone's story." At Annapolis, Andrea did not rise to the top academically, finishing near the bottom of the class of 1937, ranked 281st out of 323 graduating midshipmen. Andrea also "almost didn't graduate" because he "couldn't swim very well." In his last Christmas home while at Annapolis, he told family members he was considering "going into the Marine Corps" after graduation, perhaps in the hope of staying on solid ground. In spite of his difficulties at the Academy, "the Navigator," as he was called by his fellow Middies for his tireless efforts to learn navigation, was described heartily in the *Lucky Bag 1937,* the annual of the regiment of midshipmen:

> *Jimmie has the enviable faculty of being able to believe that everything happens for the best. A ready smile and an eternal song are the outward manifestations of his contented nature. His congeniality and his ability to provide entertainment under the most depressing circumstances make him a welcome guest in any circle. Dancing is not the least of Jimmie's abilities. Dim lights and rhythmic music never fail to allure him. Whatever the future may hold for Jimmie, we can be sure that his present course will lead him to the fullest enjoyment of life.*

Following his graduation from the Naval Academy, Andrea was assigned to the battleship *Pennsylvania* (BB-38). For three years he served as assistant engineering, division, and gun turret officer on the dreadnought, which took part in tactical exercises, battle practices, and fleet maneuvers in the Pacific and Caribbean. In early 1940, Andrea was transferred to the

hydrographic survey ship *Sumner* (AG-32), then charting areas in the Caribbean and along the west coast of South America. Crew members on *Sumner*, where Andrea soon took over as gunnery officer after being promoted
to lieutenant (j.g.), found him to be "all Academy" but also thought "you
couldn't find a nicer guy . . . also shipmate all around." In fact, Chief
Gunner's Mate John O. Hill "named one of his sons" after the friendly
young officer. In receipt of secret orders to Surabaya, Indonesia, *Sumner* was
en route in early December 1941 when she put in briefly at Pearl Harbor.
When the first bombs fell during the Japanese attack, Andrea was dressing
in his small cabin to attend morning mass. At the sound of roaring air-
planes and a loud explosion, he raced up on deck to direct the gun
crews. One of *Sumner's* .50-caliber machine guns soon scored "a direct
hit" on an enemy torpedo plane that was "making an approach" on
Battleship Row. The aircraft "disintegrated in flames and sank in fragments" and its torpedo "sunk without exploding." In rage and frustra-
tion that morning, Andrea was "seen firing his Colt .45" sidearm at an
enemy plane as it flashed by. *Sumner* spent the next year and a half survey-
ing new anchorages throughout the Pacific at places such as Samoa, New
Caledonia, Tonga, the New Hebrides, the Solomon Islands, and New
Guinea. In July 1943, Andrea was detached from *Sumner* and ordered to
new construction/destroyers—a highly prized assignment among junior
officers. He was to put into commission and serve as executive officer of
the *Fletcher*-class destroyer *Mertz* (DD-691), a new "Bath boat" about to be
launched. After a shakedown cruise off Bermuda, *Mertz* headed for Pearl
Harbor. The destroyer began convoy escort duties in March 1944 and
within weeks saw her first action: attacking in a hail of 5-inch shells and
sinking an enemy merchant ship. In May, *Mertz* returned to Pearl Harbor
to prepare for the Marianas campaign, which Andrea was to miss. On
May 25, after six months as *Mertz's* second in command, he was ordered to
Spence after attending a short technical school for prospective commanding officers.

The personable style of the new *Spence* skipper, who was "much
younger" than his predecessor and seemed to the crew "not much older
than some of us," soon became evident. Andrea let it be known that he

had a motto: "An efficient ship can be a happy ship." His actions showed that he also believed a happy ship could be an efficient ship. Among those appreciating Andrea's style of command was supply officer Al Krauchunas, who noticed that when Andrea went to the wardroom it was often for "a friendly visit" and not just "to talk shop." The new captain's impromptu visits were not restricted to officer country. In fact, Krauchunas judged their new commander to be "an enlisted man's skipper and a grand fellow." No one aboard *Spence* had any way of knowing that Andrea had in his career been warned by higher-ranking officers against becoming "too close to the crew." But that was what Torpedoman Al Rosley liked about the new skipper, finding him to be an "ordinary fellow who didn't try to rise above you." The ship newsletter was soon raving: "It would only be a waste of space to tell you that we have both the best ship and the best skipper. Who could ask for more? The height of our regard and affection is 4.0 unanimously."

Ordered stateside for a major shipyard overhaul, *Spence* departed Eniwetok in early August. Following an overnight stop at Pearl Harbor to fuel, the destroyer sailed for San Francisco, arriving August 18. After passing beneath the Golden Gate and Oakland Bay bridges, *Spence* eased into a berth at Hunter's Point Shipyard. A week later, the destroyer was nudged by a yard tug into a nearby dry dock. Once the underwater caisson closed, pumping commenced. Within two hours, *Spence* was high and dry atop keel blocks.

As the well-traveled ship underwent her first complete overhaul, a steady stream of men came and went with orders in hand and seabags on shoulders. In all, nearly 100 crew members transferred off in August and September—including half of the ship's complement of twenty officers. Departing for schools and other assignments, many of these veterans would use their experience to help commission new ships and train their young crews.

Gone were old hands such as Yeoman Al Bunin, already with more than two years of sea duty when he became a *Spence* plank owner and one of "5 percent" of the crew who put the ship into commission—now with enough sea time to request shore duty. Bunin, who "never wanted to be a hero," transferred to a naval communications office in California, where he was to safely spend the remainder of the war.

Replacing the veterans were newcomers such as Seaman 1st Class Ramon Zasadil, eighteen, of Cicero, Illinois. Fresh out of radio school, to which he had been sent after boot camp at Great Lakes in Illinois, Zasadil, who had been told "they needed radiomen in a hurry" in the fleet, had been surprised to be assigned to a warship sitting out of the water. For Zasadil, who had joined the Navy seven months earlier—two weeks before his eighteenth birthday—it would be a couple more months before he would have a chance to find his sea legs.

Those crewmen who were to remain aboard *Spence*—all had been told they would be headed back to the Pacific when the overhaul was completed—were given twenty-day leaves. They left in droves to try to put the war behind them for a time. One who traveled home was Machinist's Mate Robert Strand, the Pennsylvanian who had mailed to his parents from Purvis Bay the menu for *Spence*'s belated Thanksgiving after the Battle of Cape St. George.

Strand, now twenty-three, had a serious girlfriend at home. He and Jane Michel, who was two years his junior, had dated since 1940, the year she graduated from high school. After Bob went in the Navy in 1942, Jane, a slim brunette, had considered joining the WAVES. He discouraged her, reasoning that their military leaves would never coincide. Instead, she had gone to work in an insurance office and waited for him. Now, with Bob home on leave, many believed he would propose to Jane—but he did not. A family friend who lived next door to the Strands later told Bob's younger brother, Richard, that when he asked Bob why he and Jane were not getting married during his leave, "Bob got a funny look in his eyes and said, 'I'm not coming back.'" He expressed a foreboding that "all of the luck" *Spence* had in the war was "going to catch up" with them, and he did not want to leave Jane a widow.

Strand wrote his parents, Josephine and Alvin—a U.S. Army ammunition driver and veteran of the Battle of Argonne Forest in World War I—upon his return to *Spence*:

Guess I wasn't such good company while I was home but I had missed so darn much for the last year and a half that I had a lot of catching up to do. I really had the best 2 weeks that I've ever lived and that is no fooling. Jane was really swell and to me she is the one and only. Would have really liked to have been married but for the fact that I had to go back to the Pacific zone and that was the big reason. Should I get a decent break, I expect you will be a daddy-in-law and a mother-in-law. Of course, I have to get a break first.

With *Spence* perched on keel blocks through September, Strand, who still hoped to one day own his own bowling alley, rediscovered his enjoyment of the sport. After visiting a bowling alley in nearby Redwood City, he was taken home for dinner by the owner and his wife, who had a son away in the military. Soon, in recognition of his competitive game, Strand was being sponsored in a tournament against "all the big name bowlers from Frisco, Oakland and Los Angeles." Unfortunately, while working on shipboard machinery the day before, he injured his bowling hand, and played with a "very, very sore thumb." He still rolled a strong score of 232 in the final game and finished with a 197 average in front of a cheering crowd, he proudly wrote his parents, "pulling for the sailor against all the big-time bowlers." Thereafter, while many shipmates went drinking nightly—"it is no wonder they are broke" all the time—Strand regularly hopped a bus to Redwood City and bowled.

Flooding of the dry dock commenced on September 14 and in two hours *Spence* was again "waterborne." When the dock gates swung open, the destroyer was towed out by a tug and moored to a pier where the remaining work would be completed over the course of the next two weeks. Although a normal destroyer overhaul lasted about ninety days in peacetime, *Spence*'s overhaul took only half that long—a common occurrence during wartime.

In the final days before their departure, the young skipper gave a tour of his new ship to his wife, the former Jean Barton, with whom he had fallen "in love at first sight" while still at the Naval Academy, and their two-year-old daughter, Judith, the first of what her father hoped would be "lots of kids." The couple had met when she accompanied a girlfriend to a social function at Annapolis. In June 1939—two years after his graduation—they had returned to the Academy to be married in the chapel under tall stained-glass windows and a great dome rising higher than the Maryland state capitol. It had been a traditional military wedding in a beautiful setting, and the newlyweds exited the chapel under the raised swords of some of Andrea's former classmates who, coming back for the affair, wore dress white uniforms with matching gloves and oxfords. A reception followed at the Alumni House, with drinks and canapés.

Before sunrise on Saturday, September 30, fires were lit under *Spence*'s number three boiler. Two and a half hours later the ship was under way from the dock, "backing into the stream" at first, then "standing down the channel on various courses and speeds" heading for the ammunition depot at Mare Island. From 11:00 A.M. to 5:30 P.M. ammo was loaded aboard by a working party consisting of most of the crew not on watch. The ship then returned to Hunter's Point.

All hands knew that with live ammunition aboard, *Spence* would be leaving the shipyard shortly. Before they did, it was time for a bon voyage party. The supply division had accumulated extra goodies—such as cigarettes and candies—which were freely passed out that night in a large banquet room at the fashionable, seven-story Oakland Hotel across the bay. Tuxedoed bartenders were kept busy all evening—"everyone was half bombed"—and a local torch singer and band performed favorites, keeping the dance floor filled with married couples and singles with girlfriends and dates; officers, chiefs, and sailors alike. Clearly enjoying the evening was a beaming Jim Andrea, and the new commanding officer warmly greeted all who approached him, regardless of rank. He often took his wife's hand and adjourned to the dance floor—"such a romantic"—where they swung to popular numbers such as "I'll Buy That Dream" and "Sentimental Journey."

After a run to the Farallon Islands—27 miles outside the Golden Gate—to test-fire weapons, and several other mornings and afternoons spent maneuvering in the San Francisco Bay calibrating the magnetic compass and testing new and refurbished equipment, *Spence* was ready.

At 8:00 A.M. on October 5, the crew mustered with only a single absentee: a seaman 2nd class who would be reported as AWOL. *Spence* was under way at 9:10 A.M. for Pearl Harbor in the company of three other destroyers that had also concluded shipyard overhauls. Arriving five days later, *Spence* took on fuel and supplies, and replenished the ammunition that had been expended in gunnery drills during the crossing. For several days they conducted exercises in Hawaiian waters, during which they practiced antisubmarine warfare with a U.S. submarine.

On October 26 *Spence* and six other ships—three of them the newly commissioned escort carriers *Makin Island* (CVE-93), *Lunga Point* (CVE-94), and *Bismarck Sea* (CVE-95), each carrying approximately thirty aircraft—left Pearl Harbor bound for a fueling stop at Eniwetok in the Marshall Islands. Things were moving quickly now, and there seemed an urgency behind fleet orders and ship movements. After a short layover at Eniwetok, *Spence* and the other ships pushed on westerly.

On the second day out, *Spence* conducted a surprise abandon-ship drill. After the general quarters horn blared *oooga-oooga-oooga* came the announcement over the loudspeaker: "This is a drill, this is a drill. All hands prepare to abandon ship." It was the one order that no seaman ever wanted to hear for real, but it had to be practiced, like everything else aboard ship. It took twenty minutes for the crew to be in their proper places wearing their life jackets, and with large life rafts ready to be cast off—at which point they were "secured from drill."

On November 5, *Spence* arrived at Ulithi, a large atoll in the Caroline Islands occupied by U.S. forces with no opposition only two months earlier. *Spence* had no sooner anchored at a depth of 130 feet "over a sand and coral bottom" some 100 yards from shore when a priority message was received indicating that all ships in the lagoon were in "Condition of Readiness Typhoon II," requiring them to stay in a "high degree of readiness" to encounter a typhoon that might hit in the next twenty-four to forty-eight hours. The alert lasted three days, until a lower condition of readiness was set, which meant that a storm was "no longer imminent" but still could materialize.

Aboard *Spence* and other destroyers in the squadron, the weather alerts "impressed upon" officers and enlisted crew alike that they would be operating in a region where they had to be prepared to deal with typhoons. Commanding officers and junior watch officers alike broke out copies of Austin M. Knight's *Modern Seamanship* and Nathaniel Bowditch's *American Practical Navigator*—both classic naval titles were carried on the bridge of every U.S. Navy ship—"to reacquaint themselves" with cyclonic storms. The former title was of little help, as it contained only a single-page reference to typhoons, and stated: "At the present time the available evidence concerning the formation of tropical cyclones—they are called typhoons in the Far East—is incomplete and inconclusive." Bowditch, however, had a twenty-two-page chapter titled "Cyclonic Storms," which contained information "of great value to the uninitiated destroyer skipper so far as typhoons were concerned." Sections in Bowditch included "Fixing the Bearing of the Storm Center" and "Handling the Vessel Within the Storm Area." Of special interest were the "thumb rules regarding local indicators of a typhoon's approach," including increased and shifting winds, and falling pressure readings on the barometer. Fully developed typhoons were described as covering an area 300 miles in diameter with a calm center—the eye of the storm—up to 20 miles wide. For mariners, the "dangerous semicircle" of the storm—the right-hand side in the northern hemisphere—was the worst place to be, according to Bowditch, as the "seas within this area are violent and confused, sweeping in from all sides with overwhelming violence." The left-hand side or southernmost side was considered the "navigable semicircle."

While most crew members of *Spence* and the other destroyers at Ulithi had not been through a typhoon and had "no conception of the overwhelming destructiveness of such a storm," their officers hoped that should they have the "bad luck to encounter one" they would at least be able to recognize its approach and take appropriate measures to avoid the worst of it. While all agreed there was value in discussing typhoons, they reasoned that "only if we were operating independently did it seemed necessary to concern ourselves with avoiding bad weather." Otherwise, they believed, the fleet would provide the eyes, ears, expertise, and leadership needed to avoid a typhoon, which in the lower to middle latitudes traveled at relatively slow speeds of from 5 to 17 miles per hour along its path.

On November 10, *Spence* in company with other destroyers and several aircraft carriers headed farther westward across the Pacific. One evening during this leg of the journey, Water Tender 3rd Class Charles Wohlleb, twenty, who had been on *Spence* since May 1943 and had gone through all the exploits of the Little Beavers, finished his watch in the after fire room. Coming topside on a "pitch-black" night, he went to the engineering division's berthing compartment near the fantail—right above the twin turbine-powered propellers that drove the ship—and found fifty or sixty guys off duty, shooting the breeze and playing poker and games of checkers. "All of a sudden, the screws stopped." It was not a noise Wohlleb had often heard at sea, as it meant the ship was dead in the water. Then the ship was thrown into "full reverse" and "the whole boat shuddered." Wohlleb and the others "looked at each other," jumped up, and rushed topside to see "what the hell was going on." Scaling the ladder to the deck and emerging through a hatch, Wohlleb was alarmed to see a "huge shadow" looming close in front of *Spence*. As his eyes adjusted to the darkness, he was shocked to see an aircraft carrier—darkened per wartime regulations and not showing a single light—crossing silently in front of their bow.

The skipper had been at the conn, Wohlleb found out, and "almost hit" the carrier *Wasp* (CV-18). Andrea had ordered an emergency stop and reversed the engines to avoid ramming *Wasp,* and barely succeeded in doing so. Wohlleb had never seen anything like it, not in the eighteen months he had been aboard and all the tight spots *Spence* had been through. Their old skipper had always had a deft hand at the conn no matter the situation. Now Wohlleb "got to thinking" about Andrea, who hailed from the same Jersey town (West New York) as he did, although they had not known each other. It dawned on Wohlleb that the new skipper, whom he thought was a "nice, regular guy," had demonstrated he "just didn't have the experience." Added to this was the fact that they also had a new executive officer; two weeks earlier, shortly before leaving Pearl Harbor, Lieutenant Frank V. Andrews had taken over as second in command and navigator. That meant *Spence*'s two highest-ranking officers were new at their jobs, along with a majority of the other officers and crew.

Wohlleb and some of the old hands in the engineering division talked it over. Their young commanding officer had made his "first mistake," and it had been a close call. No matter how pleasant the new skipper, they begrudgingly agreed they would, if they could, trade him for Henry "Heinie" Armstrong—that "tough bastard" who had "gotten us through a lot" and "never did anything stupid" such as nearly hitting another ship. What in the world would happen, the sailors wondered, if things got really rough?

Nine

Early on September 27, 1944, fires were lit under *Monaghan*'s boilers. By noontime, all preparations had been made for getting under way
from Puget Sound Navy Yard to conduct engineering trials following a six-week overhaul.

In the Strait of Juan de Fuca that afternoon, engines were brought up to full
power—and in short order a "gasket blew in
boiler #2." The boiler was secured and other
tests continued, including the firing of new 20
mm and 40 mm antiaircraft weapons. By evening, *Monaghan* was dockside again, and the
shipyard was presented with a list of final fixes
to be made.

Monaghan ran another full-power test the
next day to determine any new handling

characteristics after the overhaul, such as how much the ship heeled over during turns at certain speeds and how long it took to recover from turns of varying degrees. A sister *Farragut*-class destroyer, *Dewey* (DD-349), also wrapping up an overhaul, was in the seaway at the same time undergoing similar trials. Among experienced personnel on the bridges of both ships there was the recognition that "something was wrong." Long considered top-heavy, the prewar destroyers had always been prone to steep, slow rolls even in relatively calm seas. The officers and enlisted crew who had served on *Farraguts* were "very aware of their lack of stability," and it was a "matter of constant concern." However, the ships now seemed even "more sluggish than when they first entered the shipyard" a month earlier. In turns as moderate as 10 degrees, they "lurched awkwardly and heeled over about fifteen degrees." Failing to "snap back" normally, they hung precariously to one side for "a long time" before slowly righting themselves. It made for serious questions about their seaworthiness in strong winds and high seas.

Dewey's skipper of only four days, Lieutenant Commander Charles R. Calhoun, who had previously commanded the fast minesweeper *Lamberton* (DMS-2), a converted destroyer, was so alarmed at the instability of his new ship that he decided against executing "radical turns at speeds in excess of twenty knots" because doing so seemed "imprudent." His urgent request—upon returning to the shipyard—to have the vessel's "serious stability problem" more extensively tested was "beyond the scope of local authority," and it was passed along to Washington. In response, the Navy's Bureau of Ships conceded that the *Farraguts'* "stability might have undergone some reduction" due to the added weight of new equipment installed topside, but claimed they were still "basically stable." It was also pointed out that "in light of pressures from the operating forces to deploy all available destroyers to the western Pacific as a matter of urgency," no further shipyard delays for these vessels were "feasible." Calhoun and the officers of other *Farraguts* had little choice but to accept the opinion of the Bureau of Ships that the overhauled ships were stable, and "went about the business of getting ready for sea."*

Since the launch of the *Farraguts* in 1934–35, many newer destroyers had been built—more than 250 during the war. These newer classes of destroyers were equipped with better armament and communications and more modern engineering systems. Notwithstanding the *Farraguts'* documented stability problems, the remaining seven destroyers of the class were needed in the war effort, as fleets tended never to have enough destroyers to screen and protect larger ships.

In the days preceding *Monaghan's* departure, the usual sense of excitement and anticipation was palpable throughout the ship at the promise of getting under way, no matter what the destination. When *Monaghan* departed on October 1—in company with *Dewey,* escorting the battleship *North Carolina* (BB-55) down the coast to San Pedro, California—the ship "mustered the crew on stations" that morning and found two "absentees," both young seamen who would be reported as AWOL.

From San Pedro, *Monaghan* steamed to Pearl Harbor, arriving October 10. For several weeks in Hawaiian waters, the destroyer trained her crew—many of them right out of recruit training—in gunnery, torpedo-firing runs, and antisubmarine warfare. Everyone knew what to expect next.

Gunner's Mate Joe Guio, who had written home in 1943 announc-

---

* The observations aboard *Farraguts* during power runs in September and October 1944 after major overhauls were supported by the findings of a dockside inclining test conducted at Puget Sound Navy Yard on the *Farragut*-class destroyer *Aylwin* (DD-355). This test, which determines a vessel's center of gravity and calculates how far it can heel over and still safely recover, was done a few days before *Monaghan* and *Dewey* left the shipyard. The results—not made widely known until later—found that *Aylwin's* stability "had been substantially reduced during the war years" by the added weight of new equipment, most of which had been installed on the other *Farraguts* as well. The test, most often done on only one ship in a class but used to determine the characteristics of all the other ships in the same class, confirmed that the top-heavy design for prewar destroyers had been made worse by continuous additions and modifications. That said, the *Farraguts* were certified by the Bureau of Ships—based on "theoretical computations"—to be able to recover from rolls of 70 degrees, although those who had taken them to sea thought the "idea seemed preposterous."

ing that the president of the United States was sending them to San Francisco for Christmas, now wrote a different message to his folks in West Virginia.

The Navy will get together & pay the Japs a visit in the Philippine Islands. Don't worry about me. I'll be alright. The Japs don't worry me a bit.

I didn't get to stay in the states very long. Hell, I didn't like Seattle anyway but at least that's a safe place to be, isn't it? I hope it isn't too long before this War ends. It will last a couple of years at the most, then I come home to stay for keeps. I guess you people know what is happening out here in the Pacific by what you read in the newspapers. I don't have anything more to say so I'll drop my anchor until I hear from you. Love & kisses to all.

Toward the end of October, a kind of "miracle" happened to Water Tender Joseph Candelaria, who had been aboard *Monaghan* since shortly after the attack on Pearl Harbor, and who had figured during the Battle of the Komandorski Islands in the Aleutians that he and his shipmates were all goners.

Candelaria was one of three senior water tenders eligible for transfer to three-month boiler school in Philadelphia, which would include leave to go home—but only one man could be sent. The other two candidates, Joe "Mother" McCrane and Water Tender 2nd Class William D. Weaver, were both married. Candelaria was single, and at first he said one of the other two should go. Besides, he had spent all his money on liberties in Seattle, and going home broke didn't seem like a good idea. On decision day, however, Candelaria received a letter from his hometown sweetheart, Alvina Holguin, in California. She told him she had joined the WAVES and expected to be leaving shortly for training. Candelaria thought it might be a good idea, after all, to get home. Chief Water Tender Martin Busch stood on the quarterdeck before the three men with a hat containing folded slips of paper; two said "no" and one said "go." After all reached into the hat, Candelaria unfolded the "go" slip. His luck amazed him; he had quit gambling because he "never caught a lucky break."

Just before Candelaria left the ship a couple of days later carrying his seabag and with new orders in hand, two buddies—Boilermaker 1st Class Frank A. Cain and Water Tender 3rd Class Leonard R. Bryant— tried to talk him out of leaving so they could enjoy more liberties in port together. In addition, Candelaria had been offered "over $200" to sell his "go" slip, as each of the married water tenders tried to outbid the other. As much as he could use the money, Candelaria knew he would have to work with the fellow who didn't get to go. Although he would miss his liberty buddies, he decided to punch his lucky ticket off the ship and get back to the woman he hoped one day to marry.

Candelaria said his goodbyes and boarded the whaleboat to be taken to the fleet landing. Casting a last look back at the old destroyer on which he had spent the last thirty-three months, he started his long journey "back to the States."

His shipmates, Candelaria knew, would be heading in the opposite direction: toward the war and whatever fate awaited them across the Pacific.

LIEUTENANT COMMANDER Charles Consolvo was relieved of command on October 2, 1944, with *Hull* still dockside undergoing a major overhaul at Seattle's Todd Pacific Shipyard.

Consolvo had been skipper for ten months, a period of time that placed him in line to rotate to a new assignment. He had performed well in his position, not only in the opinion of his crew but also according to Consolvo's squadron and division commander, who had rated Consolvo in his latest fitness report a 3.9 (out of 4.0) in three categories: present assignment, ability to command, and ship handling. Another comment was "Commander Consolvo is an excellent officer in every respect. He is conscientious, energetic, and sincere." When asked for his preference for his next duty station, Consolvo had written: "No change." Still, the Navy had decided it was time for him to move on, and had a special job in mind for him: Consolvo was headed to Annapolis to teach future officers.

Among those disappointed to see the popular skipper depart was Storekeeper 3rd Class Kenneth L. Drummond, twenty, of Jamesport, Missouri, where all nineteen boys in his twenty-seven-member high school graduating class were in military service. A strapping six-footer with a mop of black curly hair, Drummond joined the Navy six months after graduating from high school in June 1942. Aboard since early 1943, he considered *Hull* "a very good ship" with "high morale." He found the crew to be a "tight-knit group and very close," and agreed with the consensus that *Hull* was a "lucky ship" for having gone through so much without loss of lives. Drummond also liked his own duties: as storekeeper, he always had "plenty of food" close at hand. His battle station was on the bridge as a captain's talker over the ship's sound-powered phone system, and he liked being "aware of all that was going on" during the action. With his bridge duties, Drummond had the opportunity to get to know many of the ship's officers. They were, he thought, uniformly competent, as well as friendly toward enlisted men. The officers had all been trained as underway watch officers by Consolvo, and Drummond witnessed how they had been given plenty of time at the conn under the captain's watchful eye and instructive hand.

Drummond considered Consolvo "the very best"—an excellent ship handler who could "lay a ship against a dock with precision the way some people park a car and never bump the curb." He had been standing near Consolvo when *Hull* was shelling the beach of an enemy-held island and suddenly a rain squall came up and "visibility dropped to almost nothing." Moments later, everyone on the bridge looked up to see the bow of the battleship *New Jersey* "almost upon us and very close to ramming us amidships." Consolvo ordered full speed ahead and hard left rudder, and *Hull* swung up alongside *New Jersey* like a harbor tugboat. Everyone on the bridge "breathed a sigh of relief." *New Jersey*'s captain signaled by light: "How does our bow look?" Consolvo signaled back: "Second link in your anchor chain is rusty, sir."

An incident Drummond thought revealed Consolvo's true character happened as they were heading stateside for overhaul. Drummond was at the fantail with several other enlisted men having coffee and shining their shoes, which were lined up in front of them. When the skipper approached, everyone started to stand, but Consolvo said, "No, sit down." Consolvo seated himself alongside them, picked up a rag and a scuffed shoe, and began polishing. The skipper, who had been a "private in the U.S. Army for a year before receiving his appointment to Annapolis," didn't leave until the enlisted men's shoes were all shined.

Drummond appreciated the way Consolvo "took care of his crew" in ways large and small. During a stopover at Pearl Harbor some months earlier, Consolvo had purchased black baseball caps for all the crew. The caps "kept the sun off your forehead and out of your eyes" and made it easier to wear the sound-powered phone headsets. While the caps "made a lot more sense in the Pacific than the regulation white sailor hats," *Hull*'s crew did get some "strange looks from other ships" when they were spotted in the baseball caps.

In the course of going through the change-of-command procedure, Consolvo took his relief on a tour of *Hull*, introducing him to officers and enlisted men alike and endeavoring to explain the ship's little quirks and his tricks for handling them. Normally, they would have spent time operating at sea, with Consolvo showing how the vessel handled, but there was no opportunity to do so with the yard work still going on. Likewise, they could not hold the usual general drills—battle stations, fire, collision, abandon ship, and such—with all the civilian workers aboard. The new commanding officer would have to do that on his own when the yard finished its work—conducting the first full-power runs after the overhaul, and testing all the new equipment installed and repairs made. Consolvo tried to cover everything else the regulations required, and whatever else he could think of that would be useful.

During the tour, Consolvo introduced his relief to Sonarman Pat Douhan, explaining that the petty officer was "changing his rate" to yeoman because he was doing so much clerical work these days, including typing the ship's daily log. The new commanding officer said abruptly, "I don't go for this change-in-rate business. Too many men want to do it." Consolvo looked at the younger officer as if taking his measure. "Douhan didn't ask for this," said Consolvo, a discernible edge in his voice. "I asked him to change rates."

At 1:40 P.M., the change of command was complete, with Consolvo "detached as commanding officer"—and not long after, Douhan, bemused and bewildered, was typing up the new skipper's first set of general orders to the crew, in which he decreed there would be "no profanity aboard this ship." Knowing that the general obscenity and blasphemy for which sailors were well known would not stop on demand, Douhan wondered how long it would be before someone was written up for cussing, which would turn a few heads.

From where he stood on the foredeck, Storekeeper Drummond was amazed by the new commanding officer's first speech to the crew. The first words out of his mouth were about how they were "going out to win the war" and "make history together." The new skipper, young-looking for a destroyer captain, went on: "We're going out there to fight, and we might die together." Eyeing some of the ship's veterans exchanging furtive glances, Drummond thought *Hull's* good luck might have just run out.

When he took over *Hull*, Lieutenant Commander James A. Marks, twenty-nine, of Washington, D.C., where his father held a high-ranking staff position in the Treasury Department, was among "the most junior [commanding officers] in destroyers," being a member of the Naval Academy Class of 1938. Always a good student, Marks ranked 52nd in his class of 438 graduates. While excelling in academics at Annapolis, Marks, "short and slight" with dark hair and an olive complexion, engaged in many extracurricular activities—including soccer, wrestling, and tennis—but found his calling in music, playing clarinet and becoming leader of the academy's swing band, the NA Ten. While some classmates judged him to be "very serious and very regulation," other peers found an agreeable side to him, as reflected in the *Lucky Bag 1938*:

> *Valedictorian of his class in high school and captain of Cadets, Jim entered the Academy with a high set of standards to which he has never been false. His wide range of talents and his infectious enthusiasm have made him a mainstay of the*

NA-10 as well as a savior of no mean order. In athletics his success has been only moderate—a shortcoming explained perhaps by his frequent attendance at hops. However, it is for his ability to bring a smile to even the most bewildered face at the end of a long drill and his unfailing willingness to let others benefit from his prowess at academics that Jim is most valued as a classmate and a friend. With his willingness to work, his much appreciated ability to get the word, and his warm sense of humor, Jim is certain to meet with success in the Fleet.

While he had obtained a command for which he was rather young, Marks did have experience in destroyers, most recently as executive officer of the newly launched *Allen M. Sumner*–class destroyer *Brush* (DD-745). As soon as *Brush* arrived in Pearl Harbor—after training off both U.S. coasts—ready for action in the Pacific, however, Marks had been detached to command *Hull*. Prior to that, he spent four years (1939–43) as a junior officer aboard *Trippe* (DD-403), a *Benham*-class destroyer assigned to Atlantic convoy duty. Fresh from the Academy, Marks had been assigned to the battleship *Colorado* (BB-45) as an assistant engineering officer. One night in New York harbor, *Colorado* had been in a collision with another ship, which "pierced" the battleship's hull, crushing to death several crewmen in the engineering spaces. It had fallen to Marks to go below and identify the dead sailors, a sad task he "never forgot."

Lieutenant (j.g.) C. Donald Watkins, twenty-two, of Columbus, Ohio, a friendly, soft-spoken 1942 graduate of the Carnegie Institute of Technology who had gone to Midshipman School at Notre Dame and been assigned to *Hull* in September 1943, could see that things under the new commanding officer were going to be different for the officers as well as enlisted men. Watkins understood it could not be easy to command a warship and shoulder all the related responsibilities, and he also realized that a "stellar captain and excellent ship handler" such as Consolvo would be a "hard act to follow." In Marks, however, Watkins came to see a "remote" man who brought many of "his problems" on himself by being demonstratively "asocial." All of *Hull*'s officers were reservists—recent civilians from fields as diverse as business, banking, and law—and from the beginning Marks showed no interest in getting to know any of them. Watkins found it revealing that Marks "did not eat in the wardroom with the other officers" and took his meals alone in his cabin. While such a self-imposed distancing of a ship captain from his junior officers was commonplace on larger vessels such as cruisers and battleships, where regulations and formalities were generally more rigidly enforced, it was rare on destroyers and other smaller ships. On *Hull,* it went in the direction of increasing the new commanding officer's remoteness.

Proving that his first speech was not an aberration, Marks repeatedly made it clear that he was "in a hurry to get out" into combat in the Pacific, while most of the veteran crew were happy in port in Seattle, where many *Hull* wives had taken up temporary residence. To Watkins, Marks acted as if he was "afraid the war would be over" before he could take *Hull* into action. To get *Hull* released sooner from the shipyard, Marks began "closing up all the open work orders whether they were finished or not," also not endearing himself to his veteran crewmen, who after a long combat cruise were happy to be home.

Within three days of assuming command, Marks held his first captain's mast—a nonjudicial disciplinary hearing aboard ship wherein the captain hears the evidence and either dismisses a case or imposes punishment on an offender. It was the first captain's mast held aboard *Hull* in months. Ten days later, Marks held his second captain's mast. In all, more than twenty enlisted sailors were brought up on charges ranging from smoking in a barracks ashore to being late returning to the ship to wearing an "improper uniform with cuffs rolled up and hat on back of head." The improper-uniform charge, more than the others, showed that a new day had dawned on *Hull,* which like most destroyers had enjoyed the relaxed dress code of the dungaree Navy. The initial punishments handed down by Marks included warnings, extra duties, and losses of liberties. The regularity of captain's masts was unlike anything *Hull*'s crew had known before. Marks, showing that he intended to lead with an authoritarian hand, held his third and fourth masts during his first thirty days. He took away from a radioman 2nd class his next five liberties, effectively keeping him from going ashore for two weeks, for "use of foul language," and he sentenced a young seaman—whose offense was not documented in the log—to "20 days confinement on bread and water" with a "full ration every third day." The enlisted men judged their "hard-fisted" new skipper a poor replacement for a guy who "just couldn't be beat." Consolvo was remembered as being "all for his crew and the crew all for him," one veteran crewman wrote to friends, "but our new skipper is just the opposite." Soon Marks was being called "every unlegal name in the dictionary."

At 3:00 P.M. on October 9, the crew was called to quarters for the issuance of life jackets, which took twenty minutes. The new kapok life jackets, complete with whistles and lights—ordered by Boatswain's Mate Ray Schultz—were the subject of much conversation. In the wardroom, officers discussed how long they would keep a man afloat before becoming waterlogged—the consensus was "three or four days"—while among the deck and engineering divisions there was concern about their bulkiness when performing certain tasks. In those cases, sailors were told, they could keep the life jackets "available at GQ stations or on their bunks" in berthing compartments when off duty. When it was suggested that men wearing the life jackets "could sleep in the water" because the high necklines would hold their heads up, most thought the notion of ever having to "stay afloat for that long" seemed "totally unrealistic."

The next day, *Hull* pulled away from the dock for the first time in nearly two months. With Marks at the conn, they headed into the Strait of Juan de Fuca "on various courses and speeds attempting [a] full power run."

Manning the helm was Chief Quartermaster Archie G. DeRyckere, twenty-four, a rangy, six-foot-two, good-natured native of Laurel, Montana, not far from Billings. It took strength and coordination to handle the wheel of a destroyer and keep the bow on an exact compass heading—particularly in rough seas—and DeRyckere, a solid 225-pounder, had the requisite long, powerful arms and wide shoulders. While being transported in January 1941 on the battleship *West Virginia* to

Pearl Harbor, where he would pick up *Hull,* DeRyckere had taken offense at being cursed by another sailor and lit into the guy, "giving him a good fight" until it was broken up by an officer, and only then learning that he had held his own against the heavyweight boxing champion of Battleship Division 3. DeRyckere had learned the sport in a youth boxing club run by a former pro boxer, and after nearly four years in the fleet he could still boast that he had "never been whipped."

Traditionally the master of the quarterdeck—the location behind the bridge where the helm is situated and most navigation is performed—a ship's chief quartermaster is responsible for the maintenance, correction, and preparation of navigational charts and instruments, as well as the training of the ship's lookouts and helmsmen. It was a job for a keen and precise mind, and DeRyckere—his lack of formal education belying his natural intelligence—had what it took in that department, too. A skilled helmsmen and excellent navigator, he had recently been promoted to chief petty officer after taking a test filled with questions taken directly from Bowditch's *American Practical Navigator,* which DeRyckere had to know inside and out to pass.

That day in the "relatively calm water" of the seaway, Marks took *Hull* up to 30 knots, then called for the rudder to be "thrown hard over in one direction," at which point the ship "rolled over at least 50 degrees." Although some 20 degrees less than the maximum roll *Farragut*-class ships were certified by the Bureau of Ships to be able to recover from, it was extremely alarming. At the helm, DeRyckere released the wheel to help bring her back upright, then went with hard opposite rudder to counter the roll. DeRyckere held his breath as the ship "sat back up" very slowly. With years of experience at *Hull's* helm, DeRyckere was familiar with the ship's top-heavy characteristics—once in 1943 when he steered across a cruiser's wake *Hull* had laid over so far that the whaleboat, secured at the edge of the main deck, dipped into the ocean like a ladle into a big bowl of gravy. Even DeRyckere was shocked, however, by the suddenness and steepness of the roll that day in the Strait of Juan de Fuca, and also by the slowness of the ship's recovery. He surmised that the added weight of all the new heavy equipment installed topside during the overhaul was the culprit. *Hull* had received "more alterations during the recent overhaul" than had the other *Farragut*-class vessels, and it was the only one to have had the plate-glass bridge windows replaced with small portholes, a modification that would "add slightly to the topside weight of the ship." It occurred to DeRyckere that the increased top-heaviness should be reported to officials at the shipyard. However, he had been in the Navy long enough to know that it was not his place to do so.

While it was "quite evident" to Marks that *Hull*'s stability was "very poor" compared with the new destroyer on which he had recently served, he deemed it "within satisfactory limits." Upon their return to the shipyard, the new commanding officer made no complaints about his ship's stability or readiness, going forward with preparations to deploy to the Pacific.

One evening shortly before *Hull* was due to depart, a group of her enlisted crew were at their "hang out"—Seattle's Club Maynards— "having the time of our lives" when a shipmate hurried in exclaiming that he had just visited a fortune-teller. He claimed to have been warned of a dire future for *Hull*: foretelling that the ship was "going to be sunk on the 23rd of December or before." The sailors laughed their "heads off," but with the worsening morale on what had once been a happy and proud ship, word rapidly spread among the crew about the terrible prophecy, and "it stuck in our minds all the time."

At 7:00 A.M. on October 17, *Hull* departed Seattle bound for Pearl Harbor. The destroyer left behind an unusually large number of AWOL crewmen, "a sure sign of poor morale." In all, twenty enlisted men "jumped ship," choosing to be reported for missing their ship's movement in wartime, a serious offense. In some cases, they had revealed their plans in advance to shipmates, who "tried to talk them out of it." Their main reason had to do with the prediction of the fortune-teller, which seemed especially believable given the crew's growing lack of confidence in their new commanding officer. The prevailing fear, in fact, was that Marks would do something to "lose the ship." Surprisingly, half of those who missed *Hull*'s departure were rated petty officers: gunner's mate, water tender, fire-control man, torpedoman, steward's mate. Most of the missing men turned themselves in soon afterward, aware that they would be dealt with punitively—busted in rank and sentenced to time in the brig. None of them would ever rejoin *Hull*'s crew.

As soon as *Hull* steamed clear of land, Marks ordered a "simulated depth charge attack." The following day and each day thereafter gunnery practice was held for the 5-inch, 40 mm, and 20 mm gun crews. However, due to the cancellation of work orders, they had left without the newly installed torpedo tubes operating correctly. Don Watkins, *Hull*'s torpedo officer, found they "would not train," which was "a dangerous situation" in the event *Hull* made an enemy contact during the crossing. Furthermore, it would not go over well for a ship right out of a major stateside overhaul to have to be scheduled for repairs at the crowded fleet repair facilities upon arrival at Pearl.

One of the wives left behind in Seattle was Portia (Elam) Kreidler, a vivacious, twenty-two-year-old brunette whose husband, Sonarman 1st Class John Kreidler, twenty-three, had transferred aboard *Hull* a month earlier. The couple, both of whom had graduated from high school in Yakima, Washington, where they had met, had been married less than three months. Prior to reporting to *Hull*, Kreidler had taken a long leave, and the couple enjoyed an idyllic honeymoon, spending time at a "hunting cabin up in the hills," then settling in Navy housing at Port Orchard, not far from the shipyard. Mostly it had been about being together after not seeing each other for two years. Before he had left the first time for the South Pacific, they had discussed getting married, but Kreidler was "not too sure he was going to survive."

The blond, blue-eyed Kreidler, who had enlisted a month after the Pearl Harbor attack, had served on the patrol craft *PC-476* beginning in July 1942 and for the next two years "participated in operations in the forward areas under hazardous conditions," including the nighttime evacuation of twenty-nine civilians—of which a dozen were Catholic nuns—from the shores of enemy-held Bougainville, and another mission "attacking with PT boats" in the Guadalcanal area a "resupply mission of Japanese destroyers."

When they left Seattle, *Hull*'s crew believed they would "not be back until the war was over," although Douhan and others who had been aboard a while remembered thinking the same thing when they left for the Aleutians in early 1943. Nevertheless, the word was passed that it could be a long wait for the wives who had relocated to Seattle to spend time with their husbands, and most of them soon packed up and moved back closer to family.

Portia Kreidler returned to San Francisco, where her parents owned an apartment building. Going back to the same city was Greil Gerstley's wife, Eleanore, who had moved to Seattle after their September wedding. Also heading back to California was Pat Douhan's wife of nine months, Kathleen (Lassley). These young women had something else in common with ten other *Hull* wives: although some of them did not yet know it, all thirteen were in their first trimester of pregnancy.

Arriving at Pearl Harbor on October 23, *Hull* put on a "really embarrassing" show in the harbor of the Pacific Fleet's home port. For Boatswain's Mate Ray Schultz, it was the most humiliating arrival in port that he had ever experienced in all his years in the Navy. It was not such a surprise to Schultz or anyone aboard *Hull* because Marks had already demonstrated a lack of ship-handling ability. With Marks at the conn, Schultz considered "no dock safe," and the destroyer had plenty of nicks and dents in her new camouflage paint job to show for it. But coming into a major fleet anchorage so sloppily after being commanded by one of the smoothest and best-known ship handlers in the destroyer Navy made it sting that much more. Schultz stood at the bow with his deck force, prepared to secure a line to the buoy. Each time Marks tried to "take the buoy and missed," the destroyer had to make a wide circle and line up for another pass. After several failed attempts, Schultz told the deck phone talker to advise the captain they could put the whaleboat in the water and hook the buoy by hand. After the call went to the bridge, Marks shouted down from the bridge: "I don't want any more smart remarks from the peanut gallery!" After a half dozen more misses, however, word came from the bridge to launch the whaleboat. Schultz did, and had the boat motor up to the bow. The hook rope was lowered and secured to the buoy. Schultz signaled the winch operator, and the ship and buoy finally were brought together. *Hull*'s achievement was acknowledged by blaring sirens and whistles from scores of ships in the harbor.

By telephoning a friend on the staff of Commander Destroyers Pacific, Don Watkins was able to arrange for repairs to the torpedo tubes. *Hull* pulled alongside the tender *Yosemite* (AD-19) for several days' worth of work.

Then for several weeks, *Hull* spent days at a time training off Hawaii. During this period, countless crewmen tried to get off *Hull* through transfers to school or elsewhere, sometimes succeeding after "getting into it with the captain," who expressed his own desire to "get rid" of any malcontents from "Consolvo's crew," which in general he viewed as "spoiled." A chief machinist's mate had a loud row with Marks, and he was soon gone, as "everyone else would have been if they had their way." For Douhan, whose own hope for a transfer never materialized— and neither did his change of rate to yeoman, thanks to the new skipper—it was "tough" seeing the morale of a good ship "destroyed by a guy like Marks."

On November 8, Greil Gerstley became the ship's new executive officer and navigator, replacing Lieutenant Maury M. Strauss, who had been aboard since a year before Pearl Harbor and was respected and well liked. Gerstley, now a lieutenant, was well qualified to become second in command, having benefited from Consolvo's training and served as senior watch officer. His fellow officers considered Gerstley a "good ship handler." The "gentlemanly" Ivy Leaguer was known for "never saying a bad thing about anyone," including Marks, who had given him a stern dressing-down one night on the quarterdeck for having renamed a star on the navigational chart Eleanore —"after my wife," Gerstley had explained to the "not amused" skipper, who ordered him to "go back to the star's real name."

By this time, there was "a lot of grumbling" about Marks in the wardroom as well as the crew quarters. The officers' discontent had to do not only with his martinet-like personality but also with his deficiencies at ship handling. In the view of Lloyd Rust, the Texas lawyer who was now *Hull*'s combat information center (CIC) officer, whether a commanding officer won a popularity contest wasn't as important as whether he was "capable of doing the job." To Rust and others, Marks had shown he was "incapable . . . in every way."

For *Hull*'s sonar officer, Lieutenant (j.g.) Edwin B. Brooks Jr., twenty-three, a Virginian and graduate of the University of Richmond with a degree in economics, artful ship handling was a "very big thing." Aboard *Hull* since Christmas 1943, Brooks, five foot ten with dark features and a beaming smile, had enjoyed his time understudying Consolvo as he learned to become an officer of the deck (OOD) when under way—his "main ambition" as a naval officer. Like other *Hull* officers, Brooks' major criticism of Marks was over his inept ship handling. A proper "southern gentleman" to his core, Brooks was not a profane man, yet he was not above "calling Marks a real bastard."

At the conn, Marks' mistakes and failings were so elementary that there was universal disbelief he had spent much time at the conn of destroyers, which given their narrow beam could be tricky to handle, particularly in a harbor at slow speeds or at sea in heavy weather. It was speculated that his previous captains had kept him off the conn as a junior officer (or even in his short five-month stint as executive officer) after he proved to be such a "bad driver." In a wardroom where everyone had always good-naturedly "joked about everyone's idiosyncrasies," there was now "an awful lot of serious talk about whether Marks was competent."

On November 11, *Hull* left Pearl Harbor in a task force bound for the western Pacific. After a brief stop at Eniwetok for fueling, the ships made for Ulithi, home of the U.S. Third Fleet, arriving on the first day of December 1944.

The next day, Storekeeper Drummond wrote to his mother in Missouri:

My only hope is that I get a transfer. . . . The ship sure has changed since last time we were out here. I don't think it will ever be like that again. We lost our old Captain, and believe me the new one isn't like our old one.

As *Hull*'s new commanding officer hoped, the destroyer and her crew were about to make naval history together, although not the kind he sought.

Ten

With a protected lagoon 20 miles long and 10 miles wide, capable of holding hundreds of ships, Ulithi was well positioned to serve as a staging area for the upcoming naval operations in support of the liberation of the Philippines, 850 miles to the west. Since being occupied by U.S. forces in September 1944, Ulithi had turned into the largest and most secret naval base in the world; for the next several months it would be identified in news stories only as "a Pacific base."

At 8:05 A.M. on November 30, 1944, while anchored in the Ulithi lagoon, *Monaghan* held a brief change-of-command ceremony. Departing was Commander Waldemar Wendt, who had been in command for eleven months. Newly arriving was an officer who had graduated from

Annapolis five years after Wendt: Lieutenant Commander Floyd Bruce Garrett Jr., twenty-nine, of Clarendon, Arkansas (near Little Rock), where his mother—Laura Redding Garrett, the "first female graduate of the University of Arkansas" and a lifelong teacher—still held class in a one-room schoolhouse. When, at age seventeen, he asked for permission to join the Navy, his mother initially declined in favor of his continuing his education. She changed her mind when he came home one night admittedly "skunky drunk." Deciding that the Navy might teach him some discipline, she signed the necessary papers. After a year in the fleet, the seaman 2nd class was honorably discharged "to study for the U.S. Naval Academy," which accepted him in 1934 at age nineteen. The "short, slight and boyish" Garrett—at five foot six and 115 pounds the "smallest and probably one of the youngest looking" in his class—paid attention to academics, ending up ranked 123rd in his class, which graduated 438 new ensigns in June 1938. One classmate remembered Garrett's "quiet dignity and winning smile." His biography in the *Lucky Bag 1938* stated:

> *Small in size, but full of fight and determination, Brucie came all the way from Little Rock to learn this naval trade. He has become the most seagoing fellow in the class, and some of the yarns he spins would turn the "Old Navy" green with envy. Always ready to have a good time, but serious enough to stand well in the upper third, his knowledge of the academic side of life has made him more than helpful as a roommate. His cheerful, level point of view is always dependable. His activities have been limited to holding down the radiator and complaining about the food. Seriously, we would have been lost without him. Our suite could never be complete without Brucie's helpful, encouraging and determined companionship.*

Garrett's first sea duty was aboard the aircraft carrier *Ranger* (CV-4), commanded by Captain John S. McCain, who would become a decorated wartime carrier admiral. After ten months on *Ranger*, Garrett was the subject of an exemplary fitness report written by McCain, who added: "Ensign Garrett is wide awake, energetic and capable. He is one of the finest and most promising young officers it has been my pleasure to have serve under me."

Ordered to the battleship *New York* (BB-34) in early 1939, Garrett made an unusual request the following year: to leave the ranks of young line officers—in training to one day command a ship, traditionally the most coveted position in the Navy—and be sent to the next session of the Navy's Finance and Supply School, and upon completing that course to be transferred to the Supply Corps. Garrett was sent to the three-month school in Philadelphia. However, shortly before graduation he asked to return to line duty following the school. Showing indecision about his career path, he soon followed that request with another: "I now request that my request for return to line duty be cancelled." Four days later, Garrett was informed by the Bureau of Navigation, which handed out assignments for the Navy's line officers, that it was "impractical to approve" his latest request. He was ordered to report to the destroyer *John D. Edwards* (DD-216)—an obsolete four-piper (ship with four stacks) in service since 1920—then in the Asiatic Fleet on China station, which was considered a backwater assignment but which would heat up at the outbreak of war. In February 1942, *Edwards* participated in the Battle of the Java Sea—up to that time the largest naval surface engagement since the Battle of Jutland in World War I—in which ten U.S. ships and four Japanese ships were sunk.

In January 1943, Garrett volunteered for ordnance engineering school "because I would like to have a class of duty wherein poor eyesight is not such a great handicap." If granted, the request likely would have resulted in his removal from lightly armored destroyers and future service in cruisers and battleships. *Edwards'* commanding officer, Lieutenant Commander William J. Giles Jr., although a close friend of Garrett's, disapproved the request before sending it up the chain of command because he considered Garrett's experience in destroyers too valuable to lose. Giles, who had been awarded a Silver Star for "an outstanding piece of navigation under fire" in bringing *Edwards* and three other destroyers safely through the treacherous Bali Strait with the enemy in pursuit during fighting in the Java Sea, explained:

Experience possessed by this officer as an actual participant in the Southwest Pa-cific Combat Zone during the Defense of Java, and the qualifications for the com-mand of a destroyer, which he unquestionably now possesses, make the retention of this officer in destroyer duty practically mandatory. Although near-sighted, this defect has in no way interfered with the high standard of performance of the duties of Officer of the Deck, Gunnery Officer, Navigator, and Executive Officer, in all of which offices he has served since the outbreak of war.

Garrett's request for ordnance school was denied, and he remained second in command of *Edwards* until the destroyer was transferred to the Atlantic in June 1943, at which time he was ordered to the Bethlehem Steel Co. shipyard at Terminal Island, California, to become executive officer of the newly launched *Fletcher*-class destroyer *Cowell* (DD-547). That same month his wife, Virginia (Corbin), presented him with their first child and his namesake: Floyd Bruce Garrett III. Capable of "sud-den and strong emotion," Garrett had met his future wife at a costume ball, where she wore a veil that hid all but her eyes, and danced with her the entire evening. The next day he wrote his mother, vowing: "I am going to marry the girl who has those eyes."

Assigned to a carrier group in the Pacific, *Cowell* took part in the landings at the Gilbert Islands in November 1943, and screened carriers launching air strikes at Hollandia, Kavieng, Truk, Wake Island, Guam, Iwo Jima, and elsewhere. In November 1944, facing his own rotation to a new assignment, *Cowell*'s commanding officer, Commander Charles W. Parker, urgently wired higher authorities requesting that Garrett "be retained on board" *Cowell* due to his "experience and qualification for command this class destroyer." While it seemed logical to Parker that Garrett should take command of *Cowell*—a ship he knew well—the Navy had other plans, and Parker's request was "answered by orders" two days later detaching Garrett from *Cowell* to command *Monaghan*. Both ships were then anchored at Ulithi.

Along with James Marks of *Hull*, Garrett was among the first wave of young destroyer skippers from Annapolis class of 1938. As it happened, a total of five classmates within weeks of each other joined Destroyer

Squadron One—composed of the seven remaining *Farragut*-class ships—forming up at Ulithi under the command of the "intelligent, demanding and precise" Captain Preston V. Mercer (Annapolis 1924), who looked "more like a professor" than a career naval officer but was a "thoroughly seasoned professional" about the Navy in general and "destroyers in particular." Mercer inspected each of the destroyers in his command for "watertight integrity" and found them "up to a proper standard." He made sure, too, that all his young first-time commanding officers were "very much aware of the lack of stability" of the *Farragut*-class destroyers.*

Within hours of assuming command, Garrett was at the conn of *Monaghan* patrolling around Ulithi for enemy submarines. During patrols and training exercises over the next ten days, the new skipper began to gain "the confidence and respect" of the crew for his "kind and considerate" ways of making "our duties more pleasant" whenever possible. Exuding "southern charm and personality" in dealing with officers and enlisted men alike, he also showed he knew his way around destroyers. Whatever doubts he had once harbored about ascending to command of his own ship, Bruce Garrett clearly took "great pride" in his beginning tour as a destroyer skipper.

AT DAWN on December 11, an armada nearly a hundred strong—large flattops crowded with warplanes, big battlewagons, menacing cruisers, and greyhound-like destroyers—sortied from Ulithi in the morning twilight of "crimson, yellow and green upon the eastern sky," steaming

---

* The squadron's other new destroyer commanders from Annapolis class of 1938 were William K. Rogers, *Aylwin* (DD-355), "a man's man . . . with a sense of humor that never lets him down and a modesty that is real"; Charles C. Hartigan Jr., *Farragut* (DD-348), an academy soccer player known for enjoying "life in general and a good time best of all"; and C. Raymond Calhoun, *Dewey* (DD-349), the "good-natured, easygoing son of a naval officer." The other two commanding officers in the squadron were classmates of *Spence*'s James Andrea from the class of 1937: Burton H. Shupper, *Macdonough* (DD-351), an academy wrestler and "Long Island's pride and our joy," and Stanley M. Zimny, *Dale* (DD-353), a handball champion and "loyal, energetic and unfailing pal."

hard through "caps of silvery . . . cresting waves" for the Philippines. "The greatest fleet that ever sailed the seas" was commanded by Admiral William Halsey, "neat as a pin" with slicked-down graying hair, pressed khakis, and shoes that "shone like brown mirrors." The admiral watched his warships returning to sea from his chair on the port side of his flag bridge aboard one of the newest battleships, *New Jersey* (BB-62). "As was his habit," Halsey's posture was "head up, chin out [and] back straight."

The U.S. Navy was "fresh from its greatest triumph" two months earlier at the Battle of Leyte Gulf, where, thirty-four months after the Imperial Japanese Navy struck so suddenly and devastatingly on December 7, 1941, "Japan's capacity to wage another fleet battle" had been "terminated."* Revered as the Navy's most famous fighting admiral, Halsey had, however, received widespread condemnation rather than credit for his actions at Leyte.

Halsey had sported a bloodlust since observing from the flag bridge of the carrier *Enterprise* the horrific smoking ruins of Pearl Harbor the day after the surprise attack. He spoke regularly of his loathing for the Japanese—complete with denigrating racial slurs—to an American military and public that had vowed to "remember Pearl Harbor." His well-publicized hit-and-run carrier raids in the Gilbert and Marshall islands in 1942, as well as the launching of Colonel Jimmy Doolittle's flight of B-26 Army bombers from the deck of the carrier *Hornet* on their mission to bomb Tokyo, helped the American public become "aware of the formidable Bull Halsey and his exploits." Although hailed as "Pearl Harbor avenged" in some U.S. newspapers, these early strikes were "essentially pinpricks" that had "no discernible effect on enemy operations." However, they were the first offensive punches thrown in the Pacific, and as such "morale rose appreciably in the U.S. armed forces and among the American public." Halsey's reputation grew as the "first

---

* The Battle of Leyte Gulf consisted of four sea battles fought between October 23 and 26, 1944: the Battle of Sibuyan Sea, the Battle of Surigao Strait, the Battle of Cape Engano, and the Battle of Samar. In all, the Japanese lost at Leyte four aircraft carriers, three battleships, ten cruisers, nine destroyers, and more than 10,000 men.

American victor over the Japanese." In October 1942, at a time of crisis when American and Japanese forces were stalemated at Guadalcanal, Halsey was put in command of the South Pacific Area (combined air, sea, and ground forces). Within a month, he "completely reversed the course of the conflict" on Guadalcanal, throwing the Japanese "on the defensive from which they never recovered"—and the war-torn island was eventually secured by U.S. forces in February 1943. By then, the bullish, take-no-prisoners, highly quotable admiral, who received his fourth star for his "successful turning back of the [Japanese] attempt to take Guadalcanal," had become "a legendary figure in the American press" and a popular symbol of wartime military leadership.*

At Leyte, Halsey, who had assumed command of the Third Fleet in June 1944, had been assigned to keep the Japanese from steaming through the San Bernardino Strait and attacking General Douglas MacArthur's invasion convoy, including loaded transports and troop ships. Instead, Halsey had taken his fleet—"sixty-five ships strong"— hundreds of miles away to chase what he believed to be a powerful enemy carrier force but was in fact four battle-scarred carriers depleted of all but a few dozen aircraft, along with a handful of escorts. Much to his consternation, Halsey had missed the historic naval battles at Midway and Coral Sea; as the war progressed he had become increasingly obsessed with wiping out "the last of the enemy's once mighty carrier force." Regarding as "childish" his assignment "to guard statically" the San Bernardino Strait, Halsey "glimpsed the prospect of a moment of glory" in sinking the enemy carriers and rushed "recklessly toward it." In his flag plot, Halsey had placed a finger down on a chart marked with the location of the Japanese force and told his staff: "We will go north and put these Jap carriers down for keeps." He opted to do so in spite of expressed opposition from three of his task group commanders. One of

---

* In 1943, Lieutenant (j.g.) John F. Kennedy stood at the rail of a ship, preparing to disembark in the Solomon Islands after a month at sea. On orders from Halsey, a large sign had been erected on a hillside near the entrance to Purvis Bay. Kennedy, like every other newcomer to the theater of operations, reflected on its bellicose welcoming message: "Kill Japs! Kill Japs! Kill more Japs! If you do your job well, you will help to kill the little yellow bastards. Halsey."

them, Rear Admiral Gerald F. Bogan (Annapolis, 1916), thought Halsey was making "one hell of a mistake" and picked up his ship's TBS phone to personally deliver the message that he feared a Japanese fleet was "heading right for" the unguarded strait. Bogan, who found himself speaking to the "rather impatient voice" of a Halsey staff officer, was told, "Yes, yes, we have that information." Another of Halsey's top commanders, Admiral Willis A. Lee on the battleship *Washington* (BB-56), an officer of "alert mind and keen analytical sense whose advice was often sought on strategy," reviewed a "mass of intelligence" that reached him and concluded that the Japanese force must be "a decoy with little or no striking power" attempting to lure the Third Fleet away. Lee sent by flashing light a signal to Halsey expressing his views. After not hearing back, Lee sent Halsey a similar message by TBS phone.* The vexations of Bogan, Lee, and others proved accurate: with the Third Fleet far removed from the San Bernardino Strait, a major Japanese naval force with a "formidable aggregation of fire power"—including the new battleships *Yamato* and *Musashi,* the world's largest fighting ships, possessing 18-inch deck guns—with "nothing but empty air and ocean between" them and Leyte Gulf, sailed through unmolested, heading straight for MacArthur's invasion force. During the resultant enemy attack against overwhelmed elements of the less-powerful Seventh Fleet—whose commander had expected Halsey's Third Fleet to have a powerful battle line (designated Task Force 34) in place blocking San Bernardino Strait—Pacific Fleet Commander Admiral Chester Nimitz, monitoring the battle from 3,000 miles away in Hawaii, had wired the missing Halsey. The dispatch handed to Halsey on the flag bridge of *New Jersey* read: "Where is Task Force Thirty-four.

---

* Another Third Fleet officer who thought they were chasing decoys was Captain Arleigh Burke, then chief of staff to one of Halsey's carrier task force commanders, Vice Admiral Marc A. Mitscher. Burke, the former Little Beavers squadron leader, tried convincing Mitscher that since the Japanese could "not bomb the Third Fleet out of the way," they might well be trying to "bait" a trap to "draw it away." Mitscher thought Burke might be right, but concluded there was "nothing worse than having a subordinate butt in and criticize a plan that was being executed," and decided not "to bother" Halsey on his flagship. "He's busy enough. He's got a lot of things on his mind."

The world wonders." The last three words were "padding to confuse enemy decoders" and should have been deleted before delivery. However, to Halsey, who knew that a pitched naval battle was being waged in Leyte Gulf as he raced in the opposite direction because he had ignored repeated calls for help, it appeared that Nimitz—in adding "the world wonders"—was openly "taunting him with heavy-handed sarcasm." The tough-talking admiral was "stunned as if struck in the face" and went "pale with anger." He pulled off his cap, threw it to the deck, and before the disbelieving eyes of his staff officers "broke into sobs." His chief of staff, Rear Admiral Robert B. Carney, rushed over, grabbed Halsey by the shoulders, and shouted, "Stop it! What the hell's the matter with you? Pull yourself together!" So "mad he couldn't talk," Halsey handed the message to Carney. Then he snatched back the dispatch, threw it to the floor, and stomped on it. "What right does Chester have to send me a God-damn message like that?" (Halsey would "not until weeks later" learn that the message Nimitz had authorized was merely meant to ask the question "Where is Task Force 34?")* As "furious" as he was at what he perceived to be a "gratuitous insult" from Nimitz, Halsey realized he had no choice but to turn around short of destroying the enemy carriers. Caught 400 miles out of position, Halsey rushed back with his fleet, but too late to give his fullest support when it was most needed.

Halsey's actions had raised the formidable ire of the Navy's top admiral, Chief of Naval Operations Ernest J. King, who in the midst of the sea battle paced "like a tiger . . . up and down in a towering rage" in his Washington office. Following in-progress dispatches from the Philippines, King vented his anger to a visitor, Rear Admiral Joseph J. Clark: "[Halsey] has left the strait of San Bernardino open for the Japanese to strike the transports at Leyte." The two admirals agreed Halsey had

---

* Nimitz's short message was not solely informational in nature. Although "strongly opposed to interfering with a commander at the scene of action," Nimitz sent the message only after it became clear that Halsey's force was "nowhere near" San Bernardino Strait. He meant the message "as a nudge" to Halsey to make the "wisest use" of his firepower. In such "extraordinary circumstances," Nimitz felt justified in "interfering with the man on the scene."

given the enemy fleet a "golden opportunity to wreak havoc on Mac-Arthur's invasion forces." King and Clark were not alone in criticizing Halsey, as there was "persistent grumbling" in Washington about Halsey's "careless ways." Some officers who had served under Halsey even opined that he was "not sufficiently skilled to command a fleet." Audacious by nature, he was "prone to ad hoc rather than detailed planning" and "guilty of sloppy techniques," which resulted in his sending "often vague dispatches" to his commanders and ships. While these traits were overlooked when the goal had been simply to hang on at Guadalcanal or unleash a fast carrier strike against an enemy island outpost in the early days of the war, by late 1944 the "complex offensive" was at its height and such "inefficiency was intolerable."

The nearly bungled outcome at Leyte—and with it MacArthur's long-awaited return to the Philippines, where a final victory would not only free the oppressed Filipinos and "shorten the war" but also "guarantee Japan's defeat as a great power"—had been won with equal parts of luck (the Japanese fleet turned around "within forty miles of the invasion force") and heroism by a thin line of outgunned U.S. ships and their crews. In Navy wartime parlance, Halsey's foul-up became derisively known as the "Battle of Bull's Run," a pun on his nickname and the Civil War battle.*

For his part, MacArthur was a fan of Halsey's for his resourcefulness and aggressiveness in battle—the general found "the bugaboo of many sailors, the fear of losing ships, was completely alien to [Halsey's] conception of sea action." However, MacArthur, whose Leyte invasion had been jeopardized by Halsey, was greatly upset with the admiral's "failure to execute his mission of covering" the operation. While MacArthur sent an official "gushing" telegram to the commander of the Third

---

* Four of the six U.S. ships sunk at the Battle of Leyte Gulf were lost as a result of Halsey "leaving the straight unguarded." Those ships were the escort carrier *Gambier Bay* (CVE-73), the *Fletcher*-class destroyers *Johnston* (DD-557) and *Hoel* (DD-533), and the destroyer escort *Samuel B. Roberts* (DE-413). American casualties at Leyte totaled 2,800 killed, wounded, or missing. Halsey's failure to block the San Bernardino Strait also permitted the Japanese force that had sailed through the unguarded strait to escape by the same route.

Fleet—"Everyone here has a feel of complete confidence and inspiration when you go into action in our support"—the general privately seethed at Halsey, unleashing a "verbal castigation" of him. MacArthur "charged" Halsey with "threatening the destruction" of the invasion force by leaving unguarded the San Bernardino Strait. MacArthur's chief of staff, Lieutenant General Richard Sutherland, noted that MacArthur "repeatedly stated that [Halsey] should be relieved and would welcome his relief as he no longer has confidence in him." Although MacArthur took no steps to have Halsey relieved of command—and castigated his own staff officers one night over dinner for their nonstop criticism of Halsey with "He's still a fighting admiral in my book"—Leyte Gulf had come between two of the Pacific's top military commanders.

Now Halsey and the Third Fleet were headed back to the Philippines for a return engagement: MacArthur's invasion at Mindoro—250 miles northwest of Leyte and "under the shadow of Luzon" and its beehive of enemy airfields. The invasion of Luzon, the largest and northernmost island in the chain and the last enemy stronghold in the Philippines, where Japan had "thousands of superb ground forces," was scheduled to start after that, but MacArthur first wanted a foothold on Mindoro. MacArthur had planned to land at Mindoro on December 5, but ten days before D-Day, Halsey, although he "hated" to do so, had requested a postponement. With his fleet exhausted from continuous action in October (including attacks on Okinawa and Formosa) and throughout November in support of Leyte ground operations while fighting off the latest and most desperate weapon of the Japanese—kamikaze planes making suicide dives into U.S. ships, including six of Halsey's aircraft carriers in the past month—Halsey sought "an adequate rest period." His own flagship had spent all but ten of the last ninety-five days at sea, much of it in combat operations, and Halsey himself felt "tired in mind, body and nerves." He had accepted the decision that given the amount of carrier air support required by the U.S. Sixth Army, he would be unable to "break off supporting the Leyte operation" and strike the Japanese home islands before the close of 1944, which he "longed to do." It was a "bitter disappointment" for Halsey that there would be "no wings over Tokyo until next year." As soon as "MacArthur's obliging reply was decoded"—the Mindoro invasion was rescheduled for December 15—Halsey's ships had turned their prows toward Ulithi, where the admiral knew they would also have an "unexpected opportunity for maintenance."

The men of the Third Fleet had been rejuvenated by their two weeks at Ulithi. Everyone, including Halsey, who had turned sixty-two, "rested and tried to relax." Men who had been "too long at sea and faced death too often" spent afternoons playing sports and getting sunburned on Mogmog, a tiny atoll that had been turned into a thatched-hut recreational area, complete with a swimming beach, basketball courts, horseshoe pitches, baseball diamonds, and a football field. Sailors were allotted two bottles of beer daily—usually paid for from their ships' crew-welfare funds—and for officers there was a bar where bourbon and Scotch cost twenty cents a shot.

Before departing, men caught up on their letter writing; for too many, it would be their last missives home. Aboard *Spence*, which after rendezvousing with the Third Fleet off the Philippines in mid-November had been screening for a carrier group, Machinist's Mate Robert Strand, the bowler who hoped to marry his sweetheart, Jane, and own his hometown bowling alley one day, wrote to his parents in Pennsylvania. He made a point to do so—after "sitting in the sun all afternoon"—even though he had written home a day earlier. "Well, just brace yourself for it will be the last shock like this for awhile." He admitted to having been "a bad boy" and gambling. "But lucky!!!!"

Wish everyone a Merry Christmas for me and hope you'll have a lot of fun. I know I'll enjoy myself with a big banquet, Navy style. Don't worry, it will really be good chow on that day anyway. . . . You know I'll be thinking of you so very much and missing you all. Bye for now—take good care of yourselves and just keep writing even though your letters are being delayed.

As always,
Your loving son, Bob

En route to the Philippines, the Third Fleet conducted exercises to bring ships and crews back to fighting form. The fleet was divided into three main task groups, each commanded by a rear admiral, then into smaller units. Halsey had with him seven *Essex*-class (attack) and six light (or escort) aircraft carriers with 540 fighter planes, 150 dive-bombers, and 140 torpedo bombers, eight battleships, fifteen light and heavy cruisers, and about fifty destroyers. Their mission would be the "neutralization and destruction of hostile air strength on Luzon." The danger in failure was understood by Halsey, who knew the passages of MacArthur's invasion force "through restricted waters and between hostile island air bases" to land troops at Mindoro was "admittedly hazardous" unless Japanese air power could be "knocked out and kept out for the critical approach and unloading period." There could be no abandoning the invasion force to chase other targets. Halsey's intent—and the promise he had communicated to MacArthur and his staff "by conference at Leyte"—was to "support the Mindoro operation properly," which meant keeping enemy aircraft on Luzon from "intercepting our transports."

The morning of December 13 was used for fueling ships from waiting fleet oilers—topping off tanks that were already nearly full—and that afternoon, leaving behind the logistics group, Halsey ordered a "high speed run" started for Luzon in an effort to secure "tactical surprise." At dawn, his carriers launched their attacks. Every known or suspected enemy airfield on Luzon had been assigned to a specific carrier for its aircraft to hit. Tactical surprise "with its accompanying dividends" was achieved as the planes struck their targets on Luzon. "Very few Japanese planes were found airborne" that morning, and immediate air supremacy was achieved. Carrying out a new operation dubbed the "Big Blue Blanket"—named after the oversized blankets used on the bench by Annapolis football players—designed to meet the threat of kamikazes, continuous combat air patrols overflew Luzon airfields day and night to prevent enemy planes from taking off. Other methods tried for the first time to neutralize the kamikaze threat—most "masterminded" by Vice Admiral John S. McCain, who a month earlier had relieved Vice Admiral Marc Mitscher as Halsey's top task force commander—also worked, such as placing picket destroyers 60 miles away to provide "early warning of impending air attack" to be met by a wave of fighters patrolling at low altitude.

MacArthur's forces landed on the southern beaches of Mindoro on December 15, finding resistance minimal. Third Fleet air attacks conducted over the next three days netted 270 enemy planes (more than 200 destroyed on the ground), and "not a single bogey" from Luzon was able to deliver a strike on MacArthur's invasion forces.* In its first deployment, the Big Blue Blanket had been an unqualified success in "paralyzing the enemy's air effort over Luzon." Other results by marauding carrier planes: thirty-three enemy cargo vessels caught and sunk in shipping lanes, locomotives and truck convoys wrecked, and fuel and ammunition dumps blasted, along with antiaircraft batteries and airfields. Halsey's own losses those three days: twenty-seven planes and no ships.

With the Mindoro operation ongoing as troops and supplies poured ashore, air cover was still required. The operations during the last three days, however, had consumed much fuel and ammunition. Halsey planned to replenish his ships on the seventeenth and return for three more days of air strikes on Luzon no later than the nineteenth, when "the congestion of supplies on the Mindoro beachhead and of ships offshore would demand all available air cover."

As soon as the last of the carrier planes were recovered on the evening of December 16, Halsey ordered a change of course for a rendezvous with Third Fleet's replenishment group in the Philippine Sea some 400 miles east of Luzon. The location was selected as the "nearest spot

---

* In the days prior to the landings at Mindoro, MacArthur's invasion convoy was attacked by numerous Japanese suicide planes from bases on Formosa and south of Mindoro. The cruiser *Nashville* was hit, killing 133 men and injuring 190, and a destroyer was severely damaged, causing both ships to return to Leyte. After the assault troops landed on the beach, a handful of kamikazes destroyed two tank-landing ships. What the toll in casualties and ships would have been had the hundreds of enemy planes on Luzon not been covered by "a huge aerial umbrella" unfolded by the Third Fleet is "impossible to define," but military historians agree it would have been much higher.

to Luzon outside of Japanese fighter-plane radius," while remaining close enough to the launch position so the carriers could "get back in time to meet their strike deadline." Halsey was "driven" to continue to give MacArthur "his fullest support," and considered it "imperative" that all his ships be fueled and resupplied rapidly so the fleet could get back in time to resume air operations as promised.

Fueling could not come too soon for many destroyers, some "dangerously low" after three days of air strikes. Steaming with aircraft carriers, which seldom slowed when conducting flight operations, required destroyers to execute full-speed maneuvers, resulting in high fuel consumption. It "became standard practice" for destroyers, being "shortest in fuel capacity" with their limited storage compared with larger ships, to be fueled every few days.

Steaming in excess of 20 knots, the Third Fleet soon left behind the Japanese on Luzon. No one on the darkened ships heading into the night on a zigzag course in an increasing swell and wind had any idea they were about to engage in a battle for survival with an onrushing enemy of a different nature.

Eleven

U.S. Army Air Corps captain Reid A. Bryson, twenty-four, a meteorologist based on Saipan with the Twentieth Air Force, had been tracking for the past ten days the steady westward movement of what he had marked on his weather map as a "potential typhoon." When the latest weather report from Guam came in on December 17, 1944, he knew his worst fears had been realized. A tall, slender, fresh-faced midwesterner with a blond crew cut that made him appear even younger, Bryson had only months earlier received his first "real-world forecasting experience," and with it a painful lesson: typhoons had to be watched "very carefully" and avoided by fleets and ships at all times—even if doing so held up the "progress of operations" and the war.

A 1941 graduate of Ohio's Denison University with a degree in geology, Bryson had gone from being a geology graduate student at the University of Wisconsin to "meteorological cadet" in early 1942 in the Army Air Corps program at the University of Chicago. Similar training courses for military weathermen were taught at the Massachusetts Institute of Technology (MIT), the University of California at Los Angeles, and New York University. Upon his graduation nine months later, Bryson received his commission and became an instructor in the program. The following fall (1943) he was sent to study at the Institute of Tropical Meteorology at the University of Puerto Rico, and after a few months he became a teacher there, too. With still "no practical forecasting experience," he was assigned in summer 1944 to a seven-member weather group that would be handling forecasting for a wing of the Twentieth Air Force in the Pacific. They first gathered at Hamilton Air Field in California, then were flown to Pearl Harbor, where they were to await deployment to Saipan in the Mariana Islands—some 1,500 miles east of the Philippines and an equal distance southeast of the Japanese mainland. Saipan was still in the process of being secured by U.S. forces prior to establishing a major air base from which long-range bombing runs against mainland Japan were soon to be flown.

While in Hawaii, some of the Army Air Corps meteorologists had been put to work at the Navy's Pacific Fleet Weather Center, which had been "practicing forecasting for some time." One of their first mornings at Fleet Weather, Bryson and his friend William Plumley, an Army captain who had also taught in the Chicago program, found themselves working with an aerologist (what the Navy calls its weathermen) who had been one of their students. Their assignment was to make a weather forecast for a carrier-based strike at Marcus Island the following day. They looked at the "nearly blank" weather map, then at each other, and again at the map, feeling "a little like Magellan setting out into the western sea." At that point in the war, a typical Pacific weather map had data for the western parts of North America, the chain of Alaskan islands, a handful of U.S.-held islands west of Hawaii, a "few low quality Russian and Chinese bits and pieces on the western rim of the map," and a smattering of ship, submarine, and aircraft reports spread across an expanse of ocean covering roughly one-third of the Earth's surface. On the map they were analyzing, Bryson noted "seven areas larger than the United States that had no data whatsoever." Notwithstanding the limited data, one thing was immediately clear: southeast of Marcus Island (aka Minami-Io-Jima), neighbor to Iwo Jima, there was a "cyclonic storm."

By definition, a typhoon in the Pacific (or hurricane in the Atlantic) had to meet a specified wind force based on an early-nineteenth-century scale created by British admiral Sir Francis Beaufort, who had charted the effects on the canvas sails that could be carried by a fully rigged frigate in different wind conditions. Up to that time, reports of wind conditions were subjective by nature; one sailor's "squall" might be another's "stiff breeze." Beaufort determined that a ship's canvas was unable to withstand a wind of 75 knots or higher. The scale—ranging from 1 to 12—was modified several times and had long since become a standard for ship's log entries documenting observable wind and sea conditions. For a storm to be categorized as a typhoon the winds had to reach Beaufort 12, which equates to a wind speed of more than 73 knots or 90 miles per hour, although winds in a typhoon "may reach double that speed." Beaufort 12 wave height is listed as 46 feet or greater.

As for what looked to be a developing typhoon south of Marcus Island, the limited data showed "a cyclonic whirl of winds and dense clouds several hundred miles across that fed a vast layer of high clouds." How strong the winds were, how large the waves, and whether the storm had developed an eye "could only be imagined" by the weathermen at Pearl Harbor. In the final analysis, an accurate forecast would come down to one key finding: where was the storm headed? From their studies of meteorology, they knew that in the trade wind regions— an area of converging winds extending between two belts of high pressure located some 2,000 miles on either side of the equator—strong upper winds could work to steer a storm. They discussed how they could not predict if that would happen in this case—or in which direction the storm would head—because they lacked upper-air data for the area. Suppose there were no upper winds blowing that day? They decided to "construct an upper-air chart, somehow." From a few wind observations across a wide region at or near the 10,000-foot level and some surface data (pressures, winds, temperatures, and cloud amounts and types) combined with their "basic knowledge of atmospheric physics," they plotted an upper-level pressure map. Soon they had "a picture of the wind flow," which allowed them to project on the map the path of the storm. When they finished, their collective "hearts sank." They saw that the upper winds could be predicted to curve the storm northward—on a path toward Marcus Island, and a "rendezvous with the fleet at about strike time." The high winds and waves—they were sure by now that they were looking at a full-fledged typhoon, although none had yet experienced one—coupled with dense clouds and heavy rain made a U.S. carrier attack in the area "hazardous at best." In fact, adequately forewarned, the fleet would do well to steer clear of the entire region. They discussed their findings, "recalculated and reconsidered," and finally were convinced. The duty officer wrote out a forecast for the fleet showing the "recurved typhoon" heading for Marcus Island. Just before the forecast was to be transmitted via radio, the Pacific Fleet's senior aerologist—a Navy captain—walked into the room "drunk as a lord." He "blearily looked" at the intended message and "roared out: 'Nonsense! Typhoons don't recurve at that longitude at this season, they move straight west!'" He ordered the forecast to be changed, which it was prior to being sent to the fleet. The next day the typhoon curved and hit Marcus Island and the fleet, causing "losses of planes and brave men." Those tragic results, in which Bryson felt he had played a role in sending out the "bad forecast," as well as the cavalier way the accurate forecast "based on a combination of good science" and data had been rejected in favor of a "totally subjective, off-the-cuff opinion" by a senior officer, had made a "strong and lasting impression" on the young Air Corps weatherman. They were dealing not just with theoretical predictions here but with real-world weather forecasts that could hold countless lives in the balance.

In October 1944, a few months after the Air Corps meteorologists had set up operations on Saipan, Bryson observed on his weather map a probable typhoon to the southeast and requested an aerial reconnaissance of the storm. Deciding he had best venture "where I ask others to go," he hopped on the four-engine B-24 Liberator heavy bomber for the flight. Spotting the "radiating cloud bands" in front of them, they headed into them at 8,000 feet, an altitude that proved "lousy for roughness." Every so often, as the plane bumped and ground its way farther into the dark maelstrom, they heard "a rivet pop like a gunshot." As they entered what Bryson recognized as the "eye wall" of a typhoon, the pilot radioed over the intercom that they would "not make it in and out" safely, and turned around. Back at Saipan, the bomber's airframe was found to have sustained so much structural damage that the plane had to be scrapped and salvaged for spare parts, teaching Bryson still another lesson—this one firsthand—about the power of a typhoon.*

Knowing December was the tail end of what had been an active five-month season that had spawned some fifty Pacific typhoons, Bryson was not about to take any chances with the potential typhoon he had been tracking for ten days. Typically, the first data he had received about the storm had been sparse—east of Saipan the nearest weather station was at Eniwetok, a distance of 1,200 miles. Between Eniwetok and Saipan there had been only a few aircraft reports. Still, Bryson had noticed "a little shift in the wind" near Kwajalein—east of Eniwetok by another 800 miles—as well as heavier than usual cloud cover. This had suggested to him some type of low-pressure center—perhaps an "incipient typhoon"—south of Kwajalein. After that, each time he received a new set of observations, he looked for further evidence of a typhoon.

---

* The Army Air Corps weather facility at Saipan soon became the first to assign women's names to typhoons. Bryson and his fellow meteorologists had "all read" the 1941 novel *Storm,* by George Rippey Stewart, about a violent storm affectionately known as Maria that sweeps through California and changes the lives of many in its path. They eventually began naming the storms after their wives and girlfriends. Bryson irritated his wife, Frannie, "to no end" when he named a typhoon for her that soon "fizzled out." From 1950 to 1952, typhoons were identified by the phonetic alphabet (Able being the first storm of the season, Baker the second, etc.). In 1953, the U.S. Weather Bureau switched to women's names; since 1979, a list that also includes men's names has been used.

There were hints here and there—one aircraft report described larger than usual clouds, and another identified strong shifts in wind direction from the north to the south, suggesting to Bryson something other than the "usual rippling of the trade winds." Then on the morning of December 17 came a new observation from Guam, considered a reliable weather station. The winds at Guam had been gaining in strength with a "strong northerly component," only to "abruptly shift" to the southeast as the local weather rapidly deteriorated. To the west and southwest the sky was covered with a "streaky veiling of cirrus clouds"— which, Bryson knew, "mariners for centuries" had considered the "forerunning of a typhoon." At that point, Bryson was certain a full-blown typhoon had passed to the south of Guam.

Bryson knew that the Third Fleet was "assembled off to the west and northwest" on a track somewhere between Guam and the Philippines and that there were a lot of aircraft carriers, battleships, cruisers, destroyers, oilers, and men "out there." It was vital to determine whether the typhoon that passed south of Guam would continue on a westerly track toward the Philippines and pass south of the fleet's position, or whether it would recurve northwesterly—as the Marcus Island typhoon had done—and turn toward those ships at sea.

As newer weather reports arrived, Bryson identified a trough of low pressure approaching from the west that could cause the typhoon to recurve. Immediately he called the flight line and requested a reconnaissance flight to find the center of the typhoon. The aircraft took off within a few minutes, its mission to determine the intensity and location of the storm. By charting an exact location, Bryson would be able to tell whether the typhoon had started to recurve or not. Then he waited.

After "some hours," Bryson was handed a radio message from the plane, reporting they had "located the eye of the typhoon," giving its latitude and longitude, and estimating the surface winds at 140 knots, or 160 miles per hour. This was nearly twice the minimum wind force necessary to categorize a storm as a typhoon, and to Bryson's mind strong enough that "not only could no canvas withstand it," neither could "steel ships driven by modern power plants."

When he marked the position on his map and compared it with the previous position, he confirmed the typhoon was "recurving to the northwest." Bryson hurried over to the Teletype that connected the Army Air Corps meteorologists with the Navy's weather office on Saipan. He sat at the keyboard and typed a message warning of the typhoon and giving its exact latitude and longitude, as well as the estimated wind speeds. He stated that it appeared to be "recurving to the northwest" and "on a track toward the fleet."

Bryson received an almost instantaneous reply. "We don't believe you," a Navy aerologist typed on the other end.

Bryson was shocked despite his "previous experience with the Navy and typhoons." He typed back this was *not* a guess and that a reconnaissance aircraft was "out there in the eye of the storm" and had radioed a report.

The Teletype again clattered out a quick reply: "We still don't believe you but we'll watch."

COMMANDER GEORGE F. KOSCO, thirty-six, a big-boned, round-faced aerologist aboard *New Jersey,* had been following what appeared to him to be a tropical storm forming between Ulithi and Guam since picking up a weather broadcast from Ulithi on the morning of December 16. That evening, as the fleet rushed toward its fueling rendezvous with its logistics group, a routine weather report from Pearl Harbor mentioned "storm indications in about the same location." On his weather map, Kosco, who had been the Third Fleet's aerologist for only two weeks, thought it looked to be a "very weak" storm. He anticipated it would "move off to the northeast," and he "didn't expect any trouble" because he judged the storm to be of a "very small caliber."

One of ten children born to Slovakian immigrants in Pennsylvania coal country, Kosco had to move away from home to a neighboring town to attend high school, and earned his room and board by working in a local hotel. An older brother who knew the importance of an education because he had quit school after the sixth grade to work in the mines "found a way" to secure an appointment to Annapolis for him from a local congressman, but Kosco failed the entrance examination by three points. He studied on his own for a year while working as a plumber, and passed the next test. He was admitted to the U.S. Naval Academy in 1926. An inch shy of six feet and a solid 190 pounds, Kosco was a three-year member of the varsity boxing team, which was undefeated in dual meets for the past decade under the tutelage of coach Spike Webb, who at the end of World War I had trained a young Marine heavyweight named Gene Tunney. Ranked in the middle of his graduating class of 402 new ensigns, Kosco's biography in *Lucky Bag 1930* remembered him this way:

> *"Well, if I am not the dumbest 5*\$!?," such is the usual introduction to [his] post mortem to the monthly examination in anything in general and navigation in particular. To the uninitiated it would seem like the harbinger of a reluctant farewell, but his friends know better. George is not the dumbest, by about four hundred numbers, and we all know that even the savviest of our band frequently arrive at the impossible answer of five when multiplying four by two, superior knowledge and previous training notwithstanding. . . . While his resourcefulness and originality do not always fit the narrow groove of academic recitation, they do show excellent prospects for the future. . . . Of a cheerful disposition, always ready to lend a hand, sincere, and a hard worker—that is Georgie.*

Kosco applied for aviation training but was disqualified due to imperfect eyesight. He was ordered to the battleship *Colorado* (BB-45) as a gunnery officer. On November 4, 1931, as *Colorado* engaged in training exercises off San Pedro, California, a 5-inch deck gun exploded during firing practice. In what the Associated Press reported as "one of the worst accidents in Pacific Fleet history" up to that time, four men were killed instantly as "broken steel, hurled like shrapnel, raked the decks." Two more would die as a result of their injuries; another twenty-two men were hurt, including Kosco, who, although hit by shrapnel that imbedded in his neck, back, and buttocks, was singled out in a front-page newspaper article the next day for "giving all possible assistance to fallen comrades" before

seeking treatment. For the next six years Kosco served as a gunnery officer on several ships. In 1937, his request to attend postgraduate school—
listing ordnance engineering as his preferred course of study, with the
newly developing field of aerology his second choice—was approved. He
returned to Annapolis for a two-year program during which aerology
became his specialty. To complete his meteorology studies, Kosco was
sent with a select group of other naval officers to MIT, where he earned a
master's degree in 1940. Following duty as an aerological officer on two
aircraft carriers—six months on *Saratoga* (CV-3), operating off California,
and a year and a half aboard *Ranger* (CV-4), in the Atlantic—Kosco, by
then a commander, was assigned in fall 1942 to the Naval Research Laboratory in Washington, D.C. For the next two years he was involved in
several projects, not all weather-related, including the study of wave
propagation, training in chemical warfare, hurricane research in the Caribbean, and the establishment of naval air transportation to Africa and
Ireland. After receiving his new orders, Kosco had caught up with the
Third Fleet at Ulithi, boarding Halsey's flagship on December 2, 1944—
his first wartime assignment to the Pacific and first tour to the region as
an aerologist.

When Kosco had boxed at Annapolis, the officer representative to
the team (as well as commander of the academy's receiving ship, *Reina
Mercedes,* which served as both barracks for instructors and "prison ship
for midshipman transgressors") was Captain William Halsey Jr., a fortuitous crossing of paths early in Kosco's naval career that would "not
hurt" in his new assignment to Halsey's flag staff, a close-knit group led
by chief of staff Rear Admiral Robert B. Carney. Kosco found himself
rooming with one of the more notable members of Halsey's staff, assistant chief of staff Commander Harold E. Stassen, the former Minnesota
governor (and youngest, having been elected in 1940 at thirty-two) who
had resigned his office in 1943 to serve in the Navy.

As December 17, 1944, dawned amid stiff winds and boiling seas
some 430 miles east of the Philippines, Kosco noted that they were getting "a little bad weather" and assumed it meant the fleet was running
into the front already on his maps. The latest reports from various

stations showed southwest winds at Ulithi and east winds at Guam. Weather map analyses were made every six hours at Pearl Harbor by Pacific Fleet Weather Central, which then radioed their forecasts to fleet units. Kosco also had some "quite delayed" plane reports from the previous day. While there were long-range patrol aircraft flying in and out of the Marianas, Ulithi and elsewhere, reporting weather conditions was not their primary mission. In fact, they pointedly avoided heavy weather and only in "exceptional cases broke radio silence" to send any reports back prior to landing. As a result, the majority of aircraft advisories were at least twelve hours old before they reached individual ships. Checking the atmospheric pressure as measured hourly aboard *New Jersey*, Kosco noted that the barometer was dropping slightly—down from the normal fair-weather pressure of 29.92 inches at sea level, which it had been shortly before midnight—to a still "reassuring 29.84" inches at 8:00 A.M. As Kosco well knew, an "unsteady barometer" was listed in Bowditch's naval bible as one of the "rules for establishing the existence of a tropical cyclone and for locating its center." A barometer fall of between .12 and .15 inches could place the center of the storm at only "50 to 80 miles distance." A drop of a tenth of an inch or more could be found in a larger territory "surrounding the actual storm area." While Kosco "sort of thought" that the accumulated data suggested a "wind with cyclonic circulation," he believed the "disturbance . . . lay 450 miles east" of the fleet's position, and stuck with his prediction that it would move on to the northeast—as he believed "all normal storms" did in the area—without threatening the fueling operation. Of course, the rendezvous position selected by Halsey, while expeditious for the purpose of getting back in time for the planned air strikes, did "lay in the normal track of typhoons." The Third Fleet had been "chased out" of Ulithi on October 3 by a typhoon; Halsey well knew, as he wrote Nimitz that day, that "the same thing may happen again at any time up to the middle of December." However, no reports had been received by the Third Fleet about a new typhoon "even existing." In fact, that possibility was "not given serious thought" by Kosco.

Awaiting Halsey's carrier group, designated Task Force 38, at the

planned fueling area in the eastern half of the Philippine Sea—at latitude 14 degrees 50 minutes north and longitude 129 degrees 57 minutes
east—was the replenishment unit, commanded by Captain Jasper T.
Acuff. At forty-six and a 1921 graduate of Annapolis, Acuff was a seagoing veteran. His At-Sea Logistics Group, Third Fleet, currently included
a dozen filled-to-the-brim fleet oilers—several at a time alternated
steaming back and forth to Ulithi to refill their bunkers with Navy Special Fuel Oil, the heavy by-product of crude oil that fed the fires in ship
boilers, as well as aviation gasoline and diesel. Also attached to Acuff's
group were several escort carriers that would launch fighters to provide
protection for the fleet from air and submarine attack while it fueled, as
well as to supply Halsey's carriers with replacements for pilots and
planes lost in the latest round of air strikes. With newer and speedier
destroyers—including the *Fletcher*-class ships—assigned to Halsey's fast-
attack carrier force, the tankers, cargo ships, and escort carriers of the
logistics unit were screened by older destroyers, among them *Hull,
Monaghan,* and their *Farragut*-class squadron mates, along with ten smaller
destroyer escorts, including *Tabberer.*

With visibility holding at 8 miles, Halsey's Task Force 38 began ap-
pearing around 10:00 A.M. as ghostly silhouettes emerging over a misty
horizon. Acuff's logistics group was already in formation, deployed
along several widely spaced parallel lines. Throughout the morning,
thick, low-slung cumulus clouds closed in overhead until the last
traces of sun and sky were obliterated. The wind was gusting to more
than 30 knots with increasingly choppy seas, along with longer and
deeper swell sets that gave sailors the most discomfort—as opposed to
smaller waves caused by the wind. Some 130 ships came "steaming in
close proximity to each other," causing a "congested area" for miles
around. The fleet course was set at 040 degrees—a northeasterly
heading—and speed at 8 knots. Warships lined up astern of one of
three replenishment units, ready to take their turn at the pump like
cars at a busy gas station.

The men on the tankers and cargo ships were well drilled in what
had become for them these past months a routine chore that never lost

its potential for sudden danger. Ships running together—separated by 30 or 40 feet of churning ocean—on parallel courses at identical speeds for the hours it took to fuel up and transfer ammunition and other supplies left little room for error even in favorable weather conditions. With ships close alongside in winds and seas as unruly as this day, "collisions were constant threats."

By the time the underway replenishments—known as "unreps" in naval vernacular—began in earnest, there had been another "noticeable change in the air and sea." In the last hour, the swell had increased; its "latent power could be felt," especially aboard the destroyers and other smaller ships as they were "rhythmically lifted" to the top of each crest and then "let to settle into the trough" that existed between swells, only to be followed by another cycle. The wind, too, was picking up steadily, with gusts of 45 knots recorded on *New Jersey*. Even the bigger ships—the carriers, battleships, and cruisers—rose and fell with the seas as they took on fuel and supplies. Replacement pilots, hanging on "for dear life," were sent over from the escort carriers to the attack carriers in "swaying, swirling" chairs hung on lines between ships, and were hauled in soaking wet. Soon these transfers of personnel were cancelled for safety reasons. The aircraft sent aloft to protect the fleet were "fighting air as rough as the sea" in pelting rain, gusty winds, and low visibility, and finally were ordered to land. Before they all made it down safely the conditions became "too rough" to attempt further carrier landings, and two pilots still aloft were ordered to turn their planes upside down and bail out nearby. After asking for the unusual order to be repeated, the pilots did as instructed and were rescued by a destroyer.

Before long, anxious exchanges over ship radios announced the plight of the smaller vessels trying to receive fuel from the tankers. The destroyers were having great trouble "maintaining station" alongside their assigned oilers. Yawing in the swells, they either came too close to the tankers or suddenly veered in the opposite direction until the distance became too great for the extended hoses, which were "lashed and whipped until they were unmanageable." When a hose parted—or had to be cut adrift to keep from losing it altogether—hundreds of gallons of black oil gushed onto the destroyer's deck and superstructure, adding to the danger for the deck hands as they struggled to do their work on the slippery and rolling steel surfaces.

Members of Acuff's logistics group were "shocked" to learn that Halsey's destroyers had been "allowed to deplete their fuel" to such low levels—some reporting as little as 15 percent of capacity remaining aboard—which seemed "incredible." *Dewey*'s commanding officer, Lieutenant Commander Charles R. Calhoun, and the commander of the squadron of *Farragut*-class destroyers, Captain Preston Mercer, who used *Dewey* as his flagship, had "never known destroyers to be that low on fuel before." Why, Calhoun and Mercer asked each other on the bridge that morning, had the refueling of the destroyers screening the carriers during the past several days been "deemed inconsistent" with the carrier task force's primary objective? Everyone knew a fleet's destroyers had to be fueled every second or third day, oftentimes replenished by bigger warships rather than awaiting the arrival of oilers. Calhoun and Mercer were left with the "unhappy impression" that Halsey, as fleet commander, had been nothing less than "remiss in permitting this critical fuel situation to develop."

"I wouldn't be surprised," said Mercer, "if some of them ran out of gas before the weather allowed them to refuel."

One of those destroyers running dry was *Spence*.

ABOARD *SPENCE*, Water Tender Charles Wohlleb, who had seen his ship nearly ram an aircraft carrier one dark night the previous month and had wondered along with his shipmates how their new "nice guy" skipper, Lieutenant Commander James Andrea, would handle the ship when things got rough, was topside as they approached *New Jersey* shortly after 11:00 A.M.

Spence had been attached to the Third Fleet since rendezvousing with Halsey's carriers off Guam in mid-November. During the Mindoro strikes, *Spence* (along with other screening destroyers) had "operated at high speeds for three days" and had been detailed numerous times for "high-speed pilot rescue missions," all of which had expended most of the destroyer's 150,000-gallon fuel capacity. *Spence* had been directed to replenish her almost empty fuel tanks from *New Jersey* in the hope that the bulk of the battleship would block the wind and seas for the smaller ship and "provide a steadier platform" than could an oiler for the two hours or so it would take to fill *Spence*'s tanks. Any quantity of fuel transferred successfully to *Spence* from *New Jersey*, which had "on hand 1,878,398 gallons" that morning, would not be missed by the battleship, but could keep the destroyer from running out of fuel and losing her engines, steering, and electrical power.

At the same time *Spence* closed on *New Jersey*, another *Fletcher*-class destroyer, *Hunt* (DD-674), approached on the opposite side to take on fuel.

Off duty in the after fire room, Wohlleb was free to come on deck and watch the underway fueling from the Third Fleet's flagship, and right now there was no better show anywhere. *New Jersey*, displacing 60,000 tons and as long as three football fields placed end to end, loomed implausibly massive as *Spence* came close alongside to starboard, the leeward side of the battleship. The battlewagon sat so much higher in the water that Wohlleb had to crane his neck to look up to the main deck, and strain further to see the battleship's bridge, where for a moment he could have sworn he saw Bull Halsey himself looking down at the fueling operation.

It took some time to get lines across and pull two six-inch fueling hoses—forward and after—into place and connect them. Fueling commenced, only to be interrupted more than once when the hoses parted and had to be reconnected. Green water swamped sections of *Spence*'s main deck forward as the helmsman struggled to keep the bobbing and corkscrewing destroyer on course. *Spence* was so low on fuel she was riding unusually high in the water, making the ship more difficult to handle in the heavy seas and high winds.

Shortly after noon, a huge swell rolled in that caused both hoses to rupture as *Spence* surfed up the crest so high that they "almost landed on top" of the battleship, the bottom of her hull "almost even" with the battleship's main deck. With no ocean buffer remaining between the ships, there was a horrible sound of steel on steel as "paint on both vessels was scraped off." Wohlleb was suddenly looking directly across to *New Jersey*'s bridge—and into the whites of the eyes of her shocked officers. He saw and heard it all in slow motion: the battleship captain on the wing of his bridge raising a bullhorn, then his angry, amplified words echoing: "Get that goddamn ship outta here!"

Spence, her foredeck and bow stained black with spilled oil from the forward and aft hoses being wrenched from the fueling trunks, pulled away after receiving only "6,000 gallons of fuel oil" in her tanks. *Hunt,* connected on the other side and fighting the wind about twenty compass points off her port bow, fared better, receiving "22,265 gallons of fuel oil" before casting off.

Wohlleb, knowing the show was over, went below for chow. He did not think to blame the new skipper for his ship handling, even though they had been in the lee of *New Jersey,* which should have provided some protection for fueling. It was apparent to Wohlleb that "Mother Nature had control" of the sea that morning. Tired and hungry, the sailor was not concerned about the depleted state of *Spence*'s fuel supply nor alarmed about the worsening storm.

At twenty, Charlie Wohlleb was "too young to think of tragedies."

HALSEY HAD BEEN LUNCHING in the wardroom with his staff as *Spence* attempted to fuel from his flagship, sitting at his usual place at the head of the table. Facing a starboard watertight door that was toggled open for fresh air, he could gaze out "upon the open waters." Whenever he did, Halsey could see the tips of *Spence*'s antennae wiping in the wind.

When *Spence* had rocked wildly atop the powerful swell, Halsey had suddenly seen the "upper works and masts" of the destroyer, which momentarily turned "so fast and so menacingly" bow-on to *New Jersey* when the two vessels scraped hulls that he reflexively ducked to get out of the way, much as a boxer would slip a punch.

Halsey commented on the "violence of the action," suggesting the destroyer must be at fault. It was an opinion easy to form while riding and dining comfortably on a ship bigger than a New York City block. He was promptly informed by his staff that this was not an "isolated instance" and was told that other destroyers were having a difficult time fueling in the heavy seas. In a space of thirty minutes, several radioed reports had come in:

> *Collett* reported conditions very bad, and both hoses carried
>   away;
> *Stephen Potter* reported a parted forward fueling hose;
> *Mansfield* reported she had broken loose from her station;
> *Lyman K. Swenson* reported both hoses parted;
> *Preston* reported casualties to both hoses and lines;
> *Thatcher* reported parting one hose and being forced to cut
>   loose.

Surprised to hear of the widespread difficulties, Halsey spoke briefly with his chief of staff, seated next to him. Declaring they might have to "knock off fueling," Halsey cut short the meal and announced that a "weather evaluation conference" would be held at 1:00 P.M.

Without waiting for other staff members to stand, aerologist George Kosco left the table and hurried to the weather office on the navigation bridge several levels above. It did not escape his notice that he had to fight a stiff wind as he made his way up the outside ladders. Upon entering the office, Kosco found to "his great relief" that the morning Fleet Weather Central report had arrived from Pearl Harbor and had already been deciphered by the aerology department, and that the "data had been modified to include local weather conditions." This meant that Kosco had an updated weather map to show the admiral. Quickly reviewing the map, Kosco "right away formed the opinion" that a tropical storm was "getting a little closer" than he had expected, although he still "didn't think it was a typhoon." Before he left for the meeting, Kosco tapped a few times on the barometer in the weather office and noted that the pressure—now 29.73 inches—was "falling steadily."

When Halsey entered the wardroom promptly at 1:00 P.M., Kosco unrolled the weather map and placed it on the table that had been cleared of dishes and was now covered with a green felt cloth. The weatherman pointed to "a weak cold front stretching northwest to southeast" of their position. Kosco said he believed they were inside this "frontal zone," which would account for the strong winds, heavy seas, and moderate rain squalls.

Then the aerologist showed Halsey the "storm center to the southeast" that was advancing northwest at "12 to 15 knots." Kosco estimated it was "about 400 miles away." From "all indications and reports," he went on, it was "only a tropical storm"—the same one he had been watching for more than a day. As he had previously predicted, he still expected this storm to meet up and merge with the cold front, dissipate with the loss of warm air needed to fuel tropical storms, and "move off to the northeast" as it weakened.

When asked by Halsey where they might find better conditions for fueling, Kosco recommended a course "at right angles to the cold front" and moving a "safe distance behind it." The area he thought best for fueling in the morning was 140 miles to the northwest, which would take the fleet away from where he believed the center of the tropical storm to be located.*

Halsey liked the idea of heading to the northwest because it would shorten the return trip to Luzon. Since they were losing another day fueling, he figured they would be "cutting things pretty close." As he was determined to launch strikes on time against Luzon—"storm or no storm"—Halsey ordered fueling discontinued and for Task Force 38 to form up in a semicircle that stretched for miles and proceed on a new fleet course at 17 knots toward the rendezvous point at 17 degrees north

---

* According to later testimony by Captain Wilbur M. Lockhart, the senior aerology officer of the Pacific Fleet at Pearl Harbor, in "hindsight" this was a crucial moment and key mistake. Had the Third Fleet turned to the south or southwest at this time and proceeded at "10 knots or more," the assembled ships would have been "clear of the high winds" and the storm center would have been "at least 100 miles to the north" by the following morning (December 18, 1944). However, it was not the only ill-advised course change the fleet would make over the next eighteen hours.

and 128 degrees east, where fueling would commence at 6:00 A.M. Several destroyers so low on fuel that they could "not be expected to operate more than two days at moderate speeds without running empty" were directed to proceed to the fueling rendezvous at a slower speed in company with the tankers, and en route to avail themselves of any opportunity to fuel should conditions allow. For some destroyers, it figured to be a close call. *Maddox,* for example, reported 14.6 percent fuel capacity, with *Hickox* at 18 percent, and *Buchanan* reported that she could make the rendezvous only by "mixing diesel and black fuel oil."

Soon left in Halsey's wake and proceeding several knots slower was Acuff's logistics group. They fell farther behind as the tankers tried for the remainder of the afternoon to service the destroyers that were so desperately low of fuel. At times, they turned off course to go downwind while attempting to fuel the destroyers over the stern, a "valiant but futile" makeshift effort for which the oilers were not rigged and which resulted in many snapped hoses.

Left behind with the replenishment group, *Spence* tried to fuel twice more that day but was thwarted each time. The destroyer butted hulls with one tanker—unbelievably, *Spence's* second collision of the day, which ended further alongside fueling efforts. They next tried to fuel from astern of a tanker, which was observed by Supply Officer Al Krauchunas, the ship's paymaster and the former professional ballplayer. The attempt had to be abandoned due to a "lack of visibility" that prevented seeing the "inflated canvas ball to which a line was attached"— the line by which the oil hose at the tanker's stern was to be pulled over the destroyer's bow.

As *Spence* expended more fuel and the level dropped in her tanks— now around 15 percent capacity—the destroyer rode "high on the water like a Spanish galleon," which resulted in a loss of stability as well as providing the wind with more "sail" surface above water. The procedure for increasing stability in low-fuel situations was to take on ballast, filling empty oil tanks with seawater to provide weight and increase the ship's draft in the water—a "relatively simple" task, but one that took time and had to be monitored to ensure that no ballast water got into the tanks that were feeding the boilers because the water would put out their fires. Krauchunas learned that the new skipper had decided against ballasting after "word was received" that fleet fueling was to commence at sunrise. Deballasting, or pumping out the seawater, took time, too—up to "six hours" for a fully ballasted destroyer—and had to be completed before tanks could be filled with fuel, as *Spence* had done during the night en route to the morning fueling rendezvous. After the long day of fueling failures, Andrea did not want *Spence*, which was expected to rejoin Halsey's Luzon-bound task force after fueling, to be the cause of further delays in the morning should the storm subside enough to allow fueling, only to find their tanks filled with briny seawater. Andrea's choice was a gamble: if the storm worsened and fueling was cancelled, *Spence*, in a light state, would be ill prepared to ride out heavy seas.

By midafternoon on *New Jersey*, Kosco was reviewing new weather information, and some of it was conflicting. The afternoon weather summary from Pearl Harbor indicated the presence of a "severe cyclonic storm . . . 160 miles" northeast of the fleet, which was believed to be moving to the northwest. A delayed aircraft report from earlier that morning had the tropical storm "less than 200 miles southeast" of the fleet, in the same direction but much closer than Kosco's own estimate of "400 miles to the east." Still, to Kosco's knowledge "no typhoon warnings had been broadcast" by any ship or land-based station, and he continued to believe they were dealing with "one of those violent tropical disturbances that spring up." While he knew such storms could be "troublesome enough," he didn't consider them "capable of the devastation" that could be dealt out by typhoons. Some data he reviewed disagreed with the location set for fueling as potentially being in the path of the storm should it recurve to the northward. Some of the Third Fleet's other weathermen—every aircraft carrier had its own aerologist and fully staffed weather office—discussed the situation over the TBS radio. Listening in, Kosco came to the conclusion that it "might be a good idea to go south." With the wind "at the time from the north," he thought the fleet might have an opportunity "to make a run for it," and recommended so to Halsey. After discussing the "pros and cons" of a run to the south with his staff, Halsey cancelled the previous rendez-vous and designated a new location to the southwest at 14 degrees north, 127 degrees 30 minutes east.

That evening, Kosco studied "every report available" about clouds, seas, wind, and pressure readings in preparation for the usual 10:00 P.M. "round-table rehash" Halsey held with his staff. Some of the data even originated from intercepted Japanese reports, but most came from U.S.-held bases and ships at sea. Kosco scoured everything, searching "vainly for the evasive center of the storm," knowing that only by locat-ing it could he "predict its behavior." With a usual mix of scattered and conflicting reports, Kosco fell back on his extensive book learning and "historical data" to try to predict the storm's behavior. Unfortunately, there wasn't much of a written record for the region, as U.S. naval forces had not been operating in the far western Pacific for long.

Messages from the logistics group indicated it was not possible for Acuff's tankers and other slower ships to reach the rendezvous point by 6:00 A.M. At the staff meeting, a place roughly midway between the two earlier positions was selected—west of the fleet's position. Task Force 38, however, was closer to it than the logistics group and would reach the rendezvous well before sunrise. Both to "kill time" and "seek better weather," Halsey directed a southerly heading beginning at midnight, then "reversing course" to the northwest in two hours for the new fuel-ing rendezvous with the logistics group at 15 degrees 30 minutes north and 127 degrees 40 minutes east.

Following a "bit of shut-eye," Kosco awakened after midnight, feel-ing that "something was wrong." He went to the weather office and checked the reports. Although he continued to see nothing to cause him to think they were dealing with more than an intense tropical storm, he decided that it would be "safest" to continue on a "southerly course to clear the storm." After briefing Carney, Kosco was taken by the chief of staff to the admiral's quarters.

"What do you think of a turn to the north?" Halsey asked.

Kosco reiterated that the safest course was to continue southward and that there was a "danger" that a turn to the northwest might take the fleet "nearer the track of the storm" and cause it to "hit us." He explained that the storm was "increasing in intensity."

At that point, Halsey asked what it was like outside.

"Severe," Kosco said, "although not excessively so."

Halsey discussed his operational concerns with Carney. Continuing to the south, while it might be the safest "in a weather sense," would put them farther from Luzon. Also, spending time to search for calmer conditions in which to fuel would take up valuable time and delay the planned air strikes. Halsey expressed his determination "not to be feinted out of position" for the Luzon strikes by bad weather.

Carney, forty-nine, a methodical planner who nevertheless put value on doing the unexpected in the course of battle, knew the thinking processes of Halsey, whom he informally called "Admiral Bill," having been his chief of staff for the past eighteen months. Prior to that, Carney, a 1916 Annapolis graduate who saw action against German U-boats in World War I, had commanded the cruiser *Denver* and was twice decorated for valor in the Solomon Islands. Once, proceeding through unfamiliar waters near Bougainville, Carney took advantage of adverse weather to lay a large quantity of explosive mines along sea lanes extensively under use by the enemy while also bombarding shore installations. Like his boss, Carney knew that MacArthur was "counting on" their carrier air support for his invasion of Mindoro. The chief of staff understood that Halsey felt he should "live up to that commitment . . . rather than retreat" before bad weather. By giving the word, Halsey could have "deserted MacArthur and headed south for a hundred miles or so" and "bypassed" the approaching storm, but after being caught out of position so recently at Leyte, Halsey felt he must "stay until the last minute." In any case, it was the four-star admiral's decision to make, and nobody—Carney included—was "disposed to argue with it" once Halsey's mind was made up. Halsey decided to let stand his order to turn to the northwest at 2:00 A.M.

With "doubts about the wisdom of changing course to the northwest," Kosco returned to the aerology office. He felt as if they were engaging in "hand-to-hand combat" with the storm rather than doing any "long range planning." Getting "out of the way" seemed the best action, and for that reason he had advised continuing to the south. He put in an order to be awakened when the 3:00 A.M. Fleet Central weather report came in.

At 2:00 A.M., Task Force 38 changed course to 320 degrees. Almost immediately, the barometer dropped and the weather worsened.

The fleet was in the path of the typhoon that Army meteorologist Reid Bryson had issued a warning about thirty-six hours earlier.

WHEN KOSCO AROSE from a "short and restless catnap," he made his way to the navigation deck. Now as he climbed the outside ladders the wind tore at him and the seas were so "turbulent they made even the mighty *New Jersey* roll and sway." After checking the latest reports, Kosco decided that "things didn't look that bad" on paper. Although the barometer continued to fall (29.65 inches at 3:00 A.M.), he still believed the worsening weather was "due to a tropical storm passing."

At 3:45 A.M., Kosco made his way to Carney's cabin to brief him, and together they went to Halsey's quarters. The admiral came out of his bunk and put on robe and slippers, and the three of them went into the flag wardroom.

Kosco said all indications were that the storm was increasing in intensity and was "almost due to hit us if we continue to the northwest."

Halsey asked, "What do you recommend?"

"We turn immediately to the south," Kosco answered.

Halsey asked Kosco to conference with other group commanders and their weathermen via TBS radio to see what he could glean. The aerologist did so around 4:30 A.M., coming away with various opinions as to the storm's position and path—or even what to call it, some saying "cyclonic storm" and others "typhoon," a term Kosco had avoided because he did not feel he had a large enough "field of reports" from which to make that forecast, and did not want to make a "snap judgment" and sound a typhoon warning on a "mere hunch."* There was agreement that a course to the south would be best to avoid being overrun by the storm. It was also unanimous that no ships would be successfully fueled that morning in such heavy seas. One task group commander, Rear Admiral Gerald Bogan, who had sent his own message that morning to Halsey recommending a run to the south to avoid the storm, considered the possibility of fueling "under existing conditions very slim."

Shortly after 5:00 A.M., Halsey "reluctantly decided" to turn to the south. He ordered all ships to change to a southerly course and maintain 15 knots. "Loath to give up his last attempt" to fuel in time to carry out the planned strikes against enemy air forces on Luzón, however, Halsey directed the fleet to "commence fueling as soon as possible" after sunrise, adding that the destroyers should fuel first from astern of the oilers if necessary—a method that had been "clearly demonstrated" as "not feasible" the previous day.

Halsey's two orders—go south and fuel—were "not compatible." With heavy winds and swells from the north, it would not be possible to fuel on a southerly course. To fuel, the ships would need to turn around and head upwind—into a "collision course" with an enemy of a different kind.

AFTER A STARLESS, "blacker than pitch" night of howling winds and mounting seas, the sun did not rise at dawn on December 18 as much as it backlit a lurid yellow-gray sky, illuminating the heaving swells pushing down from the north that formed endless "mountains of waters."

---

* One aerologist who took part in the radio conference and did forecast an approaching typhoon was Lieutenant Daniel F. Rex, aboard the carrier *Hornet* (CV-12). So certain was Rex that he joined aerologists from other carriers in sending a direct message to Halsey soon after the conference. The weathermen warned if the fleet did not clear the area immediately, they would rendezvous "in about eight hours" with a typhoon. After the war, Rex became a prominent meteorologist—studying under the renowned Carl-Gustav Rossby at the University of Stockholm, where he received a Ph.D.—and serving as commander of the Naval Weather Research Facility prior to his retirement from the Navy in 1962 as a captain. Halsey was inclined "not to believe aerologists," Rex later explained.

The wind—"a hell's chorus of fury"—whined deeply and sorrowfully, as if the "sea and air were in pain." Clouds of spray, spume, and rain whipped across the surface, blending sea and sky, obliterating the horizon, and lowering visibility.

At 7:00 A.M. the Third Fleet turned northward into the brunt of the storm for fueling. Before the course change, Kosco briefed Halsey on the worsening weather, and the aerologist "voiced an opinion" that continuing south would be safer, while heading north would "probably take us back toward the storm." After discussing the situation with his staff, Halsey decided that, "rather than to retreat," they would "stay in the area" and "carry out our commitments."

On many ships, there was less thought that morning about fueling and more about how to "just stay afloat." Directed to a new course of 060 degrees and a speed of 10 knots, the ships came about in "very rough seas" against headwinds up to 50 miles per hour, and many were soon "riding as though caught in some giant washing machine." The fleet had never before attempted to fuel in such rough seas, and under these conditions Acuff decided that his "oilers [should] not attempt to fuel" the destroyers. Rather, the destroyers would be brought alongside aircraft carriers, utilizing the greater stability of the larger ships in the undulating seas. A destroyer was ordered alongside each of Rear Admiral Bogan's three big carriers, *Hornet, Hancock,* and *Lexington.* Observing from his flag bridge on *Lexington* as one yawing and bouncing destroyer approached—disappearing except for her mast into a deep trough before rising atop a high swell—Bogan "didn't like" what he saw and directed the destroyers to clear lest they collide with his carriers. Not long after, a message was sent to Halsey stating that "present conditions" did not permit fueling the destroyers by any method.

The 7:30 A.M. fuel reports of destroyers low in capacity were:

Yarnall, 20 percent;
Stockham and *Welles,* 22 percent;
Moore, 21 percent;

Taussig, 18 percent;

Colahan, Bush, Franks, Cushing, and *Wedderburn,* 15 percent;

Maddox, Hickox, and *Spence,* 10–15 percent.

A little after 8:00 A.M. Halsey ordered the fueling efforts discontinued and "regretfully notified" MacArthur by radio dispatch that he "could not meet our commitment" to strike Luzon the next day. The fleet turned south again.

Shortly afterward, radio reports came over the TBS of ships "pounding heavily in the mounting seas" and experiencing loss of steering control, engine difficulties, and men washed overboard. The smaller carriers with the fueling group, not as stable as their larger brethren, reported planes breaking loose from their tie-downs, colliding and slamming into bulkheads, and catching fire.

It had become "rather apparent" to Kosco that they were dealing not with an "ordinary tropical storm" but with one that was exhibiting "typhoon conditions," as the wind and sea continued to raise havoc with the fleet and the "ceiling rested virtually on the deck," with visibility down to 500 yards. He watched with alarm the barometer on *New Jersey* "falling very, very rapidly" from 29.59 to 29.20 inches in one hour—the "typical barometric nosedive of a typhoon." Then the "wind went around further to the northwest," backing counterclockwise—another "sure sign" to the weatherman of a typhoon.

Kosco believed the fleet would be far enough south to escape the mountainous, confused seas that he knew could cause damage to "all ships," and which any smaller ship "improperly ballasted would not be able to ride out." Still, he thought about their backtracking to the north that morning in a "last, vain attempt" to fuel. He wondered how "much difference" the short (an hour in each direction) but unfortunate detour in the wrong direction—keeping them that much closer to the center of the storm—would make in the end.

The aerologist now understood that the Third Fleet was "making a race" with a full-blown typhoon to get out of its path.

Twelve

That morning aboard *Spence*, Quartermaster 2nd Class Edward F. Traceski, twenty-two, born in Turners Falls, Massachusetts, to Polish immigrant parents, assumed the 8:00–to–noon watch in the wheelhouse. Stationed near the helmsman and with a panoramic view from the bridge's long row of brass-framed portholes, Traceski maintained the daily deck log, "recording all the happenings of the ship, course changes, speeds, and formations."

A slender five foot ten with thick brown hair and a shy smile, Traceski had come aboard *Spence* in the Boston Harbor in June 1943, in time for the trip through the Panama Canal out to the fight in the Pacific. "Every-place we stopped," he noticed, "they put on more armaments and more armaments," all of which *Spence* had used amply during those heady Little Beavers days.

Traceski, who had stood two four-hour watches on December 17, was on the bridge for several fueling attempts. For a day and a half it had been an increasingly "rough ride" in tumultuous seas—made all the worse by *Spence*'s lighter-than-normal condition with mostly empty fuel tanks. Throughout the night they had steamed on only one boiler to conserve fuel, although in order to "maintain station" in the formation of the replenishment group *Spence* had to keep her speed between 10 and 12 knots. No one below deck had gotten much rest or food; everyone's main preoccupation had been with finding someplace safe to hold on as the destroyer was tossed about in high seas.

When Traceski came on duty, he took one look into the near-zero visibility outside and knew there would be "no possible chance to fuel" this morning. Minutes later, the fleet's latest plan to fuel had been scrubbed. He was "not sure" whether the skipper had "done the right thing" in not taking on ballast during the night, as Traceski heard over the TBS other ships had done.

Also coming on watch at 8:00 A.M. was Seaman 1st Class Edward A. Miller, nineteen, of Clark, New Jersey, a former deckhand who had previously witnessed the nearly disastrous attempt to fuel *Spence* alongside a tanker, stopped after the destroyer "slammed into the side of the tanker, jolting several men off their feet" and causing some injuries.

A happy-go-lucky former Boy Scout who had been advised by an uncle who served years earlier in the Navy to do the same so he could have a "dry bunk and three hot meals a day," Miller had enlisted in January 1943 and gone to boot camp at Great Lakes in Illinois before being assigned to *Spence*. He had requested destroyer duty because his uncle told him there were fewer by-the-book regulations aboard smaller ships. From the beginning he had "loved the dungaree navy" and the camaraderie of knowing everyone aboard, not possible on carriers or battlewagons with crews numbering in the thousands.

Before going on watch, Miller went by the galley and found that the cooks had discovered it "simply out of the question" to serve hot meals.

Most of the food they tried to prepare had ended up on the deck, along with assorted pots and pans. *So much for three hots a day,* Miller thought as he downed a few sips of lukewarm coffee and stuffed a cold dinner roll in his pocket before taking off down a passageway. Even his rolling, well-balanced gait on experienced sea legs was not enough to keep him from staggering like the town drunkard, bouncing from one bulkhead to the other. When he emerged on deck heading for the outside ladder to the bridge, Miller saw the seas "were like mountains" and the winds were "blowing the tops off the waves in a foamy, horizontal spray as if a giant was blowing the froth from an overfilled beer mug." The wind-driven rain stung like needles on any exposed skin, and the salt spray was so thick it was difficult to see anything. When he reached the main battery's fire-control compartment above the bridge, he opened the hatch and quickly closed it behind him, relieved to be out of the elements. Here, in a cramped space with two other sailors and a gunnery officer, he was scheduled to spend the next four hours, ready to control the firing of *Spence*'s 5-inch deck guns in the unlikely event they had to be used in such heavy weather, which usually found submarines staying below periscope depth and enemy aircraft grounded.

That morning in the wardroom, supply and disbursing officer Al Krauchunas took part in a meeting in which the ship's officers "conversed with the captain over the problems of the day." Over inky coffee and sliced cold cuts, Jim Andrea asked to hear from each division head about any "problems they were encountering" or which they anticipated would "present themselves during the day." It was a calm and orderly meeting, and Krauchunas, when asked, gave an upbeat report on the status of the ship's supplies: they had consumed 20 percent of their provisions and had plenty of stores.

The chief engineer delivered news of a more dire nature. Lieutenant (j.g.) Lawrence D. Sundin reported they had remaining only "9,000 gallons of fuel; 5,000 gallons in forward tanks and 4,000 gallons aft"—which was approximately 6 percent of the destroyer's capacity. As long as they could continue to operate on one boiler and maintain their consumption rate of "300 to 350 gallons per hour," Sundin estimated they had enough fuel for "24 hours at a speed of eight knots." After that, they would be dead in the water. They were carrying 10,000 gallons of diesel fuel for the auxiliary generators, Sundin said, but he did not want to burn diesel unless they had to because it contained residual water, which could put out the fires in the boilers and contaminate the plant.

Sundin also pointed out that the vessel was operating below the ballasting point required by the Bureau of Ships. *Fletcher*-class destroyers were instructed to carry 200 tons or 60,000 gallons of fuel and/or water ballast—"about 40 percent of capacity"—to maintain optimal stability in any sea state. It was strongly felt by Sundin that if fueling was not going to take place that morning, the pumping of seawater into the empty fuel tanks should commence.

Krauchunas knew Andrea had not taken on ballast during the night because he wanted the ship to be ready to receive a full capacity of fuel so they could return to their screening duties with Halsey's task force. Without ballast, Krauchunas had not considered them to be "in danger of turning over, nor was there any thought along that line." In fact, "normal routine was carried on through the night." The ship had stayed in damage-control condition Baker—the "normal wartime cruising condition which left open those watertight doors necessary to permit a normal flow of traffic" between compartments and levels.

Not long after the wardroom meeting broke up, Chief Water Tender James Felty, who had turned twenty-five two days earlier, was ordered to transfer fuel oil from two forward wing tanks to an inboard tank that provided fuel for the forward boilers. The wing tanks, A-507 and A-508, were then to be filled with seawater ballast. Each held 8,000 gallons.

Born and raised in the heart of coal country in Matewan, West Virginia, Felty had joined the Navy in 1937 at age seventeen, intending to make the Navy a career. His two older brothers also "took off for the Navy" as soon as they finished high school, so determined were the Felty boys not to follow their father, and nearly every other adult male they knew, into the coal mines.

Strong and long-limbed, Felty was a steady and accomplished worker who took pride in being one of the plank owners of *Spence,* which he considered a "good ship" after earlier assignments on a fleet tanker and an old four-pipe destroyer. Felty's first action had been in the Solomons, and even working below deck in the fire rooms he had heard plenty of gunfire. He grew accustomed to being in combat without actually seeing anything; the secret was paying attention to one's own job, and there was always much to do in engineering. In the days when *Spence* fought under Arleigh Burke, they had "all four boilers on all the time" in order to produce ample steam to feed the turbines that powered the propellers to answer the calls from the bridge for added speed, which Felty learned meant "right then" and not later.

Shortly after 10:00 A.M., Felty finished transferring the fuel. A few minutes later the order came to commence ballasting, which Felty did, passing the word over the sound-powered phone to two men standing by to take soundings on the wing tanks, the tops of which were accessed through a hatch in the mess hall. Felty learned from speaking on the phone with another chief water tender in the after fire room that pumping had started back there, too, with salt water being directed into empty fuel tanks C-8 and C-9. As the ship could take on about "30,000 gallons of ballast in one and a half hours" with simultaneous pumping forward and aft, it would be two hours or so before *Spence* reached the level of 40 percent capacity required for proper stability.

About thirty minutes later, Felty, unable to reach the men at the wing tanks, decided to "go up and see what was the matter." That took him a short distance across the heaving, slippery main deck. When he reached the mess hall, he found everything okay with his men and the tanks filling without any problems. Before going back to the fire room, Felty went forward to the chief's berthing compartment in the bow and retrieved his kapok life jacket, which he slipped on and secured for another weather deck crossing.

As Felty headed down a port passageway he was thrown hard against the bulkhead when *Spence* took a big roll to port—so far over that he found himself on his hands and knees on the bulkhead listening to the din of loose gear crashing about him. The ship hung there momentarily before slowly starting to right herself, as she was designed to do. It was then that the lights flickered and went out. No emergency lights came on in the passageway. "In complete darkness," Felty worked his way down the passageway until he came to a port hatch open to the main deck. Eight or ten young sailors were gathered at the hatch, not sure whether they should stay inside the listing ship or take a chance out on the deck awash in churning green water.

Felty still intended to go out on deck to the hatch and ladder that went to the forward fire room because "it was my job to be there," figuring since he had made it across the deck once he could do so again. But now his exit was blocked by the "scared sailors," who would not move from the opening.

At approximately 10:30 a.m., supply officer Al Krauchunas left the galley after making sure that sufficient sandwiches had been prepared, since the crew would be subsisting on them until further notice. Clutching the handrail as the ship "pitched and tossed," he made his way to his disbursing office, where he checked to see that all cabinet doors, furniture, and equipment had been made secure. He decided to go back to the wardroom, a journey that would take him on deck for a short distance. He did so facing "hazardous footing and raging wind." On one roll to starboard, the whaleboat was engulfed in the flood and ripped free of the davits holding it in place. The whaleboat swept by, "narrowly missing" Krauchunas before disappearing over the side. He allowed himself the thought that it would be his responsibility to report the loss to the Bureau of Supplies and Accounts, and wondered if a checkmark in an appropriate column would suffice or whether he should write an eyewitness account.

"Slowly and steadily," Krauchunas made his way forward, overcoming the "blinding mist, wind and water" to stumble into the passageway leading to the wardroom, where he landed "completely drenched." When he found the wardroom deserted, he decided to "seek relief from the pitching and rolling" by going to his nearby stateroom and "hitting the sack." Once there, he removed his wet clothes, donned fresh dry ones, and lay in his bunk listening to the sounds of *Spence* enduring "continuous bucking and rolling," including the roll to port from which the ship had not come fully back.

Krauchunas heard a phone ring in an adjacent room. Soon the chief engineer peered into his room. "You better get topside, Al," Sundin said. "We've taken water down the stack into the engine room." Sundin hurried away, closing behind him the watertight hatch leading from officer country.

Having recently been topside and nearly washed overboard with the whaleboat, Krauchunas still thought his stateroom was a safer place to stay. But then the lights blinked and went out. A dim emergency light flickered on in the passageway, and Krauchunas, "concerned with the sudden turn of events," decided to head back to the wardroom. He stopped at the door to the wardroom and peered inside. Several officers not on duty were now sitting on lounges and in chairs, each with a "concerned look" on his face. Deciding the atmosphere was "not at all encouraging," Krauchunas turned away and started down the darkened passageway toward the main deck, "struggling to remain upright," as the ship had never straightened up again after rolling to her port side. As he passed the captain's stateroom, he heard a voice "spouting a few words of anger and disgust." Krauchunas looked in and saw the ship's medical officer, Lieutenant (j.g.) John C. Gaffney, sitting in a chair and having trouble keeping it from sliding across the compartment. Krauchunas liked Doc Gaffney and decided to stay. He sat down on the lower bunk of a double-bunk unit, and the two men exchanged "words of fear and some assurances regarding the storm."

Spence took a "sudden roll to port" and remained in this position for "agonizing moments" before slowly starting to come upright.

Krauchunas by then was on the deck, having been thrown backward off the bunk into the passageway and showered with books, ashtrays, and other loose items from the cabin. Without hesitating, he scrambled on all fours for the nearest hatch—left open or pushed open—that led onto the deck.

Then *Spence* rolled back to port just as far as before, only this time she did not stop until "almost completely upside down."

Krauchunas, grasping the railing of a ladder as a torrent of seawater gushed through the open hatch, found himself almost instantly underwater. With lungs bursting for air, Krauchunas experienced "a hopeless feeling." He wondered: *Is this the way I'm going to die?*

The wounded ship, still fighting, came back slightly, which abated the inflow of rushing water long enough to allow Krauchunas to surface and gasp four or five gulps of air. He pulled himself along the ladder closer to the opening. Now the ship began to settle again to her overturned position amid "gurgling and sucking sounds" as the sea again poured in.

Krauchunas went under, reaching out in the dark for the open hatch.

ON THE BRIDGE, quartermaster Ed Traceski noted the time—11:00 A.M.—when *Spence* lost all electrical power after seawater poured down the after fire room ventilator shaft, shorting out the ship's main electrical switchboard. Up until then, "all the reports of damage" the bridge had received involved the loss of the whaleboat and a depth charge breaking loose from one of the racks.

For most of the morning the captain had been at the conn, with the executive officer also on the bridge along with the duty OOD, Ensign William I. Sellers, the ship's sonar officer. Chief Quartermaster Harlan K. Carrigan was handling the wheel, an exhausting job in such elements when both ship and sea exerted powerful wills of their own. For hours they had listened to reports coming in over the TBS about the struggles of other ships in the storm and men washing overboard. Concerned about "the safety of his crew," Andrea had ordered all topside watches—those in exposed areas on deck—secured to prevent mishaps. Addressing the crew over the ship's intercom, Andrea directed "all personnel not authorized topside" to remain below in their compartments, and anyone with business topside "should wear life jackets."

With engineers working below to restart the ship's main steam-driven generator and restore electrical power—the cessation of which stopped not only ballasting but the electric bilge pumps used to control flooding—*Spence* continued to try to maintain her assigned station in the formation. Although still listing after rolling as much as "43 degrees to port," no distress call was broadcast. In fact, at 11:17 A.M., Andrea sent a routine voice message to his task unit commander over the TBS: "My last position 7,000 yards north of Task Unit 30.8.4," he said calmly, "course 220 degrees, formation speed."

Steaming at 12 knots, *Spence* had winds "well over 100 knots (115 mph)" off her starboard quarter and "huge" swells rolling in from dead astern as she ran southwest. The wind and unruly seas had pushed the destroyer "a little left" of the formation course that was "supposed to be steered." Eager to get back on their assigned course, Andrea ordered, "Hard right rudder."

As directed, the helmsman spun the wheel to 30 degrees right rudder, which immediately increased the list to port to 47 degrees, as noted by Traceski on the wheelhouse inclinometer.

"Caught in a trough" between swells higher than her mast, *Spence* wallowed. When the helmsman tried to correct, Traceski saw that he "didn't have steering control"—the rudder stayed jammed on hard right. *Spence* laid way over, hung there, came back only partway, and then laid over again.

Just before the combined forces of wind, sea, and instability joined for one final push and *Spence* "gradually capsized," Traceski, who was trained to make note of important shipboard events, looked at the bridge clock. It read 11:23 A.M.

Thirteen

Storekeeper Ken Drummond, who had feared *Hull*'s luck had run out as he listened to the new captain's first speech to the crew two months earlier, saw a shipmate die around 10:00 A.M. on December 18.

Drummond, who stood his regular watches in the radar shack, tucked into a corner of the wheelhouse, arrived on the bridge that morning for the 8:00-to-noon duty. Crossing the storm-tossed deck from the aft hatch where the ladder rose from the crew quarters had been "really tough . . . holding on to whatever you could find" as he worked his way forward. With the ship rolling so steeply, the radar dish on the mast swept straight down into walls of seawater, obscuring the returns on the screen. Drummond, who alternated with another radar operator every hour, put on his life jacket "because it seemed like a good idea" and, "trying to stay out of the way," stepped onto the port wing of the bridge, which was enclosed except for a small platform on each side—called wings—from which lookouts and deck officers stood to search for obstructions in the path of the vessel and to monitor underway replenishments or dockside approaches. Leaning against the bulkhead near the entrance to the bridge, Drummond held tightly to a railing as the ship pitched wildly in swells that broke forcefully over the forecastle and weather decks, shooting spray higher than the bridge. Whenever it was his turn to man the radar, Drummond came inside, where he found neither the bridge nor the wheelhouse "a great place to be," with all hands struggling to remain upright.

Out on the starboard wing, also at a railing bracing himself against the storm, was Radarman 3rd Class Billy Bob Dean, nineteen, a "good ole boy with a twang" from Texas who had never worn anything on his feet other than cowboy boots before joining the Navy. Finding that regular shoes hurt, he had received permission from Consolvo to wear his cowboy boots—and so far the new by-the-book skipper had "not become aware" of the Texan's nonregulation footwear.

On one steep roll to starboard, water came up as high as the bridge. As the raging foam cleared, Drummond looked across the bridge to see Billy Bob parallel to the deck, twisting in the wind like a flag in a gale and struggling to keep hold of the railing. When he lost his grip he fell, his body slamming against the rigging of the whaleboat before being flung overboard like a cloth doll. Drummond had little doubt the likeable Texan was dead before hitting the water. Most of the bridge personnel saw the accident, although "no one said anything." Drummond believed "all thought it was a lost cause" and there was "nothing anyone could do." The ship, fighting to maintain station in the formation, did not slow or alter course.

After all the battles and close calls since December 7, 1941, *Hull* had taken her first wartime casualty. The loss of Billy Bob Dean, washed overboard in his favorite cowboy boots, became for Drummond a harbinger of the might of nature's combined forces. At that point, the situation "really started getting bad," as the seas took on "those confused pyramidal shapes" characteristic of a typhoon with wind velocities in excess of 100 miles per hour.

Chief Quartermaster Archie DeRyckere, who had been at the helm two months earlier during full-power runs in Puget Sound and was shocked by the increased top-heaviness of the old destroyer after the latest additions to electronics and armament, had assigned two helmsmen to alternate at the wheel that morning. Given the physically exhausting work of steering in the violent seas, Quartermaster 3rd Class August R. Lindquist and Quartermaster 3rd Class John T. Horton, "two of the best helmsmen in the Navy," relieved each other every twenty minutes.

Pacing the wheelhouse and bridge as he kept a wary eye on the worsening weather—"so thick and dirty that sea and sky seemed fused"—DeRyckere was still fuming over the scuttlebutt from *Hull*'s radio shack that an admiral on an aircraft carrier had been heard on the TBS radio earlier that morning asking Halsey for permission to steer around "this typhoon" because he was concerned about the "small boys." Upon hearing the one word that filled with dread every sailor in any man's navy, DeRyckere had railed: "Typhoon! What in the hell are we doing near a typhoon?" The reported casual response from Halsey on *New Jersey* that the storm would give the crews on the smaller ships a "chance to practice seamanship" had caused DeRyckere's blood to boil.

Having recently been tested for his chief's promotional examination on Bowditch's *American Practical Navigator,* including the "Cyclonic Storms" chapter, DeRyckere was well versed on the advice given to mariners for "fixing the bearing" of a cyclonic storm center in the Northern Hemisphere: when an observer was standing facing the wind, the center would be found "approximately 10 points (112 degrees) to the observer's right." When DeRyckere went out on deck and faced the wind, he placed the storm to the west. And yet for one hour that morning the fleet had steamed *toward the typhoon* to facilitate a fueling operation that anyone in his right mind who looked outside at the tumultuous seas knew would never take place. DeRyckere understood that Bull Halsey could be aggressive and stubborn, and after being "fooled by the Japanese at Leyte" the feisty old sea dog had proved he was not the world's greatest naval tactician, but hadn't Halsey read Bowditch? Hadn't the admiral ensured something as basic as checking the wind conditions to find the storm center? Instead, he had kept the fleet "maneuvering right in front" of a *typhoon*? If there was any seamanship to practice, DeRyckere seethed, it damn well should have started with Halsey.

Boatswain's Mate Ray Schultz was equally perturbed, not only with the reported admiral-to-admiral TBS radio chat and news of the approaching typhoon but also with one James A. Marks, whose initials he had morphed into the moniker "Jack Ass Marks," which he peppered his speech with whenever talking about the unpopular commanding officer. Since his embarrassing performance at the conn upon their arrival at Pearl Harbor two months earlier, Marks had done nothing to redeem himself with the crew. In fact, when they entered Ulithi three weeks earlier, Marks had ordered the men on watch topside to change into undress whites for their arrival at the big Navy base, although the uniform of the day in the Pacific theater had long been blue dungarees for enlisted crewmen because whites provided good targets for enemy gunners. A message from an admiral ordered *Hull*'s captain to "report to headquarters immediately." After docking, Marks did as directed, and word soon spread among the crew that he "got his butt chewed off for having us in the wrong uniform."

Around 9:30 A.M., Schultz made his way across the heaving main deck to the bridge, where he asked Marks for permission to "ready the ship for heavy seas." Schultz knew as well as anyone in the Navy that the *Farragut*-class destroyers had been top-heavy from their earliest days, given their narrow beam. After the addition of topside weight through the years, they had become "too top heavy" even in normal seas. With a typhoon approaching, the experienced sailor figured anything done to reduce topside weight would be beneficial.

Marks asked what it was that the boatswain's mate proposed. Schultz explained that he would organize a special detail to "button the ship up tightly" and "stow below all the ready ammunition" kept topside.

Marks looked as if Schultz had lost his mind. "*Stow the ammo?* We're in a war zone. We can't do that."

The gung-ho captain was still looking for his first combat in the Pacific, thought Schultz, who knew full well that no enemy ships, subs, or planes would be operating in such heavy weather. Schultz next asked permission to cut away the 26-foot motor whaleboat, which was "dipping in the water" and filling up on every steep roll to starboard. Knowing that a cubic yard of seawater weighed nearly a ton, the boatswain's mate knew the added weight on the side was not helpful to a ship already struggling to right herself. Too, he thought it likely that the whaleboat could break loose, and perhaps in the process cause other damage and even injure someone.

"Permission denied," Marks said sternly. "Get off my bridge."

Never one to let an ill-conceived order slow him down, Schultz gathered his gang of deckhands and went about the ship securing loose gear, setting condition Affirm by closing the watertight hatches between many compartments, and otherwise preparing the ship to ride out the typhoon, although even the irascible Schultz was not so insolent as to stow the big ammo lockers or cut away the whaleboat without orders to do so.

Although Marks had "expected to receive orders to fuel" that morning, *Hull* was—after the cancellation of fueling operations—better off than *Spence* and a number of other destroyers. As of 8:00 A.M., *Hull* had aboard "about 124,000 gallons of fuel," representing 68 percent of capacity. Marks had been informed that morning by the ship's engineering officer that they were "above the required ballasting point" as directed by Bureau of Ships for *Farragut*-class destroyers, and therefore he had elected not to take on seawater ballast. The required level for ships of the *Farragut* class to maintain stability was 82,500 gallons, or about 45 percent of fuel capacity, although that figure was based on a 1942 inclining test and not the more current fall 1944 inclining test of the destroyer

Aylwin in the Seattle shipyard. As a result, the correct level of fuel and ballast had never been recalculated after all the wartime additions and modifications to the *Farragut*-class ships.

By 10:30 A.M., *Hull* was taking a terrific pounding. Her bow would rise out of the sea on a giant swell only to slam down moments later, which then brought the stern out of the water along with the rudder and spinning screws. At the same time, *Hull* was being knocked over repeatedly, taking "steady 70 degree rolls to starboard," recovering briefly, then laying over again. During one roll, the front davit that held the whaleboat in place sheared off, and the boat swung wildly, snapping the rear davit. Free of its restraints, the whaleboat "washed right down the deck," careening against the torpedo tubes and crashing into the after deck house.

Schultz, nearby securing some gear topside, saw what happened next in slow motion as if stuck in a horrible nightmare. A group of sailors, "about twelve of them" by his count, had emerged from the after hatch leading up from the crew quarters and were working their way forward between rolls. The absence of a covered, safe means of passage between the forward and after parts of the ship—one of the design flaws of the *Farragut*-class destroyers—required crew members to navigate across the wet and slippery main deck. The motorboat, after bouncing off the torpedo tubes and the aft deck house, now slid into the group of sailors, knocking them down like bowling pins. The boat along with the men, many of whom did not have on life jackets, were carried overboard.

When Schultz burst onto the bridge, he had a full head of steam. Finding Marks wedged into a corner of the bridge between the port bulkhead and the chest-high pelorus, an navigational instrument used to take accurate relative or compass bearing at sea—where he had been situated since the ship began rolling heavily—Schultz spat out the news about the whaleboat carrying a dozen men over the side. He felt strongly the tragedy could have been avoided had "Jack Ass Marks" let him cut away the whaleboat as he had wanted to do earlier.

Marks was pale and seemed to be "frozen in place."

Schultz, droplets of water cascading down his face, looked hard at the captain he despised and now blamed for the deaths of a dozen shipmates. "Sir," he said loudly over the cacophony of crashing swells and howling winds, "shall I have everyone man their life jackets?"

"What are you trying to do?" Marks asked. "Panic my crew?"

Schultz saw that Marks was wearing his own kapok life jacket, tied tightly across the chest. Without the slightest hesitation, Schultz went over to the ship's public address system and pressed the transmit button. The man who was most responsible for having ordered the new kapok life jackets was now going to make damn sure they were worn. Schultz said in the firm, steady voice he had used many times over the PA when reading the daily schedule each morning, a job of the duty boatswain's mate: "All hands man your life jackets. Repeat. All hands man your life jackets."

Throughout the ship, officers and enlisted men alike who didn't have their life jackets went to where they had left them. Many who were off duty and could wear the bulky life jackets did so, and those who couldn't because they were working in tight spaces kept them close at hand.

One officer on the bridge without a life jacket was Lieutenant (j.g.) G. C. Nelson, of Suffield, Connecticut, the assistant communication officer. He asked if Schultz could fetch him a life jacket.

"Sure," Schultz said. "But where's yours?"

Nelson said it was at his battle station in the gun director—just one level above the bridge—but that the captain wouldn't let him go get it.

Looking directly at Marks, still wedged in his corner no more than six feet away, Schultz hollered disparagingly: "Why don't you ask the captain for his life jacket? He's supposed to go down with the ship anyway."

Marks gave Schultz a "dirty look" but said nothing.

A few minutes later, Schultz saw that Nelson had on his life jacket.

Not long after, *Hull*'s radar went out of commission for good, leaving the destroyer blind in near-zero visibility with no way to know the whereabouts of other ships in their fueling-unit formation. A TBS message was sent informing the screen commander that *Hull*'s radar was inoperable and requesting they be kept advised of their position in the formation. A reply came back stating that *Hull* was at present "fairly well on station."

Taking in the "pandemonium on the bridge," where young sailors had begun to "huddle around like wet chickens, scared as hell," and conversations that he never dreamed he would hear on a Navy ship, DeRyckere kept an eye on the inclinometer used to measure the angle of a vessel's horizontal sway. The scale of *Hull's* inclinometer had a limit of 73 degrees, beyond which it did not register. The steep rolls to starboard for the past hour were getting close to the end of the scale.

The chief quartermaster saw that the duty helmsman was occasionally lifted clear off the deck by backspin from the wheel. Next to the helmsman stood the executive officer, Greil Gerstley, who had been directed by the captain to remain there to ensure that his commands were "obeyed for wheel and engine," which seemed unnecessary and even "silly" to DeRyckere. Keeping the cool-headed and experienced Gerstley pinned to that spot—when there were other places on the ship where the second in command might have been of greater service—was one of the few orders Marks had given. DeRyckere, who stood "ten feet away from [Marks] most of the time" that morning, thought that in an odd way Marks seemed "not to be fully in command" of his ship or crew. He issued no orders to DeRyckere, who stood by waiting for an assignment, or any of the growing number of personnel who had gathered on the bridge. Obsessed with speed and course, Marks seemed detached from all else. He listened without comment to reports brought to him by officers and chiefs about deteriorating conditions on the ship, such as "leaks in superstructure decks and compartments allowing spray to be driven in by high wind" that caused "troubles in radio, radar and electrical circuits." Through it all, Marks kept himself wedged in on the port side of the bridge, clinging to the pelorus stand with his knees and arms during the hard rolls. DeRyckere noticed that most of the frightened sailors stayed on the high side of the bridge, too, apparently because they "felt safer" there than on the side closest to the sea. He wondered why Marks hadn't ordered all the hangers-on to go below, and beyond that, why the captain wasn't issuing a succession of rapid-fire orders in the fight to save the ship.

Schultz, who had remained on the bridge, lobbied Marks to ballast the ship's high side with seawater or transfer fuel to tanks on the port side to balance the ship and lessen the steep rolls to starboard. The captain—his normally arrogant countenance replaced by a pale fright mask—shook his head distractedly, as if he had weightier matters on his mind.

From where DeRyckere stood, he saw in Marks a man who "did not listen to anyone." Every recommendation concerning "damage control" was "denied or ignored," and in effect he "made fools" out of Schultz and other experienced personnel who braved conditions topside to reach the bridge and report to him.

During one terrifying roll as he practically walked up the bulkhead of the wheelhouse, DeRyckere saw the forward stack "swallow" a huge amount of water. Knowing the result would be flooding in the critical engineering spaces below, DeRyckere for the first time began to think that the ship was "going to go down."

Chief Electrician's Mate Joseph J. Jambor fought his way up from the flooded engine room to the bridge to report there was "too much free water below"—now rising to the men's waists. Jambor pleaded with the captain to "reduce the demand" for engine speed provided by the ship's two turbines being fed with steam from only one boiler. That would allow them to concentrate maximum efforts on pumping, Jambor explained. It was difficult to converse on the bridge over the sounds of the storm, and Jambor yelled not only to be heard but because he considered the situation dire.

In the corner from which he refused to budge, Marks did not acknowledge Jambor's urgent request. Seeing that he was being ignored, Jambor spun angrily and stomped away, cursing the captain loudly enough for DeRyckere to hear the chief's "very salty language" as he passed.

DeRyckere's worst fear came closer to reality shortly after 11:00 A.M., when Marks abruptly ordered a change in course that would turn the destroyer from running perpendicular to the gigantic swells to a course parallel to the swells, setting up *Hull* to be struck broadside by the monstrous seas. Marks was reacting to a static-filled TBS report from an escort carrier somewhere in the vicinity reporting a hangar-deck blaze.

Schultz heard Marks reply on the TBS: "We'll be right there."

Up to then, the destroyer had been running downwind, considered a "good place to stay." Hearing the captain's intention to assist the carrier, both DeRyckere and Schultz thought he had lost his senses. Neither had any idea what *Hull*—her own engineering spaces flooding—could do to help. They lacked radar even to *find* the carrier and would be in danger of colliding with other ships if they set out looking for the vessel.

Standing next to the helmsman, where he had been stuck all morning, Gerstley suddenly came to life. He knew what others on the bridge also knew: *Hull* was likely to be rolled like a toy boat.

"Sir!" Gerstley yelled, his voice strangely disembodied amid the crashing surf and high-pitched winds. "Don't turn the ship!"

Marks, still not listening to others, repeated his order.

The helmsman obeyed the captain and spun the wheel.

What came next happened quickly: *Hull* turned into a deep trough between swells. Like a sailboat that had lost the wind, the destroyer was caught "in irons," stuck and unable to make headway. *Hull* awaited the knockout punch like a beaten boxer no longer capable of putting up a fight. Repeatedly, swells slammed into the port quarter of the destroyer, pushing her over to starboard, where she hung precariously until starting to recover, but not succeeding before the next swell hit.

Giving orders to the conn, Marks tried unsuccessfully "every combination of engine and rudder" to turn out of the deep trough. He attempted to "turn away from the wind" as before, and when that didn't work, he tried to bring the "ship's head into the sea." *Hull* was being "blown bodily before the wind and the sea," yawing wildly in "the trough of the sea" that was so deep swells rose ominously on both sides of the ship twice as tall as her mast.

Schultz could see that the captain was confused about what to do now. In his haste, Marks had failed to "let anything take effect to see if it worked" in trying to regain control of the ship. Schultz knew that a number of other officers trained as skilled ship handlers by Consolvo, as well as DeRyckere and other veteran chiefs, could have handled the ship better. The boatswain's mate included himself in that select group, and he knew exactly what he would be doing at that moment. Rather than ringing up increased speeds, he would have shut down the engines and let the ship "ride the sea," allowing her to find her own way. Schultz agreed with mariners who believed that a modern warship "functioning properly and handled with wisdom" and adequately ballasted should be able to ride out a typhoon. The main problem was the "incompetency" of *Hull*'s commanding officer, who seemed "in a state of shock" and clearly "did not know what he was doing." The same man who had demonstrated he could not bring the ship safely alongside a pier on a calm day now had the conn during a Pacific typhoon, with the fate of more than 300 men hanging in the balance. If Consolvo—or any of a long list of others aboard *Hull* who were better ship handlers than Marks—had been at the conn, Schultz thought, they would not be in their current dire predicament.

Schultz could stand idle no longer. He approached Gerstley and, in full view and earshot of Marks and other personnel, made a plea that doubtless had not been heard on any other U.S. Navy ship during the entire war: "He's sinking the ship! You better relieve the captain!"

DeRyckere heard Schultz and thought it was a damn fine idea. The chief quartermaster would support the change of command, even if it meant escorting the captain off the bridge to his quarters. He thought the ship "could be saved" if "action was taken" by someone other than Marks.

At first Gerstley said nothing. The debonaire Cornell graduate, who had been aboard *Hull* since mid-1943 and had excelled in his varied assignments aboard the destroyer, did not seem shocked at the unprecedented suggestion that he take command in a desperate bid to save the ship. Schultz wondered if Gerstley had been contemplating the deed. The executive officer looked at Marks, then back at Schultz. He had made up his mind. "If I take over and save the ship," Gerstley yelled against the rampaging seas and howling wind, "he'll say it was *mutiny*. If we don't all drown, he'll have me tried for mutiny and hanged."

Schultz knew Gerstley was right. "The bastard would," he hollered, looking back at Marks with utter contempt. "Even if you saved his life."

THE SEAWATER that DeRyckere saw go down the stack had flowed into the forward fire room, dousing the superheater on boiler number one. Lieutenant (j.g.) George H. Sharp, *Hull*'s engineering officer, ordered the boiler secured, which meant bleeding the steam down to a level that allowed the safety valve to cut off the pressure. That left *Hull* operating on only one of four boilers: the number two, located in the forward fire room. With seawater also pouring down the ventilation intake blowers that extended just four feet above the main deck—making the openings vulnerable to flooding even in moderate seas breaking across the deck—the forward fire and engine rooms were swamped in "two or three feet of seawater," resulting in "many tons of free water" sloshing to the low side on every roll.

Sharp, twenty-three, of Washington. D.C., the soft-spoken son of an admiral in command of the Pacific Fleet's minesweeper force, had come aboard *Hull* in July 1943, a few months after his graduation from Johns Hopkins University with a degree in mechanical engineering. Although everyone in the crew knew he was a top admiral's son, Sharp had earned a reputation for "not throwing his weight around." Precise and fair-minded, Sharp was respected by the enlisted men who worked in engineering.

Sharp ordered bilge pumping to commence immediately. In the engine room, however, they had difficulty doing so because the "basket for the main drain line was amidships" and whenever the destroyer rolled and the water pooled on one side they could not get suction from the drain. The accumulated free water added greatly to the weight on the low side, and made the rolls steeper and the ship's recovery slower.

With the forward fire and engine rooms flooded, and unsure how much longer the lone boiler could continue to power the ship, Sharp, who was having a hard time hearing whenever he phoned the bridge due to the "tremendous noise of the storm," decided it was urgent to light an after boiler. He asked for volunteers from members of the engineering division, explaining that the number one boiler was out, the forward fire room was flooding, and they needed to get an after boiler lit or else there was a danger of *Hull* losing all power and going dead in the water.

As he had done on America's first day of war when he was a civilian worker at Pearl Harbor awaiting transportation to Wake Island, Fireman Tom Stealey stepped forward to help save a ship—this time his own. Since coming aboard the worn-out *Hull* undergoing an overhaul at the Seattle shipyard and being assigned to the fire room, Stealey had learned firsthand just how hot and exhausting tending to boilers in a cavernous fire room far below deck could be. Still, the conscientious, hardworking Stealey had learned his job well, and he liked the other two men in his watch section. Now Stealey, who normally worked in the after fire room, looked at the other guys in his section and said, "Let's go get it done." The three men put on their life jackets and went on deck. Struggling not to be washed overboard, they opened the hatch to the fire room and hurried down the ladder. The last man through closed the hatch. As they descended into the fire room it "did not enter our minds that we weren't going to come back out." They knew they had to work quickly, however. They took off their life jackets and hung them on hooks nearby.

Rather than starting the boiler-lighting process from scratch, they used an emergency procedure: tying into the main line and diverting steam from the forward boiler. The oil that fed the fire in the number four boiler began to heat up, and Stealey was soon able to ignite the oil that sprayed out of the burners. Once a fire was going, the boiler started making steam. "Within twenty minutes" they had "400 pounds of pressure" in the boiler, sufficient to provide steam to the turbine running the twin screws and main generator. During the time they were tending the boiler, Stealey and the other two men were "hanging on for dear life" whenever the ship rolled, although the three sailors were "kidding and joking" with one another the way frightened young men sometimes do to steel their nerves.

Suddenly, one roll "put us over on our side"—more than 70 degrees to starboard. Rather than fighting to come back, as *Hull* had done previously, the ship begin "to settle as if being pushed down by the sea." Stealey and the other men ended up sprawled on the bulkhead; then, as the ship kept going over, they were tossed onto what had been "the ceiling but was now the floor." As they hung on to pipes and ducts built into the overhead, Stealey was convinced they had gone over "more than 90 degrees." He knew at that point the destroyer was "done for." The sounds of the wounded ship were dreadful: the groaning and creaking of bulkheads about to give way, and the banging of everything from coffeepots to footlockers to vital equipment that had ripped loose.

As *Hull* came back a bit in the opposite direction, the firemen slid back onto the bulkhead, next to the superhot boiler. Fearing a boiler explosion, which would kill them instantly and rip apart the ship, Stealey worked himself into position to shut off the valves feeding the flames, and they furiously doused the fire. After doing so, the three boiler operators were left "trembling." There were no wisecracks now. They all knew it was time to get out of the bowels of their sinking ship or go down with her.

The only hatch leading out of the fire room was on the port side, which was a lucky break. With the ship laying over to starboard, all the hatches on that side were underwater. Their escape route, however, was now high above their heads. The three sailors put on their life jackets and skimmed up pipes and ducts. When they reached the rungs of the ladder, they swung themselves hand over hand to the hatch cover, which they managed to open. They crawled into a narrow passageway about 15 feet long that led to the main deck, then scrambled on all fours in darkness until they reached the hatch to the deck.

When they came out topside they were underneath a floater net, a mesh of heavy-gauge rope about 5 inches round, coated with tar so it didn't absorb water and with cork floaters attached at intervals. All Navy ships carried floater nets, which were designed for survivors to grab on to in order to stay afloat as well as remain together, making it easier for rescuers to see them in the water and pick them up as a group rather than individually. The nets were kept on deck amidships and aft in storage bins that were easy to open—designed to spill out when a ship began to sink.

Stealey was amazed how dark it had turned. Although just past noontime, day had "turned into night." He understood the loss of daylight came from being smack in the middle of a storm with a combination of winds and seas he never would have believed possible. He estimated winds were more than 100 miles per hour, and would not have bet against some gusts being twice that speed. The swells pounding the ship into submission were higher than the flight deck of the new, large aircraft carriers—walls of blackish water 70, 80, and even 90 feet high rolled in with the destructive force of a tidal wave.

Rushing water immediately broke over Stealey, who held tightly to the netting to keep from being washed away, as was one of the men with whom he had come up from the fire room. Stealey hung on the side of the overturned ship like a fly caught in a web, trying to catch his breath, as he found it impossible to breathe facing into the wind. Rain was coming down hard, and the wind was "really whistling." The worst of it was the sea, however, "breaking over us constantly." He looked toward the bow. In the area between the two stacks, he saw a "whole bunch of dead men, at least twenty-five or thirty," most wearing life jackets; their bodies were being slammed into the side of the ship. It was obvious to Stealey they had been unable to get clear of *Hull* and had been "beat to death against the ship."

"Gotta climb out on the number two stack," Stealey yelled between deep gulps of air. "Go to the end. Jump! Get away from the ship."

The two men timed their run down the length of the wide stack until they could catch a comber going away from the ship. Then they took off.

With the hellish wind to his back, Stealey ran "faster than he ever had in his life." At the end of his sprint and without slowing down, he planted his right foot at the lip of the stack, coiled—and jumped for it.

Fourteen

First word that *Monaghan* was in trouble on the morning of December 18 came at 9:27 A.M. in a message from the ship's new captain, Bruce Garrett, with a total command experience of less than three weeks.

"I am unable to come to base course," Garrett radioed fueling group commander Captain Jasper Acuff aboard his flagship *Aylwin*, a *Farragut*-class destroyer commanded by Lieutenant Commander William K. Rogers, one of Garrett's four Annapolis classmates (1938) who had recently joined Destroyer Squadron One within weeks of one another. With the storm having thrown *Monaghan* off the assigned southerly course of 180 degrees, Garrett reported they were going in nearly the opposite direction on a "heading of 330 degrees." Acuff feared the wayward destroyer was in danger of colliding with ships of another fueling unit 3,000 yards to the northwest. He knew that several ships in the northern unit, "caught in similar circumstances," had "given up trying to maintain station" or stay in formation, and sans radar were "manned by blind men."

Acuff's fleet fueling group was divided into three units, each made up of approximately the same composition of four oilers, one light aircraft carrier and escorting vessels. *Monaghan* was in the center group, with *Hull* in the northern group and *Spence* in the southern group.

Acuff stepped to the TBS. "Use more speed," he told Garrett.

A few minutes later, Garrett reported: "Have tried full speed but it will not work." After a brief pause, he said in a louder but still calm voice: "Cannot get out."

On *Aylwin*, which was rolling violently in the heavy seas and experiencing steering difficulties of her own, Acuff and Rogers took Garrett's message to mean the typhoon "had the helm" of his ship, which doubtless had fallen into a deep trough and was now "out of control."

The last message from *Monaghan* was about 10:00 A.M., when Garrett came back on the TBS to warn an unidentified ship, "You are 1,200 yards off my port quarter. Am dead in water. Sheer off if possible."

After that, *Monaghan* fell silent.

Aboard *Monaghan*, which lost TBS communication shortly after Garrett's final message, things were anything but quiet, as the destroyer was "slammed back and forth between rumbling winds and booming waves." At 10:30 A.M., with his ship in "grave jeopardy" after losing her main generator and steering motor, Garrett decided to take on ballast.

Joe "Mother" McCrane, the New Jersey native who had been aboard ship more than two years and was now a water tender 2nd class, had sounded the fuel tanks as part of making out the morning fuel report. He had reported the ship's fuel state as "between 122,000 and 130,000 gallons" at 8:00 A.M.—about 70 percent of total capacity. This placed *Monaghan* above the established minimum of 45 percent fuel capacity set by the Bureau of Ships for *Farragut*-class destroyers, below which ballast-ing would have been required. Even so, McCrane thought "putting on ballast" would help the laboring ship, which he had noted with concern took longer to "come back from steep rolls" since leaving the Seattle shipyard three months ago after having the new radar gear and weapons installed topside. Given the worsening seas that "racked and strained" the destroyer, McCrane recommended ballasting to Chief Water Tender Martin Busch, who went and spoke to the captain. Not long after, Busch passed the word to McCrane to begin pumping water into two empty after fuel tanks, numbers ten and eleven.

As the ship's "oil king," McCrane was responsible for monitoring the more than 600 tons of black oil *Monaghan* could carry in about twenty different tanks. McCrane had his own little shack in the engineering department with laboratory equipment for testing the quality of fuel oil and boiler-feed water. Every few hours, he used the ship's two fuel oil transfer pumps to move oil around from storage tanks to the ready-service tanks (two forward and two aft) that fed the fires under the boilers, making sure the ship was kept balanced and on an even keel.

McCrane and his assistant, Water Tender 3rd Class Leonard Bryant, who seven weeks earlier had tried to talk Water Tender Joseph Candelaria out of leaving the ship to attend boiler school in Philadelphia, went below to shaft alley. In the watertight space that housed the propeller shafting that ran astern from the forward engine room, they discovered that large metal boxes holding spare parts and tools had broken loose from their racks, spilling their contents. Also, several 50-gallon drums of lubricating oil were rolling around on the metal grating, sections of which had to be raised in order to reach valves that needed to be opened to pump seawater. McCrane called the engineering compartment to ask for help. Water Tender 1st Class William Hally and Water Tender 2nd Class Roland D. Fisher soon showed up, and the four sailors started to clear off the grating—an exercise "not without plenty of excitement" with so many objects "flying around." During one steep roll, a portable air blower broke loose, heading for Fisher, who was bent over at the time. McCrane stuck out his arm to deflect the heavy blower, but it was "coming too fast." Before Fisher could get out of the way, the equipment struck him in the middle of the back, although "luckily it only knocked the wind out of him." They were finally able to open the valves, and McCrane notified the engine room to start the "fire and bilge pumps" in order to "pump seawater into the designated tanks." With communications failing throughout the ship, McCrane never received confirmation that the pumps were started as requested or that the ship took on any ballast.

By then, *Monaghan* was "rolling so heavy" that McCrane and some of the others in engineering decided to go topside and wait out the storm inside the after 5-inch gun shelter. When they reached the small enclosed space, they found several dozen men already inside, huddled together—"just plain scared." McCrane could "feel the tension in everyone." They were all there for the same reason—to be topside in the event the ship went down—although McCrane could not imagine the unimaginable: having to abandon ship in the middle of such a storm.

At 11:30 A.M., the lights went out, which everyone knew meant the auxiliary generator had failed. A few sailors—McCrane included—hurried down to their lockers to get flashlights, returning to the gun shelter about the time "the storm broke in all its fury." With the ship rolling so heavily to starboard, McCrane could not "say for sure" whether they were making speed or stopped dead in the water. The ship had been "stopped quite a few times," only to "pick up speed again." Every man was "holding on to something" to keep from falling, and "praying as hard as he could."

One level below in the steering motor room, Machinist's Mate 1st Class Lester J. Finch hollered up to say that if someone could get a course from the bridge, he would attempt to steer by using the manual steering gear. With "all communication between the forward and after part of the ship knocked out," that meant someone had to make it to the bridge and back. McCrane thought about it but "couldn't find the courage" to attempt passage across the swamped main deck. Roland Fisher, even with his sore and bruised back, "volunteered without hesitation" and ducked through the outer hatch to begin the perilous journey—a self-less act, the other men agreed, that "deserved some kind of citation for bravery."

There were now "about forty men" crammed into the small space. One sailor fervently prayed aloud; each time the ship took another 70-degree roll to starboard, he cried out, "Please bring her back, dear Lord. Don't let us down now!" Each time the ship struggled back, "shudderingly, from disaster," he added gratefully: "Thanks, dear Lord." The deep rolls and aloud prayers were repeated about "seven or eight" times before *Monaghan* went over and instead of coming back settled down to die.

With the ship flopped onto her starboard side like a beached whale, the port hatch was 20 feet above them. Several younger men scrambled up the overhead on pipes and ductwork as if scaling a jungle gym. With the only illumination from the narrow beams of flashlights, they began the "difficult job" of trying to open the hatch against the obstructive forces of wind and waves "beating up against it."

From where he stood in the shadows waiting to make his escape along with everyone else, McCrane observed how "the fellows kept their heads," with "no confusion or pushing" and "everyone trying to help the other guy." Before long, they started through the hatch one by one.

McCrane was helped by a set of strong arms from above and eased through the hatch by Gunner's Mate Joe Guio, who "with absolutely no thought of his own safety" was leaning over "pulling everyone out." The sturdy Guio, who had worked in coal mines back home in Holliday's Cove, West Virginia, fought to keep from falling overboard but continued to help his shipmates from the gun shelter until the last man.

McCrane figured he was "about the tenth one" to get out of the gun shelter. As he stood "so nervous" on the side of the hull, he began to inflate his rubber life belt by blowing into a nozzle. He had been one of the "very few" men in the gun shelter who had a life belt or jacket, as most kept them at their general quarters station and had been unable to reach them. "Waves were breaking steadily" over the ship, each time "carrying fellows right off." McCrane did not know how he was managing to stay on, and he no sooner had that thought than a huge swell knocked him off.

When McCrane landed in the sea amid swells "seventy feet from trough to crest," he "lost all sense of direction," although he knew he was being taken deep by what felt like "a whirl-pool." The force that was dragging him down had taken other men, too, and McCrane felt them "knocking up against" him. He flailed his arms and legs, trying to "beat my way to the surface." As he started to rise, other men grabbed at him in desperation. McCrane surfaced atop a gigantic swell that took him up and up and then placed him "right on the side of the torpedo tubes." He scrambled to the highest point of the ship, which appeared to be the shelf for the 20 mm guns. He had just about reached that perch when another wave "took me and wrapped me around the antenna." He spun around the antenna several times before being "thrown out into the sea again." To avoid being dragged underwater once more, he swam as fast as he could on the surface, but soon he realized he had to slow down or his "strength wouldn't last very long." He found it "impossible to keep from swallowing the salt water and oil." He looked around to see if he could spot the ship or any of his shipmates, but visibility was "almost zero" and he saw nothing. After what seemed like "an eternity," an exhausted McCrane heard Joe Guio yelling for him to turn around. McCrane swung around to see a life raft, grabbed on to it, "and thanked God that it came when it did." Guio and other men were hanging on around the outside of the raft, which they hadn't been able to climb into because they kept being "beaten off by the sea."

One of the men with the raft was Evan Fenn, the rough-and-tumble cowboy from Arizona by way of Utah who was now a water tender 3rd class assigned to the forward fire room as a "water-gauge watcher," monitoring pressure levels on the steam boilers. He had stood the 4:00-to-8:00 watch that morning, and upon being relieved went to get chow and found in the galley "everything a mess with dishes flying all over the place." He washed down a cold sandwich with a cup of tepid coffee, then went up on deck, where he "waited to time the waves" until he found an opening to run back aft to the berthing compartment. In the head, he washed up and shaved, then climbed up to his bunk at the top of the trilevel tier and tried to go to sleep. Eventually, his mattress went out from under him on a hard roll, spilling him onto the deck. When the compartment went dark, Fenn decided to don his life jacket and head topside. He arrived on the fantail shortly before the ship went over for the last time. He became "tangled up in antenna wires" and had a "helluva time getting loose." When he worked himself free, Fenn was blown overboard. He was one of the first to reach the raft—the only one released from *Monaghan,* as all the others had been lost off the deck earlier in the storm. The raft had been freed from its own entanglement with great effort by the ever-present Joe Guio. With one foot badly injured, Guio was still able to get the raft into the water. Fenn, like others that day, owed his life to their "big husky" shipmate Guio.

Although none of them had seen the destroyer sink because the "spray was so heavy" and visibility so low, they all were convinced *Monaghan* went down quickly. Seaman 1st Class Doil T. Carpenter, twenty-four, of Pasadena, California, told of being in the flooded steering motor room and hearing from a phone talker that the overhead in both the forward engine and fire room "began to rip loose from the bulkheads five or ten minutes before she capsized." Carpenter also heard that the "bridge ripped off" just before the end, which could explain why no bridge personnel—including the captain, Bruce Garrett, whom Fenn "never once saw" in the short time he was in command of *Monaghan*—were seen after the ship went over. The majority of the survivors, in fact, had been topside near the fantail when the ship rolled, with only "one or two of them" from on deck amidships and none from the spaces or compartments below.

The thirteen men with the raft were the only ones known to have made it off the sinking ship. Gone were 256 shipmates, including Chief Martin Busch, who had lived to tell of his harrowing escape from the capsized battleship *Oklahoma* during the attack on Pearl Harbor only to perish three years later aboard a destroyer in a typhoon.

SPENCE SUPPLY OFFICER Al Krauchunas, who was taking the measure of death even as he fought for his life, made it through the flooded passageway underwater and came up "gagging on oil and salt water."

In the darkness as the sea had poured in around him, Krauchunas "relived an incident" from his childhood that he had not thought about in a long time. As a boy of six, he had walked into a store and removed a box of animal crackers from a shelf, placed it under his coat, and walked out without paying. He "regretted now" the dishonest deed. He also thought about the "essential records" from the supply department, which were in a waterproof box at his general quarters station in the decoding room. They consisted of pay accounts, cash book, receipts and vouchers. He had believed that in any emergency he would have time to grab the box and take it with him. How else would he justify to the Navy Department his monthly purchases and disbursements?

Krauchunas surfaced a few feet from the overturned *Spence,* which showed only her "freshly painted red bottom." The proud warrior ship that had been a "bulwark of strength and stability" had turned turtle.

When the crush of water swept him back against the ship, Krauchunas heard disembodied voices, "screaming and curdling yells," attesting to "the chaos and bedlam" taking place inside watertight compartments still containing air. Krauchunas had come out near the crew's galley, which he knew had been filled with enlisted men, as were other spaces after the skipper, James Andrea, had earlier that morning ordered all hands not on duty "to stay below" for their own safety. Now the sounds of the trapped men snapped Krauchunas out of his "shocked daze."

Treading water to stay afloat, Krauchunas was not wearing his life jacket because "no thought of the ship capsizing had come to mind." Concerned about getting caught in the "suction of the ship should it sink," he started swimming away. Although continually bounced and spun around by the cascading seas and by winds he estimated at 100 miles per hour, Krauchunas reached a floater net adrift 20 yards away; two dozen men were hanging on to the 25-by-25-foot net. When they began to drift toward the sinking ship, the men kicked and paddled to "push the net" in the opposite direction. After some effort they were far enough away that they lost sight of *Spence* in the "gusts of rain and violent wind spray." They all, however, heard the muffled explosion.

As did everyone at the floater net, Water Tender 3rd Class Charles Wohlleb, who the day before had watched *Spence*'s aborted attempt to fuel alongside Halsey's flagship, *New Jersey,* knew that the "rumbling blast" was a boiler exploding. He understood, too, that anyone still alive in the fire room would have been instantly scalded to death when the boiler blew. Knowing the great explosive force would certainly have broken the narrow-beam vessel in half amidships, Wohlleb imagined for the first but not last time the forward and after sections of the ravaged destroyer sinking slowly to the depths below, taking untold men with her.

Wohlleb had been in the after fire room prior to coming topside with two shipmates, Water Tender Cecil Miller and Boilermaker Franklin Horkey, neither of whom Wohlleb saw again after all three were cast into the water from the alcove where they had been when the ship went over. The New Jersey native pictured the men who had been on duty in the fire room tending to the ship's only lit boiler: Frank Thompson, operating the oil burners; Norman Small, the six-foot-two Nebraska farm boy, watching the steam pressure gauges; Claude "Roy" Turner, monitoring the boiler's water level; and Layton Slaughter, the 1st class in charge, speaking to Horkey over the sound-powered phones. To a man they had been Wohlleb's buddies, and now they were gone.

One of the last to reach the floater net was Seaman 1st Class James P. Heater, twenty, of Auburn, Washington, who had been in the after crew compartment when the ship went over, and managed to swim "out from under the ship." Although "pretty conscious at the time," Heater, who was not wearing a life jacket, was having difficulty breathing due to his lungs being "full of water." Wohlleb, one of the few men wearing a flotation device—a brown canvas life belt, inflated by "two new CO_2

cartridges," which he had kept on all morning because he had had a "funny feeling"—had Heater straddle him, then wrapped his arms around him.

Also finding his way to the floater net was Quartermaster Edward Traceski, who when the ship went over had "climbed the wall" inside the bridge to the nearest exit and "just fell out into the drink." With the rain stinging like needles on his face and arms, Traceski realized how foolish he was not to have been wearing his life jacket. Luckily, he was pushed by the wind and sea to the floater net, which he "just clung to."

Torpedoman Al Rosley, who had served aboard *Spence* since her commissioning, had been standing the 8:00-to-noon watch in the torpedo shack located behind the bridge and one level above the main deck. For most of the morning, as the ship "rolled over and came back," Rosley had been hanging on, keenly aware that *Spence* was riding unusually high and "bobbing like a cork" in the worst sea conditions he had ever known. After the ship turned into the trough, Rosley judged it wasn't "very long before we went over." He pulled off the sound-powered headset he was wearing and opened the hatch above his head. Immediately, water started coming in. Fighting his way against the inflow of the sea, he barely made it out. Standing on the side of the ship, he jumped. While he was still only 10 or 15 feet away, *Spence* "rolled the rest of the way over." Rosley was "blown way aft of the ship" by the wind and sea and found "the pressure so great" he had to cup his hand over his mouth and nose to breathe deeply. Unfortunately, his life jacket was still down below in his locker, but he was a strong swimmer, having grown up in western Maryland frolicking with his nine siblings in myriad lakes and rivers. He was pulled under several times but popped up still swimming. Willing himself to keep kicking and stroking, he swam his way right into the floater net, caught like a sardine—and gratefully so.

For the rest of the afternoon, frightened men kept themselves lashed to the tarred rope and cork floaters, with legs and arms hooked inside the foot-square openings in the heavy webbing, desperately trying not to lose hold in the raging storm that churned the floater net into a "huge round mass of rope and rubber blocks." As nightfall came, the high wind and driving rain subsided. Although the seas remained unruly, above them "the most beautiful South Pacific evening came into being." The men were soon drifting silently under the stars.

Wohlleb realized the man he was still holding had stopped breathing. He tried without success to find a pulse at the neck and wrist, then placed his hand over Heater's heart, which was still.

"Mr. Krauchunas, I don't think he's alive," Wohlleb said.

The other two officers in the group were in bad shape, which left the supply officer in charge. He had already inventoried the survival gear and supplies, finding two 5-gallon kegs of water, flares, K rations, a signal mirror, a dye marker, a hatchet, and some small bottles of medicinal whiskey.

"You'll have to let him go, Charlie," Krauchunas said.

Wohlleb did, watching his shipmate "sink slowly in the dark sea."

AT THE END it was the wind and not the sea that sealed *Hull*'s fate.

Shortly after noontime, the ship—still being broadsided by great swells, and thereafter locked in a series of deep troughs—had rolled to "about 70 degrees" and begun to right herself when the wind "increased to an unbelievable high point," which bridge personnel estimated at 125 miles per hour. The force of the wind "laid the ship steadily back over" to starboard and "held her down in the water" at an angle of "80 degrees or more," allowing the seas to come flowing into her upper structures and down her stacks.

The bridge quickly flooded, and men scrambled for the exit on the port side. When Boatswain's Mate Ray Schultz crawled out onto the wing of the bridge, he found the skipper, who appeared "to be in shock," already sitting there. It figured, Schultz thought, that Marks would be among the first out of a sinking ship and not the last man, as naval tradition dictated for the commanding officer.

Not long after declining Schultz's request to relieve Marks of his command in an effort to save the ship, executive officer Greil Gerstley was thrown headlong into a bank of equipment during a bad roll. Gerstley came up with the fingers on one hand "broken and bent back." "Would you see if you could get me something," he calmly asked Schultz, who had been trying to reach the first aid kit to find a splint and tape when the ship went over and "didn't come back."

Gerstley now was halfway through the exit from the bridge, not asking for help but "pleading with his eyes" to several men standing nearby on the side of the ship. Two sailors reached down, gripped the executive officer under his arms, and raised him up the rest of the way.

Quartermaster Archie DeRyckere assisted Gerstley in reaching the highest point of the ship: a searchlight platform on the port wing of the bridge. The chief sat down next to Gerstley, who was holding his shattered hand in front of him. The officer, shouting to be heard over the bedlam in which they had landed, said he wasn't sure he would be able to swim.

"When we go down, will you help me?" Gerstley asked.

"Yes, sir," said DeRyckere, pulling tight the straps on his life jacket.

From their perch, DeRyckere saw sailors below in the water—wearing life jackets but unable to get clear of the ship—being battered against the "guns and appendages" that "kept hitting them." The horrific scene made him determined to stay with the ship as long as possible. That point soon arrived: *Hull* "just sunk underneath" them like a diving submarine. The suction pulled DeRyckere so deep he thought his eardrums would burst. As he struggled to regain the surface, he heard a boiler explode, then the howl of the wind as he broke the surface. Gerstley was nowhere to be seen; neither was Schultz, Marks, or anyone else.

Prior to *Hull* going over for the last time, Radarman Michael "Frenchy" Franchak, who would also earn the nickname "Moose" because of his physical strength, had been "sweating it out" in the crew's galley, one level below the main deck just forward of the forward stack. Off watch since 4:00 A.M., he had stayed there rather than going back aft to the berthing compartment. Shortly before noontime, Franchak decided to head to his general quarters station on the bridge to fetch his life jacket. When he reached the radar shack, he had difficulty opening the door because there were so many men crowded into the small space—none of whom had any business there except for radar operators. Wanting to be above deck but having seen numerous men already washed overboard, they had all ducked inside the radar shack for cover. Franchak went to where he had left his life jacket, and found it gone. He saw his life jacket on Storekeeper 3rd Class Arnold Niss, of Chicago. Because Niss was "one of the best sailors on board," Franchak told him he could keep it. Franchak found himself another one, which had the name Torkildson stenciled on the back. Knowing that Yeoman 3rd Class Keith Torkildson was "one of the men who had already been washed over the side," Franchak slipped on the life jacket. Not wanting to stay in the crowded radar shack and with the ship riding like a roller coaster, he crawled through the bridge back to the chart room. On the way he noticed the "terrific pounding" Gerstley had taken, suffering what looked to Franchak like a compound fracture of the hand and perhaps lower arm.

The chart room was filled with frightened sailors, and Franchak sat in a corner on a bucket—one of many set out to catch the water coming in through the overhead. Franchak realized he was getting so "wet from above" that he put on a steel battle helmet for protection. He wasn't "annoyed for long" by the "plunking of drops" on the helmet because a few seconds later "the ship turned very quiet" and rolled over on her side.

The nearest exit was quickly nearly underwater, with just a thin horizontal slit showing. "Panic ensued" as "60 or 80 men made one dash for that opening." As those in front reached the narrow opening, the "ship descended," pushing them in the opposite direction. It was then Franchak made his move to get out, during which he "kicked and stepped over bodies." The ship seemed to rise on a crest, causing the water to recede slightly from the exit, leaving a larger escape hole for Franchak. With his "chin barely above water all the way," he grabbed on to "halyards, pipe stanchions, and whatever," using all his strength to pull himself toward the opening. He came out near the platform for the forward 20 mm guns. His first impression was that the stacks were "making an awful hissing sound," and then he was doused with "steaming hot oil" backing up the stacks from the fire room.

Looking down into the sea, the first person Franchak saw was the captain, "the most hated officer on board ship." Marks had two life jackets—one he was wearing and one he was holding tightly. The sight made Franchak "angry enough to go after him," so he jumped. Immediately pulled underwater, he became disoriented and never saw Marks again.

Franchak surfaced to a "panic-stricken scene." A life raft had been freed forward, and once in the water it was soon overcrowded. Just as the men "all got settled," the sea "hammered them against" a gun mount. To Franchak, it was "like breaking wooden matches." Long after the screams ended, the sea kept battering the broken bodies against the ship.

Making it to another raft, Franchak got hold of the rope handle and then was hit in the back of the head by what felt like a "loose timber" but was actually a swell breaking over him. Suddenly, he and the raft filled with men were "about thirty feet underwater," "spinning like a top in a whirlpool." Not until the swell completely subsided were the men and raft "shot up like they were on an elevator" to the surface. By then, though, some men had let go. The first swell, in fact, took most of the men on the raft. Franchak and the others now knew the score—"inhale as much air into your chest and cheeks" as possible before being immersed. When the next swell came it was "just as strong," but now the men "had a little experience" and their "chances to survive were better."

Storekeeper Ken Drummond, who had considered the loss of Billy Bob Dean off the wing of the bridge a forerunner of things to come for *Hull* and her crew, had been on the port wing of the bridge when the ship went over to stay. For some time before, he had been reassuring Boatswain's Mate 2nd Class Robert E. Parker that the ship was "not going to sink." A former Golden Gloves boxer nicknamed "Punchy," Parker was the "toughest guy on the ship," but he couldn't swim and was afraid of the water. He sat on the deck beside Drummond, holding the storekeeper's leg like a frightened child, asking repeatedly, "What are we going to do?"

When *Hull* went over, Drummond found himself standing on the side of the ship looking aft at some guys "sitting on the side," although there was "too much spray and wind" to recognize anyone. Then a wave hit Drummond and washed him overboard. The sea was "very cold and dark," and he "assumed" he was a "goner." One overriding thought crossed his mind at that moment: *This is really going to upset my mother.*

Lieutenant (j.g.) Lloyd Rust, the seagoing Texas lawyer, had gone on duty at 8:00 A.M. in CIC (Combat Information Center), one level below the bridge. There were no windows in CIC—the communications and electronics hub of the ship—so Rust "couldn't see anything." But he was wearing a voice-powered phone headset and had been hearing reports all morning from various locations on the ship. When the "worst thing happened" and Marks "put the ship in a trough"—"the last place you want to be in a storm"—Rust heard thereafter "all kinds of conflicting orders" coming from the bridge. It was clear to him that Marks was "quite upset" and "not doing any good" with his orders, which "kept changing back and forth." If the captain had done any one of several things and "stuck to it long enough," Rust thought, the ship might have made it. But Marks, who had "not been good at seamanship" since the day he took command of *Hull,* proved to be an "incompetent ship handler" to the end.

The phone talker on the bridge kept reporting degrees of rolls as registered on the inclinometer. Around 11:30 A.M. he exclaimed, "It hit the stop on that one!" Rust knew that meant *Hull* had rolled in excess of 72 degrees, which was as high as the inclinometer went before "hitting the peg"—and beyond what was considered a recoverable roll for the decade-old destroyer. The ship "came back from that one," but not long afterward, *Hull* went on her side.

Arriving early to relieve Rust in CIC was Lieutenant (j.g.) Don Watkins, the Carnegie Institute of Technology graduate who had come aboard as a new ensign more than a year before. Admittedly "not known for being early," Watkins had been looking for something to do other than stay in the wardroom, where furniture and other loose items were sliding from bulkhead to bulkhead. Being early that morning "probably saved" his life, as most of the off-duty officers Watkins left in the wardroom when he headed for CIC "didn't make it" off the ship.

There was only one exit out of CIC. When it was obvious the ship wasn't going to recover, Rust ordered the hatch to a weather deck opened. Water "rushed in right away," and he told all the men to get out. Two or three times the ship was lifted by a swell, which caused the water to drain out of the room. Then when the ship dropped off the crest and went the other way, CIC filled again. Each time it drained, several men went out with the flow of water. The two officers, Rust and Watkins, were the last to exit. By the time they did, CIC was again nearly filled with seawater.

When Watkins came out, he found quite a few enlisted men standing on the side of the ship, as well as engineering officer George Sharp, the admiral's son. When *Hull* went over, Sharp had been on deck heading from the bridge to the engine room, where they had been "answering bells all morning," maintaining a speed of "17 knots up until about 11 A.M." Sharp suspected that when the ship had "lurched to starboard at about 11:30 A.M.," the longitudinal bulkheads between three main fuel tanks amidships were "possibly carried away." He knew that would cause many tons of oil in those tanks to flow in the direction the ship was rolling, thereby increasing the list and making recovery from rolls more difficult. He had planned to inspect the tanks, and if they were damaged to shift the fuel elsewhere, but he and *Hull* had run out of time.

To Sharp and the others on the side of the ship it was clear that *Hull* was going to go down soon, as the vessel was "taking on water through both stacks," which were spewing back hot oil, creating a scenario for a cataclysmic boiler explosion at any moment. Sharp began organizing the men to tie themselves together and go into the water as a group.

Watkins began to assist, but on the next wave he was washed off. He came up treading water next to Fireman 2nd Class Roderick Mackenzie, twenty, of Los Angeles, California. Mackenzie was clearly having a difficult time keeping himself afloat, and Watkins quickly saw why.

"Where's your life jacket?" asked the officer.

"Couldn't get it," Mackenzie said.

"Hang on to me."

The kapok life jacket wasn't designed to support two men, and they "weren't doing too well together." When a raft passed nearby, Watkins suggested Mackenzie "swim over and get on it." As the raft was already full, Watkins decided to stay on his own. Mackenzie made it to the raft and was pulled aboard. Watkins waved goodbye to the sailor.

When Rust came out of CIC he got "hung up and cut pretty badly" on the mast and its rigging before he was able to get clear of the ship. Once in the water he saw a life raft nearby with men hanging on it. He made for the raft, and when he reached it he joined in kicking and paddling in order to move farther away from the ship.

Rust heard the rumbling before he saw the approaching swell. Looking up, he realized they were about to be hit by "a wave 70 feet or better." When the swell struck, the raft and men tumbled end over end into a seemingly endless pit. Rust was kept underwater a long time before his life jacket finally started pulling him toward the surface. He came up under the raft, which had wooden slats across the bottom. He "almost drowned trying to get out from under the raft" and past all the exhausted men hanging on to the rope handles on the side. Deciding he would be better off alone, Rust pushed off and moved away.

Rust had learned that each swell could be heard a couple of seconds before it hit, giving him time to take a deep breath before "it swallowed me up." The waves were "so much and powerful" that they kept him tumbling underwater near the limit of his stamina. Just when he thought he was going to drown, he could feel the water rushing by as his life jacket brought him back up to the surface. He then had only enough time to take a couple of breaths before he heard the next swell coming. This went on for hours—all afternoon, in fact.

Rust frankly wasn't sure where he was finding the endurance to take such a beating.

A nagging thought kept coming to the newly licensed barrister who had not yet set foot in a court of law as an attorney. Lloyd Rust knew he might have to be "resigned to the fact" that he was going to die.

Fifteen

It was several hours before the thirteen *Monaghan* survivors were able to fix the lashings for the wooden "latticework" bottom of the oblong raft and "let the wounded climb in." Before that, each huge swell that hit ripped the men away from the handles and lines they were clutching, and scattered everyone in different directions. They kept having to "fish around to help the wounded back," which left everyone "tired and weak."

One by one the others followed the injured men into the hard-shell raft, designed to hold eight men. In a "shaky condition" to start with, the raft filled with water as "waves were breaking over [it] . . . continuously." Afraid that the bottom might fall out from all the weight, the able-bodied men tried to float so that their life jackets would take some of the weight.

Water Tender Joe McCrane, the senior enlisted man present—not one of the ship's eighteen officers had survived—was in charge. Already affectionately known as "Mother McCrane" for the concern he showed others, McCrane was a man in the right place at the right time. Once all were inside the raft, McCrane "started to organize" by first finding out "who was hurt and how badly." Gunner's Mate Joe Guio and Ship's Cook 1st Class Will Ben Holland, of McMinnville, Tennessee, were the most seriously injured. Guio had "a large piece of the bottom of his foot hanging off" and Holland had "a big hole in the top of his head."

Checking the supplies in the raft, McCrane found two kegs of water, assorted canned rations, and a tin filled with medical supplies. They had no flares and only one oar—everything else had broken loose in the storm. Trying to get water from a keg turned into a difficult job, as the spigot was stuck. When they got it open, everyone took a few sips. The water tasted stagnant, as though it hadn't been changed for some time, and men complained of being sickened by it.

As darkness settled over them the rain had stopped and the wind had died down considerably. Large waves continued to break over the raft, providing "force-feedings of seawater" and leaving everyone "very cold and very miserable." They decided they might as well try to settle down for the night and "pray to be picked up" in the morning.

Guio, who had had most of his clothing torn off, was shivering uncontrollably. "I don't think I'm going to make it," he said softly.

McCrane pulled the injured man close and wrapped his arms around him to keep him as warm as possible.

Guio, having lost a lot of blood, kept drifting in and out. At one point he awoke groggily and asked McCrane if he could see anything.

"The stars," McCrane answered.

"I can't see a thing," Guio said. He thanked McCrane for trying to keep him warm, and others for helping him, too. Guio then laid his head back on McCrane's shoulder and "went to sleep." About half an hour later, McCrane tried to awaken Guio, only to find that he was dead. McCrane told the others, then decided to "hold for a little longer" the man who had rescued from the sinking ship so many others, McCrane included, and who freed from *Monaghan*'s deck—an effort that caused his own serious injury—the raft upon which their lives depended.

Twenty minutes later, the men on the raft held their first burial at sea. They recited the Lord's Prayer as the popular and heroic Joe Guio was lowered over the side. The rest of the night was "spent very quietly" with everyone "just absorbed in their own thoughts."

At daybreak, McCrane gave each man a small ration of food and a cup of water. The malted milk tablets were "very good tasting," but the biscuits were so hard they had to be soaked in water before they could be eaten. When McCrane opened a can of Spam to divvy up, they were suddenly surrounded by sharks, which made the men "plenty scared." While the sharks disappeared for long stretches of time, they would quickly reappear whenever a new can was opened. At one point, seven sharks were counted, "just circling, and waiting." The men talked nervously about how sharks must have a strong sense of smell to find food, and all agreed they would never again let anyone tell them that "sharks go after you only if you are bleeding."

McCrane and the group's next most senior man, Machinist's Mate 2nd Class Robert J. Darden, twenty-eight, of Jacksonville, North Carolina, busied themselves treating the injured—applying sulfa powder and ointment to wounds and bandaging them. Their efforts seemed "fruitless," for as fast as a wound was treated "it had to be put back into salt water."

That morning they sighted several planes, which "seemed to be going everywhere" but over the raft. Everyone was "on edge" and so sore from cuts, bruises, burns (from the sun and salt water) and other injuries that "even if we brushed against each other it was painful."

One of the water kegs was lost that day when the lashing failed to hold it to the raft and it drifted away without anyone noticing. Not knowing how much longer it would be before they were rescued, Mc-Crane reduced the daily water rations. Thirst became a major problem. McCrane and Darden had a "difficult time" with Seaman 1st Class Bruce Campbell, of Texas, who insisted on drinking salt water. They "slapped him and threatened to throw him to the sharks," but it was "all in vain."

As evening approached, McCrane observed that "some of the boys began to crack under the strain." One man bit another on the shoulder, and someone else untied the life jackets that had been secured to the outside of the raft as they were "too water soaked to be worn" all the time. Someone also "unscrewed the top to the first aid supply," and as a result almost all of the medical supplies were lost.

That night, Campbell and Gunner's Mate 2nd Class Dayton Genest, of California, passed away, and the men on the raft held two more burials. McCrane was perplexed, as neither Campbell nor Genest was seriously injured and he didn't know why they died, although he speculated that in Campbell's case it could have been from drinking salt water.

On the morning of December 20 they spotted a second raft and decided they would try to reach it. The men paddled with their hands and with the one oar they had. The swells were "still so mountainous" that it seemed "impossible to ever reach the other raft," which could be seen only when it was atop a swell, before disappearing into a deep trough.

Finally, they worked their way to within 10 feet of the smaller raft, which "no one was on." Seaman 2nd Class Melroy Morrison, of South Dakota, and Fireman 1st Class Louis Shalkowski, of Rhode Island, swam over and tried to "push it toward us as we tried to row to them." The currents, however, kept them apart. The distance between the rafts increased until the raft with Morrison and Shalkowski disappeared from sight in the swells. Neither man was seen again.

Their raft was in "pretty bad condition" at that point. Seaman Doil Carpenter and McCrane repaired it as best they could, and managed to raise the wooden bottom so the men could be "farther out of the water."

That night—their second spent on the raft—"most of the fellows had really lost their heads," with many of them thinking "they saw land and houses." Radioman 2nd Class Louis Spence, of Texas, jumped overboard and kept swimming around the raft. They other men tried to get him back, but he ignored their pleas. He finally climbed back on the raft and said he had filled the water keg with fresh water while everyone was dozing.

McCrane had Carpenter check the keg to see if Spence had dumped the water or "put salt water in it." Carpenter found it almost all gone. Everyone started yelling at Spence, who said if they didn't leave him alone he would "whistle and have the Indians surround all you guys."

"Go ahead and whistle," McCrane said.

Spence did.

McCrane waited a while before he asked, "Where are the Indians?"

"Don't worry, they'll be here."

Spence announced he was going for another "short swim" and would be "right back." Several men tried to grab him, but he pulled away from them. The water was "calmer than it had been," and Spence would swim about 10 feet away and then swim back to the raft. He did this a number of times before finally swimming out of sight. They heard him yell, and tried to row toward the sound of his voice, but in the dark couldn't get a fix on him. Spence abruptly stopped yelling—as if "attacked by a shark or drowned," everyone decided. Spence was never seen again.

After all the excitement, McCrane turned around to find that Holland, one of the most seriously injured, had died. They had their fourth burial that night, committing Holland's body to the deep. Now down to six—"all in bad condition"—they had lost a total of seven men from the raft.

On the morning of their third day, the sea started to get choppy again. McCrane felt that things were "beginning to look pretty grim," but he still endeavored to "keep up his own spirits" as well as those of the other men. They sighted "more planes and a big task force," but they were far off. With McCrane leading them, they "prayed like never before."

Someone spotted an onion floating about 25 feet away, and they paddled over. They had almost reached it when a shark about 8 feet long nosed up next to the onion. The men decided "to let the shark have it."

Darden saw something floating next to the raft and picked it up. It was a flat piece of wood with a quarter-inch chain held on by a nail. The men decided to try to make a fishing line and "catch ourselves a small shark." Someone remembered hearing that liquid from a shark was "good drinking water." Darden used Spam for bait and got a bite, but there was nothing on the line and the bait was gone. He tried again, holding the line down about a foot below the surface. A 5-foot shark came up from below. Darden pulled the line in slowly until the bait was out of the water. The shark followed until its head was up near the side of the raft.

Water Tender 3rd Class James T. Story, of Grant, Oklahoma, took his penknife and plunged it into the shark's head. The shark seemed "little fazed," and swam off. Meanwhile, Evan Fenn and Fireman 1st Class William Kramer tried catching by hand the small fish that swam around with the sharks, but were unsuccessful. The men gave up fishing.

The planes and task force in the distance seemed to be getting closer. McCrane tied a white skivvy shirt on one end of the oar and the men took turns waving it in the air. They also used cans of Spam and biscuits to try to reflect the sun at the ships and planes. After several hours, two planes came "right over the raft about 200 feet above." The men did everything possible to attract attention but were almost sure that the pilots hadn't seen them. After they were well past, however, the planes suddenly turned back toward their position. One plane signaled the men by "nosing over in a steep power dive" that ended with a "deafening roar" as the plane buzzed low overhead.

That was when the men knew for sure they had been sighted. They were so happy as to be "almost speechless." There were no cheers, yells, or whistles. "Thinking of nothing better to do than to thank God," they all recited a raspy "prayer of thanks."

The two planes, which had flown off a Third Fleet aircraft carrier, circled and dropped dye markers close by in the water. They then climbed with full throttle, dipped their wings in farewell, and departed.

Ten minutes later, the men on the raft saw "the most wonderful sight in the world": a U.S. destroyer "steaming at full speed right at us." At 11:41 A.M. on December 21, *Brown* (DD-546) picked up the six *Monaghan* survivors, who had been in the water for three days and nights.

The last to leave the raft was Joe McCrane, as "befitted the senior in command." He took a final look back before grabbing on to the line and being hoisted up the side of the destroyer. It occurred to him that to his "dying day," he would "never forget the finest shipmates that a man could have." And he would always remember "the Mighty M, a gallant ship" that had played a "brilliant role" in her country's "fight for victory." He understood, too, the sad irony that it had taken a monstrous storm at sea to do what the Japanese had been "trying to do since December 7, 1941, at Pearl Harbor."

Sixteen

When Fireman Tom Stealey ran the length of *Hull*'s stack and jumped off, he went "right into a roll." Falling deeply into the swirling sea, he thought his "lungs would give out" just as he popped up to the surface. Buffeted by fierce winds and rain, he managed "one big breath" before dropping into a "deep funnel" that took him down again. Stealey "started flapping" his arms and legs in a desperate bid to reach the surface. He came up just when he was again near the end of his endurance, "right on top of a wave" that took him "sailing 30 or 40 feet through the air" to the crest of another swell, forcing his now nearly limp body to roll and twist and turn as if trapped in a frenetic water ballet.

Stealey lost his shoes and pants; he had his

kapok life jacket "stripped off," too, but was able to snag it and put it back on. When a 5-gallon can passed nearby, he grabbed the handle, thinking the container might provide added flotation. The next funnel he dropped into "beat the living hell" out of him. When he surfaced, he was holding the handle without the can attached. He saw a bench float by but decided to let it go, fearing that he might be "beaten to death" if he tried to hold on to it.

After an hour or so he spotted the first person he had seen since entering the water. The visibility had begun to "clear up a little," and although "the big waves were still there," he wasn't dropping into as many funnels. The two men waved, then swam toward each other. "Circles of water," however, kept them separated. Stealey had the idea to swim along with the circular currents. When he did, the next time he came around they were closer. With each man "swimming with the circles," they eventually came together. Although Stealey did not know the guy and they never traded names, they used the straps on their life jackets to tie themselves together. By evening, they had "thirteen guys hooked up."

Throughout the long night no one saw any lights or ships or heard anything other than the angry wind and sea. It was as if they were all alone in the middle of the ocean. Shortly after sunrise, "one guy went out of his head all of a sudden." Untying himself before he could be stopped, he took off and disappeared from sight.

The next day, a single dorsal fin appeared in the water.

The shark looked to be 10 to 12 feet long, with a large and conspicuously rounded fin that had a white tip. The fin aimed straight for the men. With circular eyes inset into a rounded snout, the shark hurtled toward the men like a maasive gray torpedo. It first bumped a man in the center of the line in a hard body check, then twisted around, eyes bulging and jaws snapping, and seized the man at about waist level. Amid violent splashing, the shark "pulled him down so fast he made no outcry."

Surfacing some yards away, the shark released the bloody and now shrieking man like a dog dropping a bone. Whirling back toward the line of men, the shark—teeth glinting in the sun—hit furiously again, grabbing and pulling down a second man.

Panic had set in and everyone was hollering. Men desperately worked to untie themselves from the line and as soon as they were free started swimming away. Stealey did likewise—nearly "running on top of the water getting the hell out." There were more terrible cries behind him. Although Stealey did not look back, he knew that the shark was either finishing off the first two men or striking new victims. Certain that "a mess of sharks" would be attracted by the sounds and smells of men dying in the water, Stealey swam as hard and fast as he had back in the days when he was winning ribbons in high school swim meets in California.

Eventually, Stealey found himself with three men from the larger group. Once again they tied themselves together. Anxiously they searched the horizon and skies in all directions for would-be rescuers, but there was "nobody in sight, no ships, no nothing." The hopelessness of their situation "started to get to everybody." A couple of the men talked about nobody looking for them and being left for dead. Stealey changed the subject to home and family and the first things he wanted to eat when he was rescued. Steak and potatoes sounded good to Stealey, who tried to recall the exact tastes and textures.

Stealey turned to the man next to him to ask a question. To his shock, he saw that the guy was "dead, just like that." Whether from "exposure, exertion, or a little bit of everything," Stealey had no way of knowing. Before nightfall, a second man passed away as suddenly, leaving Stealey and a young seaman he knew only as Smitty, who said he was from Gridley, California, about 30 miles "down the road" from Stealey's hometown of Stockton.

In the wee hours of the morning, Stealey, who had managed to drop off to sleep in his high-collared life jacket, awakened to the sight of the sweeping beam of a ship's searchlight off in the distance, obviously looking for men in the water. Observing the sweep of the light, Stealey realized the ship was going to pass too far away to spot them.

"They're gonna miss us," Stealey said. "We gotta swim toward them."

He and Smitty started swimming. After a few minutes Smitty stopped, saying he was too tired to keep going. Stealey "egged him on." In another ten minutes Smitty was finished. "You know I'm here," the exhausted man said. "When you get picked up, send someone to get me."

Stealey swam on.

The next thing he knew, he awoke after daybreak "in torture," lying on top of his life jacket with no memory of taking it or anything else off. He was "bare-assed" and "little fish were nibbling" on his legs. His skin was blistered from the sun, and his eyes, swollen from the salt water, were nearly shut. His lips were swollen, too, and his ears ached. His life jacket was so waterlogged that when he put it back on he hung so low in the water that the sea came up to just below his chin.

There was no ship to be seen, and Smitty was gone, too.

Literally inches from death, Stealey saw a vision that looked like one of those postcards of a cascading waterfall at a national park, only this one flowed with pineapple juice. It looked wet, cool, and delicious, and he was so very thirsty—and yet, somewhere in the parched recesses of his memory, he recalled that he "didn't even like pineapple juice."

It was about then that Stealey decided to commit suicide. He knew one way he did not want to go: to be "eaten by a shark" like his shipmates. He started drinking salt water—"as much as I could get down." His first feeling was unexpected: he felt better at having something in his stomach for the first time in days. If he kept drinking salt water, he figured, it would be only a matter of time before he died. He hoped it would be like going to sleep, as some of the other men had so quietly expired. He began to make his peace: saying goodbye to his wife, Ida, and telling her that he was sorry, and how he had tried to get out of this mess but couldn't.

It was a clear day, sunny with a few puffy white clouds drifting across blue skies. Off in the distance, Stealey spotted smoke. Not believing his eyes, he looked away, then back again. The column of smoke was still there. Soon he saw a ship "starting to come up out of the water," getting bigger as it came closer, heading directly for him. *Oh, man*, Stealey said to himself, *I sure hope it's American.*

At a quarter of a mile away, the ship began to turn.

Stealey started hollering and waving.

Something hit him "in the back of the head." His first thought was that a shark had bumped him. Turning around quickly, he saw a piece of board that was yellow on both sides. He had no idea what it was or where it had come from, but it seemed as if it had dropped from heaven. He frantically waved the board above his head.

The ship "came back around," heading for his position.

As the vessel approached, Stealey heard rifle shots. He hadn't yet seen the ship's flag and thought only that it must be a Japanese destroyer and that they intended to shoot him in the water rather than rescue him.

The ship came close and all engines stopped.

The next thing Stealey knew, a "monkey's fist"—a type of knot tied at the end of a line to serve as a weight, making it easier to throw—was headed his way. He grabbed the line and was pulled in by deckhands as other sailors kept firing at the sharks that had been circling Stealey.

It was 8:25 A.M. on December 20 when Stealey was rescued by the U.S. destroyer *Cogswell* (DD-651). He had been in the water for forty-four hours.

Stealey was too weak to pull himself up the ship's cargo net, and several men jumped into the water to assist him. Unable to walk on his own—"no muscle control in my legs"—he was carried to the infirmary, where he was found to be in "unbelievably good shape." After a freshwater shower, during which he had to be held up to keep from toppling over, Stealey was asked by a ship's cook what he would like to eat.

"I'd sure like a nice steak. With potatoes, if you don't mind."

"Fix you right up, buddy."

SONARMAN PAT DOUHAN, whose hope for a transfer off *Hull* never happened, was in the after deckhouse with some twenty other men when the destroyer rolled over to starboard "to the point of no return." Up until then the men had been praying, as well as "cussing the captain" for not keeping the ship "headed into the swells or going away from them." Instead, James Marks had placed them broadside to the wind and mountainous swells—"at the mercy of the elements."

What followed was "pure panic" as water poured in through the opening where the starboard hatch cover had earlier blown off. In an effort to stay above the rising water, several men were "hanging on to the dogs" (heavy latches) that secured the port hatch cover. As the port hatch was "our only way out of a sinking ship," Douhan, who was holding on to the hatch handle and had one foot planted inside an overhead vent, did the only thing he could "in order to save any of us." He methodically began "kicking my shipmates' hands off" the latches. As the men let go, they fell into the rushing water below. It was a haunting scene, one that Douhan would "never get over." Finally able to open the hatch, he had begun to wriggle out when the heavy hatch cover came down on his back, pinning him. While several men squeezed out past without stopping to pull the hatch cover off Douhan, someone finally did stop and free him. He stepped out onto the side of the sinking ship without a life jacket—his was on the bridge, where he had spent most of the last eighteen hours until heading down to the berthing compartment that morning to hit the sack. It had been a "foolish thought," as no sooner had he climbed into his bunk than he was "thrown out by a very hard roll to starboard."

For now, the luck of the Irish was still with Douhan: he spotted a spare kapok life jacket "tied to a gun mount." There was only one, and he was convinced that a "higher power up above"—and possibly his "praying mother"—had left it there for him. He put on the life jacket and was immediately "swept up amidship" near the 40 mm gun bays. As he grabbed one of the gun railings to keep from being washed overboard, he saw Chief Machinist's Mate Archie L. Vaughan, a peacetime *Hull* crew member and Pearl Harbor survivor, "pinned against the bulkhead by a large broken life raft with no way to free himself." Douhan, who considered Vaughan "a great guy," wanted to reach the chief but with the "high winds and seas" was unable even to get close to him. Moments later, as Douhan and several other men were swept off the side of the ship, the trapped Vaughan held a fist high in the air, exhorting his shipmates: "Fight on!"

Once clear of the ship, Douhan found himself in a large group of men trying to grab hold of several wooden-bottom life rafts, which "kept being swept up into the air." Each time a raft came down, several men would attempt to climb inside, with limited success. Douhan was able to grab hold of one raft connected by a length of line to another raft. When the two rafts were swept high in the air they "came together with such force" that a number of men caught between them had their "heads popped open like popcorn." Douhan had enough of life rafts and let go. About that time a huge swell separated him from the others. He was "all alone in heavy seas," with rain coming down so hard he had to shield his face from what felt like countless needle pricks.

Shortly afterward, Douhan heard a muffled underwater explosion, and felt "a lot of pressure on the lower part" of his body. He surmised that the sinking *Hull* had gone "deep enough to cause the boilers to explode."

While floating alone, Douhan was suddenly pulled down so deep that he thought his "ears would explode." His first thought was that he was being sucked into the screws of one of the large ships in the task force; the low visibility would have made it "impossible to see any ship," even one passing nearby. He envisioned being "chopped up by the screws." To his surprise, he suddenly shot to the surface. When the same thing happened again, Douhan realized that the huge swells breaking over him were driving him down. He began to anticipate the next swell, taking a deep breath, then tucking into a roll and "riding with it." Even though the "depth was about all he could stand," he developed a workable technique.

Douhan had time to think as well as react. He was alone in the ocean in the middle of a typhoon, not knowing how many—or even "if any"—of his shipmates had survived, nor whether he realistically stood a chance of being rescued. With great trepidation, he wondered: *When will the sharks show up?* While there were lots of things to "keep me guessing," there was something he "never lost sight of": he had to make it because he had left his "beautiful wife expecting our first child." He did not intend for his Kay (Kathleen), one of thirteen pregnant *Hull* wives, "to be a widow."

After drifting alone for several hours, Douhan saw that the waves and winds started to subside and darkness was coming on. A few hours later, Douhan spotted a searchlight, which gave him hope for rescue. But the light disappeared over the horizon, "never to appear again." In the middle of the night something hit him in the back of the neck. He froze, "thinking of sharks." When he turned around he found an ordinary broom. No doubt off the ship, it was like "finding an old lost friend"—something material to prove he had once been on a destroyer with nearly 300 other men. He discovered he could rest his feet on the broom's shoulder and that it was a relief from hanging in the life jacket, which caused numbness in his legs and feet.

Around what he estimated to be midnight, Douhan spotted a "little light." He yelled out, and received a response: "Who's there?"

"Pat Douhan."

"Get over here, Douhan."

He found fourteen of his *Hull* shipmates in a partly broken-up raft. The line that secured the wooden slats to the raft had sagged, dropping the bottom down 3 or 4 feet. The men, most wearing life jackets, were standing inside the raft in water up to their shoulders. All the raft's supplies had broken loose in the storm, and there wasn't as much as a sip of water or a morsel of food; nevertheless, Douhan was thankful for having made it to the raft and finding "some company."

The only officer in the raft was Lieutenant (j.g.) Edwin Brooks, the proper and "self-assured" Virginian with an economics degree from the University of Richmond who served as *Hull*'s sonar officer. Brooks, too, had been floating alone in his kapok life jacket until earlier that evening, when he happened across the raft filled with men.

When *Hull* went over, Brooks had jumped into the water and the first wave pushed him away from the ship. As *Hull* "started to settle," Brooks was "pulled under by the suction" and thought "it was all over." The rush of water pulling at him turned out to be a "compartment filling" rather than the ship sinking. Brooks came up inside a gun turret alongside the bodies of several men "killed by being thrown back against the ship" and having "their heads caved in." He climbed out of the turret with the help of two enlisted men who "pulled me up with them," and jumped again into the water. For eight hours he rode out the worst part of the storm alone, with his head "more than half the time underwater in heavy seas." He ingested a lot of salt water, and by the time he reached the raft Brooks was "not in very good shape."

Also on the raft was Gunner's Mate 1st Class John Valverde, twenty-five, of San Francisco, who spent his early childhood in the city by the bay and then was on his own from age fourteen, "picking fruit in the country" until he was old enough to enlist in the Navy. A first-generation American of Spanish descent, the stocky Valverde had been aboard *Hull* since a year before the Pearl Harbor attack, during which he used bolt cutters to break open the ammunition lockers in order to "get the machine guns going" against the attacking aircraft. Through three years of war, Valverde's favorite skipper had been Consolvo, and his least favorite was Marks, who "wasn't qualified for a seagoing command."

When *Hull* started filling with water, Valverde had emerged from below deck near the crew's galley with Chief Yeoman Robert H. Ellis. The two men climbed over the side, where they hung on until a swell threw them into the air "right over the ship and clear to the other side." In the water together, Ellis asked calmly, "What do you think, John?" Valverde replied, "Don't give up! Just hang on!" The problem was there "wasn't anything to hold on to," and they were "sucked down" so deep that Valverde's "ears were bursting." When he surfaced, Valverde came up under a life raft, hitting his head on the wooden bottom. When he worked his way out from underneath, another raft struck him in the chest. After being "sucked back under two or three more times," Val-verde, who never saw his shipmate Bob Ellis again, was eventually washed clear of the sinking ship. For the next few hours, Valverde at times saw "a whole bunch of men" and at other times was alone. Once he was surprised to see the captain "floating by in his life jacket holding a seaman's knife in his hand." There was no question in Valverde's mind that the unpopular Marks was holding the long-bladed knife at the ready for one reason: as "protection from the crew." *Deservedly so,* thought Valverde, who knew an old-time gunner's mate who had threatened to "kill Marks with a machine gun" and was quickly transferred before *Hull* left the shipyard in Seattle. As they drifted past each other, neither Marks nor Valverde spoke to the other. Valverde, however, thought it would be a good idea to "keep away from the captain because of that knife," and so he put distance between them.

With the light of day, the men in the raft found that they faced an-other enemy. Dorsal fins appeared for the first time. As he counted the number of sharks circling the raft, Douhan could make out their torpedo-like body shapes, beady eyes, and conical snouts. He reckoned they were about 12 to 14 feet long and "looking for something to eat." With the men hanging in their life jackets inside the raft, the sharks could easily have struck at "our legs any time." They continued making constant passes at the overcrowded raft, which "made a big target," only to turn away at the last moment. For the time being, however, the sharks seemed to be biding their time.

That night, two sailors became violent and started thrashing around. They had both been seen drinking ocean water. The first to go was one of the new recruits who had picked up the ship some weeks earlier in Pearl Harbor and whom "no one really knew." He was stripped, a short prayer was said, and his body was pushed away from the raft. His clothes and life jacket were given to Radioman 2nd Class Lester C. Mullins, who had escaped the sinking *Hull* by squeezing through a small porthole. To do so he had had to remove his life jacket and clothing, and he had been suffering in the elements: his skin was burned by the sun, and he shivered in the cold at night. Mullins died later that night. After another prayer his body, too, was allowed to drift away. Although the men "did not know if the sharks got" the bodies, pushing their dead buddies out into shark-infested waters "didn't set very well" with Fireman 2nd Class Edward J. Price, twenty-one, of Topeka, Kansas, who also had escaped *Hull* by climbing out a porthole, or with any of the others on the raft. But it was something they all knew "had to be done."

Not long after, one of the men thought he saw something on the horizon. He right away wanted to "start calling out," but Douhan and other veterans picked up a sweet odor—rather like incense, and reminiscent of a scent they had detected while patrolling close to Japanese-held islands. Everyone was told to keep quiet because they might be drifting near an enemy island or even a Japanese submarine at the surface charging its batteries. All agreed that staying in the raft with hope of rescue was better than being taken prisoner by the Japanese—possibly to be executed.

The next morning Douhan and several others decided to keep someone on watch at all times to look for "some kind of life and rescue." The long, hot day was uneventful except for the "sharks still circling and making passes at us," something that never became routine. Also, several men became delirious. One wanted to "go down to the galley and get a sandwich," while another said he was going to borrow his brother's "Model T Ford and bring back some 7UP" for everyone.

When a delirious Brooks was seen to take a swig of sea water, Fireman 1st Class Nicholas Nagurney "pounced on and rammed his finger" down the officer's throat to make him regurgitate. In the process, Brooks bit Nagurney's finger. It was soon Nagurney's turn to have "strange delusions"—he swam a few yards away, intent upon finding out "how deep the water is under the raft." Before he could get back, he was bitten by a shark, which tore a thin slab off the top of the right forearm. Back in the raft, Nagurney's bloody arm—with a row of half-inch-deep teeth marks on the underside—was wrapped with a piece of torn shirt. Jolted back to his senses, Nagurney summed up his after-noon: "I guess I'm the only guy that's ever been bit by a shark and an officer the same day."

The next morning, someone hollered, "Task force on the horizon!" At first everyone thought the guy was "a little out of it," but he kept in-sisting. Douhan "rubbed the salt water crust" out of his eyes and, "sure enough, saw ships on the horizon." About nearly the same time, two planes crossed overhead a "little on the high side." Even with "all our waving," the men began to think they hadn't been seen. But then the planes came back over nearly at sea level and wiggled their wings to let the men in the raft know they had been spotted. It was enough "to make us cry."

Soon the men saw a column of black smoke. Being destroyer sail-ors, they knew what it meant: a tin can was "lighting off extra boilers" to speed to "our rescue." In short order, the destroyer *Brown* (DD-546)—which an hour later would also pick up the six *Monaghan* survivors—pulled up next to the raft and threw over a line. At that moment, the sharks—"knowing they were going to lose their dinner," surmised Douhan—went into a frenzy. Sharpshooters on *Brown* fired accurate volleys at the sharks to keep them away as the raft was hauled in.

Starting at 10:46 A.M. on December 21, the survivors were helped one by one up the side of the destroyer by their rescuers. Having not seen anyone else or other rafts or debris since *Hull* went over, the men had no way of knowing whether or not they were the only members of their ship's company of 258 officers and enlisted personnel still alive.

The thirteen *Hull* crewmen had been in the water for seventy hours.

Seventeen

Tabberer, the new Houston-built destroyer escort under the command of Lieutenant Commander Henry Plage, the tall Georgian and former retail credit company employee who had proven to be a natural-born leader, sortied in mid-October 1944 from Pearl Harbor in company with the escort carrier *Anzio* (CVE-57) and several other vessels.

En route westward to Eniwetok, the ship's dog, Tabby, stopped coming around the galley for food. The word went out, and a stem-to-stern search was conducted. It was Plage who broke the news over the ship's address system, explaining that "poor little Tabby must have been washed overboard." In a letter to his wife, Plage wrote: "The gang is pretty blue about it."

Ship's Cook Paul "Cookie" Phillips, the wiry former amateur boxing champion, had been one of the first to notice Tabby was missing, and he took the loss of everyone's favorite pooch as a bad omen. Ironically, not long afterward, Phillips found himself involved in his first fistfight aboard ship. The altercation was with Boatswain's Mate 1st Class Louis A. Purvis, twenty-four, of Chatham, New Jersey. Purvis, the leading petty officer in charge of *Tabberer*'s deck force, was a "rugged character" and "smart-aleck tough guy" known for "throwing his weight and power around" to keep his young seamen in line.

One morning Phillips went through the mess hall after chow had already been served, and he was surprised to find hot food still on the steam table. He asked one of the messmen standing by why the spread hadn't yet been picked up. "Waitin' on Purvis. He's always late."

"Purvis eats by eight o'clock or doesn't eat," Phillips said.

The messmen were clearing the steam table when Purvis came down the ladder. Told what was happening, Purvis said, "I run this mess hall. I eat when I want." It was true that Purvis provided seamen from his deck force to clean up the mess hall, but otherwise this was Cookie's turf.

"This is my food," Phillips told Purvis. "I cook it in my galley. These are my messmen. If you ain't here on time, Purvis, you don't eat."

With that, Phillips went past Purvis, headed for the galley.

Purvis, the bigger man by 30 pounds, slugged Phillips, knocking him up against the bulkhead. Before Phillips could cover up, Purvis clocked him again. At that point, Phillips' cooks wanted to stop the fight, but Purvis' seamen wouldn't let them. Phillips, who had always had a good left jab, then went to work with lightning punches and dancing feet, as he had learned from long days sparring in the gym. Soon, Purvis' men were trying to stop the fight, only now the cooks wouldn't allow it. When Phillips had Purvis whipped and defenseless against the bulkhead, he looked at the seamen. "Purvis is out now. If he had me like this, he'd work me over and scar me up. I'm just gonna put him down." With that, Phillips dropped Purvis to the deck with one last punch and walked away. Phillips went directly to sick bay to have his cuts treated.

"My God, what happened to you, Cookie?"

"Don't worry, Doc. Someone in worse shape will be here soon."

A few days later, Purvis, his face bruised and battered, gathered his deck force and told them they could fight anyone aboard ship except for one man. "Look what Cookie Phillips did to me," Purvis said. "Stay away from him." After that, Purvis, who never asked for a rematch, started eating on time. Before long, he and Phillips even became "good buddies." Purvis would ask if Phillips needed extra help in the galley. If so, the boatswain's mate would send a seaman or two around to help out.

Upon joining the Third Fleet at Eniwetok, *Tabberer's* task group was deployed to the fleet's fueling area southeast of Luzon to conduct anti-submarine sweeps. They immediately found themselves on the edge of a typhoon and it got "pretty rough." Plage was pleased, however, to see that his ship and crew "rode it okay," although they "saved money on the chow bill." Some of the new guys, fighting seasickness, "weren't very hungry," and men who could eat had to settle for sandwiches when the heavy seas made it impossible to prepare hot meals in the galley.

When things turned calm and peaceful for several hours one morning, Plage took the opportunity to go below for a quick shower. He had a feeling that as soon as he stepped into the shower "the bridge would call," so he put his hand on the water valve and waited a minute. When his cabin phone remained quiet, he turned on the water and soaped up, and then the "damn phone rang." It was the OOD reporting a routine change of speed, as he was required to do. Following that interruption, Plage figured he would be able to enjoy "the cool water." All of a sudden his cabin's emergency buzzer went off and the general quarters alarm began to clang, calling all hands to their battle stations. Plage jumped from the shower soaking wet and soapy. With no robe handy, he grabbed his foul-weather coat and went running. By the time he reached topside he had fastened only the top hook. With the "rest of the coat flying in the wind," he stepped onto the bridge "in all my glory." His immediate problem was trying to "maintain discipline," which he found difficult to do when "everyone is laughing."

On November 18, the *Anzio* task group registered its first kill after being alerted to the presence of a Japanese submarine in the area in a message from the military intelligence unit, which was reading Japan's secret war code. *Anzio* launched aircraft for an extensive search, which resulted in a radar contact on a surfaced submarine. After a fourteen-hour chase, *Tabberer*'s sister ships *Lawrence C. Taylor* (DE-415) and *Melvin R. Nawman* (DE-416) carried out a coordinated depth charge attack, sinking the Japanese fleet sub *I-41* with the resultant loss of her 114-man crew. Two weeks earlier, *I-41* had torpedoed off San Bernardino Strait the new light cruiser *Reno* (CL-96), which had to be towed 1,500 miles to Ulithi for emergency repairs in order to steam under her own power back to the States for extensive work, ending her wartime service.

On November 29, *Tabberer* pulled into Ulithi, where her depleted stores and ammunition were to be replenished. They had been at sea so long their fresh foods had run out, and main courses had consisted of canned "Vienna sausage, Spam and corned beef hash." Unlike many ships where the lowest-ranking men did most of the heavy lifting, Plage required every officer and enlisted man not on watch to take part in boarding supplies and stowing them. To make sure that "everybody who eats loads stores," Plage positioned himself in a bird's-eye seat on the ship's fantail, overseeing each work party. For many in the crew, it was such "evenhanded fairness" that made Plage so popular.

On December 5, *Tabberer* "finally got some fresh food on board"— "meats, oranges, apples, potatoes, lettuce, cabbage." In a letter home, Plage wrote: "It has been two months since we've had an orange on board." He soon noted that with the improved chow, morale was "back up tremendously."

The crew had come together in ways that pleased Plage. A new supply officer, Ensign Travis E. Nelson, twenty-one, of Bryan, Texas, had started Sunday services that "really have taken hold." At sea they were held in the mess hall with a "different person teaching or giving a short talk," and soon the place was packed with everyone not on watch. "I am very pleased to see the crew enjoy a simple church service so much,"

Plage wrote to his wife. "It means a lot and adds so much to life aboard ship." Every Sunday, several hymns were sung, "of course with no organ or piano," but Plage noted that "we have some good voices aboard." In fact, the crew soon organized a small "choir or glee club" made up of "nine or ten fellows who gather on the fantail at sunset and practice" for the following Sunday. After working on the hymns, they "usually drift off into most any kind of song," and the rest of the crew that came out to listen—again, practically everyone not on watch—"joins in." Often Plage would be on the bridge as the sun went down, listening to the a cappella songfest. After each song, "there is a dead silence for a minute or two" before someone would start "singing some song quietly" and others chimed in. These sunset events were soon being called "happy hour," with all hands, from the "kids who are naturally homesick" to the married men "just plain longing to get back to their wives and families," enjoying the musical reprieve from wartime.

Meeting *Tabberer* at Ulithi was Lieutenant Howard Korth, the former Notre Dame football player and the ship's senior watch officer. Korth had gotten off the ship at Pearl Harbor in mid-October to attend fire-control school. Finishing first in his class, he had been spoken to about an instructor's job. Korth had enjoyed Pearl Harbor, where he met up with a number of former Notre Dame classmates—including several with whom he had played football for the Fighting Irish—as they came through on ships and other military assignments. Since Korth was assigned to a ship deployed to a combat area, however, it was decided he should "first return" to *Tabberer* before being considered for any new shore assignment. Korth was not disappointed, since he considered *Tabberer* a "fine ship" and was pleased to be serving under Henry Plage—a "first-class guy all the way around" and "someone you could depend on."

Shortly after sunrise on December 10, *Tabberer* weighed anchor and departed Ulithi with *Anzio*'s hunter-killer antisubmarine group, consisting of four other destroyer escorts. Headed back to the fleet's fueling area, they were scheduled to arrive a few days before the Third Fleet's main body was expected to commence fueling operations on December 17.

On December 15, *Tabberer* came alongside *Anzio* and received "18,582 gallons of fuel oil." Two days later, when the fleet's first fueling operation was cancelled by Halsey due to the worsening weather, *Tabberer* still had "79,256 gallons of fuel on hand," some 260 tons—approximately 75 percent of the ship's fuel capacity.

As did the rest of the fleet, *Tabberer*'s small group—given the familiar assignment of protecting the warships and tankers when they were most vulnerable to submarine attack, which was as they maintained a straight course and steady speed during fueling—turned to a northwest course in the early morning hours of the eighteenth, steaming at between 12 and 15 knots for the next fueling rendezvous at 6:00 A.M.

When the fleet's course was switched to the south only to change again after sunrise to a northerly one—into the wind in an effort to fuel in spite of the heavy weather—Plage noted that the barometer on the bridge, which measured 29.58 at 7 A.M., thereafter began falling rapidly. That meant one thing to Plage: a typhoon. It was no surprise to him when shortly after 8:00 A.M. fueling was cancelled and all ships were ordered to proceed south.

At 10:30 A.M., steering on *Tabberer* became difficult due to the increasing wind and sea. The wind out of the north was measured on the bridge as "force 12"—equating to "above 75 miles per hour," at that time the highest category on the Beaufort wind scale. The official "seaman's description of the wind," often at variance with the terminology used in U.S. Weather Bureau forecasts—for example, a "calm" wind (less than one mile per hour) to a seaman was called a "light" wind by weather forecasters—was in this situation identical. Force 12 on the Beaufort scale meant the same thing to everyone: a hurricane, which in the Pacific was called a typhoon.

Due to their steering problems and the "close proximity of numerous other ships" on the same southerly course, Plage decided to head for a short time on a course of 90 degrees until they could "get clear of other ships." Unfortunately, the turn "put the vessel in the trough of the sea," but Plage judged it not immediately dangerous since the steepness of the ship's rolling at that point did not exceed 40 degrees. When *Tabberer* came clear of other ships, Plage brought her back around to the base course of 160 degrees only to find that they were unable to maintain the course because they kept falling back into the trough. He made repeated attempts to keep the ship headed downwind using "various speeds up to 18 knots with full rudder" and even "ahead full on one engine while backing with the other engine," but nothing worked. They were stuck on a giant roller-coaster ride: rising high with each swell that broadsided them, only to drop into the next deep trough. Realizing they would have to ride out the storm with wind and sea on the port beam, Plage decided to shift as much fuel oil as possible to the port tanks to compensate for the starboard list.

At 12:30 P.M., when other ships and sailors not far away were meeting their tragic end, *Tabberer* was "riding quite well" at 10 knots, "rolling up to 55 degrees" in winds estimated at "over 100 knots." Even as the wind shifted rapidly to the west and then to the south, however, the ship stayed stuck in the line of troughs despite all attempts to get out. During the "greatest ferocity" of the typhoon, *Tabberer*'s rolls reached 72 degrees, although each time she recovered "rapidly with no hesitation." In a later official report, Plage would judge "this type of vessel very seaworthy in rough sea," stating that the new class of destroyer escorts could "withstand rolls in excess of 72 degrees without danger of capsizing."

Sonarman Frank Burbage, the New Jersey teenager who had been impressed with the "high morale" of the crew under Plage ever since the ship's shakedown cruise to Bermuda, had the 8:00-to-noon watch on the bridge. From his vantage point, Burbage decided that "those big waves at the beach at Asbury Park" were nothing more than "ripples compared to the size" of the ones washing over the bow. He also observed Plage, standing nearby at the conn, "handling the ship magnificently" without looking "nervous or taking a hopeless attitude." In fact, after one deep roll, Plage asked the man at the pitometer how far they went over. Told 60 degrees, "the captain jokingly replied, 'She'll take 20 more.'" What scared Burbage "more than when I used to come home with my new Sunday pants torn" was the "persistent pounding of the waves on the bow." One minute they would be rising on the crest of a swell, and then the deck "would give out from under us" and the bow would crash down into the trough and the "whole ship would shake and tremble." Burbage knew that such pounding on the hull could eventually "crack the seams." Which, he began to wonder, would last longer: the ship or the "watery hell of the typhoon"?

Radioman 3rd Class George Pacanovsky, nineteen, of New York City, was on duty that morning in the radio shack adjacent to the bridge. He had graduated from P.S. 29 half a year earlier and joined the Navy the following month. After boot camp in Newport, Rhode Island, and two radio technician schools in Chicago and Bainbridge, Maryland, he had met *Tabberer* at Ulithi in early November. He had gone through a long period of severe seasickness before "finding my sea legs." In fact, it seemed to him he had "stopped throwing up just in time for the typhoon," during which everyone on board was "walking the bulkheads" to such a degree that work parties later went around to "wash off the footprints." While Pacanovsky was still a seagoing novice, he saw enough that morning to have confidence that Plage was "in full control." What impressed the young sailor was the way the captain "didn't try to fight the storm" but slowed the ship down and "let the typhoon bounce us around." The pounding on the ship "would have been worse if we were going faster."

When Plage received a report that some 5-inch shells and cartridges in the forward magazine had "broken loose," he held off ordering anyone into the magazine because it was "almost a certainty that someone would get crushed trying to secure" the 54-pound shells. While the shells wouldn't explode on their own, it "created a delicate situation" because of the possibility one of them might hit the primer on a 27-pound cartridge filled with gunpowder. Plage's concern was soon relieved when he learned that two volunteers went in the magazine and secured the shells and gunpowder. It was the kind of thing a "close-knit crew" did for one another, believed one of the volunteers, Gunner's

Mate Tom Bellino, the young Idaho dairyman who, although he kept getting busted in rank for minor infractions, was among the many crewmen who "loved the skipper."

Plage was informed of three radar contacts from 2,000 to 5,000 yards ahead. They were identified as two destroyers, *Hickok* and *Benham,* and the destroyer escort *Waterman* (DE-740). Raising them on the TBS, Plage learned that all three were "in troughs steaming at 3 knots and without steering control due to the wind and sea." Plage quickly ordered a reduction in speed to 3 knots to avoid colliding with the other ships.

While "just looking at the tremendous seas" could "encourage exaggeration" as to their size—"they looked like vertical mountains bearing down on us"—Plage came up with a way to gauge their height. A destroyer escort of the same class as *Tabberer* was adjacent to them in formation and occasionally visible through the storm. He knew the top of her mast was "93 feet above the waterline." When both ships were in parallel troughs, the tip of the other ship's mast would disappear behind the crest of a wave. As for frequency, Plage timed "nine seconds for a complete cycle—crest to crest." When it came to wind speed, *Tabberer's* anemometer eventually "blew away at 100 knots." Plage heard a report on the TBS that one ship's anemometer blew off at 130 knots.

So much solid water broke over the main deck that the "life nets all floated out of their storage racks," even those floater nets twenty feet above the deck. *Tabberer's* whaleboat was demolished while secured to its davits. On one steep roll to starboard, water went through the topside ventilator intakes to the engine room. As the water flowed in near an electrical panel, quick-thinking engineers rigged a canvas sleeve to guide the water away from the electrical equipment.

The barometer began to rise by 1:15 P.M. Just as everyone started to hope the storm was passing and the worst was over, at 1:51 P.M., *Tabberer* took a "quick 60-degree roll to starboard," and the pressure on the mainmast caused one of the insulators on deck that secured a guy wire to crumble, allowing the supporting wire about three inches of slack. The mast—at the top of which were attached "such precious

devices" for communication and navigation as radar, radio, TBS, and
IFF (electronic identification equipment to identify *Tabberer* to other U.S.
ships) antennae—began to sway back and forth. A work party tried to
take up the slack, but the men were unable to do so due to the force of
the wind. Soon a second and then a third insulator gave way. With the
mast now "swaying about eight feet," the weld at the bottom soon
broke.

Plage knew the mast could topple at any moment. His fear was that
when it snapped off it might open a hole in the main deck through
which seawater could pour inside. Not only could no one predict when
the mast would go or the amount of damage that would be caused, "no
one could do anything about it." Plage considered it "all in the laps of
the gods."

The "torment lasted" until 6:28 P.M., when the mast finally buckled
during a 50-degree roll. Snapping in half, the top portion with all the
electronic gear fell into the sea over the starboard side, while the base
remained tenuously attached to the ship by its tangled-up guy wires.

Plage stopped all engines. A damage control team with axes and a
cutting torch was standing by. He sent the men out with orders to cut
loose the remaining portion of the mast, which they did while secured
by lifelines to the superstructure. At 7:03 P.M., the remaining mast fell
over the side "without even denting the ship or scratching the paint."
While the dismasting resulted in lost communications and navigational
systems, the severity of the ship's rolls was noticeably less without the
drag of the dangling mast.

At almost the same time, steering control was regained. Finally
"freed of the irons" that had held her locked in one trough after another, *Tabberer,* whose power plant had "operated without serious casualty" all day, turned south for the new fueling rendezvous point radioed
some hours earlier and scheduled for sunrise. "As deaf as a stone and
blind as a bat" without radar or radio in this "highly populated slice of
ocean," *Tabberer* was provided "courses to steer and speeds to make" by
blinker signals from a nearby destroyer.

Proceeding at 8 knots and making slow headway, *Tabberer* "pitched and pounded" through a dark night filled with "weird sounds of shriek-ing winds" of gale force. Although the typhoon had "left the scene"—rushing westward before finally curving northward—bolts of lightning electrified the sky, followed by booming claps of thunder.

Chief Radioman Ralph E. Tucker, twenty-seven, of Somerville, Massachusetts, had been hunkered down all day in *Tabberer's* radio shack trying to take his mind off the "terrific rolling" of the ship by reading Bob Hope's book *I Never Left Home,* the comic's account of enter-taining members of the armed forces during the war. Admittedly very frightened, the barrel-chested and normally happy-go-lucky Tucker would not realize the book was funny "until two days after the storm."

At 9:50 P.M., Tucker was on an upper deck rigging an emergency ra-dio antenna near the forward stack to restore communications when he heard a shout and noticed a small light shining in the water.

"Man overboard!" Tucker yelled. "Light off starboard beam!"

The radioman was close enough to the bridge to be heard by Plage, who feared "one of our men" had gone over the side. He directed the duty boatswain's mate to sound the man-overboard alarm, which was soon clanging loudly followed by the "terrifying word" over the pub-lic address system that no sailor ever wanted to hear, let alone at night in storm-tossed seas: "All hands, man stations to rescue man over-board."

Someone from the bridge hollered to Tucker asking what had happened. The chief answered that it was "not one of our men" who had fallen overboard but possibly someone from "another ship."

Peering through the line of portholes on the bridge, Plage scanned the darkness. He spotted the light off the starboard beam, blinking on and off as it "rose and sank among the waves." He had the 24-inch searchlight turned on and sighted a "waving man" in a life jacket.

Plage ordered the helmsman to a new course. In picking up survi-vors, the usual procedure—and one that *Tabberer's* crew had practiced countless times—was to maneuver the ship downwind, turn the bow into the wind, and proceed upwind, as a ship did when approaching a mooring buoy. The equipment needed for a rescue comprised two cargo nets thrown over the side, life rings with long lines attached, and other lines with monkey's fists tied at the end. Also needed were a couple of capable swimmers in life jackets with safety lines attached. As the ship neared the survivor, lines were thrown to the man; in the event he was unable to reach them, they were taken to him by a swimmer. The man would then be hauled in. Two men were stationed on the heavy-webbed cargo nets to help him onto the deck, where other sailors waited to assist.

With *Tabberer* in position downwind, Plage discovered that as their forward momentum slowed in the face of the gale-force winds he lost steering control, and the "large seas and strong wind" kept pushing the ship away from the man in the water. Plage realized that normal procedures wouldn't work in this sea state. Electing to take a "calculated and highly dangerous risk," he drove the ship upwind on the windward side and about 50 yards from the man in the water, and turned broadside to the heavy seas so that the waves and wind would push them sideways toward the drifting man. The seas accommodated the maneuver, taking the ship "in their grip at once." As *Tabberer* rolled heavily with the onrushing swells—so far over that the edge of the main deck dipped underwater—she was hurried toward the man like a "huge hunk of tumbleweed" blowing across the Texas Panhandle. Plage knew it would be tricky work bringing the man aboard before the careening vessel overran him. It certainly wasn't the way he or anyone else had been taught to conduct at-sea rescues.

All the ship's lights were now blazing so as not to lose sight of the life-jacketed man, and sailors on deck stood at the ready to take part in the rescue operation. With Plage at the conn, the senior officer on the weather deck was the ship's executive officer, Lieutenant Robert M. Surdam, the upstate New Yorker and bank president's son who had previously served in the Atlantic on the destroyer *Warrington*, which had gone down three months earlier in the Caribbean during a typhoon with the loss of more than 250 of Surdam's former shipmates. In recently recommending Surdam for an appointment to the next class at the prestigious Naval War College staff course, Plage had stated that his second in command "demonstrated outstanding ability and initiative in carrying out his duties aboard this vessel."

When they were close enough to the man in the water, Surdam shouted for him to grab the line about to be thrown and loop it under his arms. When the rolling ship was almost alongside, the line "flashed through the air." The man reached out and caught it. Just then, the ship rocked hard and was pushed by the sea toward him. The vessel went sideways so quickly that it seemed impossible not to overrun the man. Deckhands quickly took in the slack on the line as "green water and white lather swirled over the edge of the deck and retreated." When the water receded, the man in the life jacket was lying unconscious on the deck. He was quickly carried below to be examined by the destroyer escort division's medical officer, Lieutenant Frank W. Cleary, twenty-seven, of Burlingame, California. Cleary, a 1943 graduate of the McGill Faculty of Medicine in Montreal, happened to be aboard *Tabberer* for this patrol—his first sea voyage—rather than on one of the squadron's other five ships.

As soon as Plage received word on the bridge that the man in the water had been revived in sick bay and was claiming to be from a destroyer that had capsized during the typhoon, he rushed below to get firsthand information.

When *Hull* quartermaster August Lindquist awakened in sick bay, the first thing he wanted to know was how many of his shipmates from the sunken ship had been picked up. Told that no one knew anything about any lost ship, Lindquist, who had been at the helm of *Hull* when the destroyer went over, had been as shocked as Plage was upon arriving below and hearing from Lindquist the dramatic details of *Hull*'s capsizing.

"There are probably more men in the water," said a weary Lindquist. "Probably right in this vicinity, sir."

"We'll look for them," promised Plage, who told the sailor to rest.

Lindquist, who had been in the water approximately ten hours and was the first survivor rescued from any of the three lost destroyers, did as he was ordered and went to sleep. When he awakened it was past midnight on December 19—his twenty-fourth birthday.

As soon as Plage returned to the bridge, he ordered extra lookouts topside and searchlights turned on in spite of the danger of being spotted by an enemy submarine, a possibility they were keenly aware of, as they were now "only 150 to 200 miles off the coast of Luzon," where they had been "hunting submarines for the past month or so." He undertook a retiring search—"start at the center and expand as you go out." Plage figured they could "get the most men that way," with less danger of passing up anybody. Steaming on various courses and speeds, *Tabberer* commenced a systematic search for further survivors. As noted in the ship's log, results were forthwith:

2215 Recovered second man.
2230 Recovered third man.
2245 Recovered fourth and fifth man.
2255 Recovered sixth and seventh man.
2315 Recovered eighth, ninth and tenth man.
2320 Recovered eleventh man.

Near enough to *Tabberer* to see "the loom" of her searchlights sweeping back and forth over the horizon, the destroyer *Dewey*, commanded by Charles R. Calhoun, the Annapolis classmate of both James Marks of *Hull* and Bruce Garrett of *Monaghan*, was steaming on course for the fleet's fueling rendezvous the following morning. On *Dewey*'s bridge, it occurred to Calhoun that another ship—he thought at the time that the light "must be from *Tabberer*"—"might have found someone in the water." He turned his vessel in that direction and "headed over to see what was happening," knowing that if a rescue operation was under way, "*Dewey* could be of some assistance."

The new course put the ship "more directly into the sea," and *Dewey* began to "pitch and pound heavily." It was a motion that all hands "had come to know only too well during the past two days." Nevertheless, Calhoun picked up the phone and called one level below to the sea cabin of the destroyer squadron commander, Captain Preston Mercer, so he would "understand the reason for the slamming." Mercer responded that he thought *Dewey*, whose forward stack had failed during the storm and was still "draped over the starboard side," should "turn back to our previous heading," as there was "some risk that the pounding" might aggravate any other damage *Dewey* had incurred in the storm. Mercer made his decision in spite of being told that a light signal from another ship—identified as *Tabberer*—indicated she was picking up survivors in the water. Calhoun heeded Mercer's "advice and abandoned" lending assistance to the rescue operations. In a darkened state with no searchlights shining and no extra lookouts topside, *Dewey* knifed through blackish waters filled with shipwrecked men without spotting or picking up a single one.

At 1:10 A.M., ten minutes after *Dewey* turned away, heading for the Third Fleet rendezvous as ordered, *Tabberer*, operating on her own and with her own storm damage for Plage to be concerned about, picked up a twelfth survivor. A little over an hour later the thirteenth man was brought aboard. The survivors, all off *Hull* and diagnosed by the doctor as suffering "exhaustion from overexposure," had been found over a swath of ocean covering "some 25 square miles."

Taking charge of the deck crew during rescue operations was senior watch officer Howard Korth, who stripped off his uniform and dove into the water four times to assist men up to the cargo net—at times from as far as 75 feet away. He worked in tandem with Boatswain's Mate Louis Purvis, also a strong swimmer. Once, coming back with a weakened young sailor who looked to Korth "about 15 or 16 years old and trying to grow his first moustache," they became caught in "tremendous suction" off *Tabberer*'s bow and were "banged against it" and dragged down. Korth couldn't keep his grasp on the young sailor. When the officer surfaced, he realized he had "lost him"—the sailor with the baby face.

Purvis was nowhere to be seen, and Korth, who finally made it to the cargo net, thought he might be gone, too. Soon, however, a sputtering Purvis surfaced on the opposite side of the ship. He had been drawn completely under the hull and had to slip out of his life jacket, tethered to a lifeline that became fouled on the sonar dome, in order to surface and be hauled aboard like a half-drowned cat. After being "pumped out a bit," Purvis recovered. He then observed to the delight of all, "Dammit, I bet I'm the first sailor to be keelhauled in 200 years," referring to the old, outlawed form of punishment whereby malcontents were attached to lines or chains and hauled under ships from one side to the other a prescribed number of times—a "painful and often fatal experience."

During the night, Plage stopped the ship every ten minutes, and the ventilating blowers in the bridge structure were turned off to reduce the noise level. All lights were also turned off, and everyone topside searched for life jacket lights and listened for whistles and shouts. With the navigational equipment inoperable, the careful box search had to be conducted entirely by dead reckoning. This required constant course changes to the north, east, south, and west at "so many minutes per leg."

At 3:10 A.M., with the emergency radio finally working thanks to the jury-rigged antenna set up by Chief Tucker, *Tabberer* transmitted the first news of the loss of *Hull,* the number of survivors being pulled from the water, and their approximate location.

The fourth *Hull* survivor picked up by *Tabberer* had been Storekeeper Ken Drummond, who when he was washed overboard from the overturned ship had thought about how much his death was going to upset his mother. After being pulled down in the water, then "popping out of the water about 20 feet into the air," Drummond landed "butt first" in a life raft. Dropping next to him within a few seconds had been Boatswain's Mate Chief Ray Schultz, who turned to Drummond and said, "You don't look too good." Schultz added calmly, "Don't believe the storm's as bad." With that, another swell hit the raft, sending both men "flying off in different directions." Alone until almost dark, Drummond bumped into Chief Radioman Francis "Burt" Martin. They discussed the situation and decided to tie themselves together for the night. As they undid their life jackets and were in the process of hooking them together, another swell hit, separating the two men. Drummond did

The destroyer *Spence* (DD-512), commissioned in 1943, winner of eight battle stars. U.S. NAVY

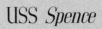

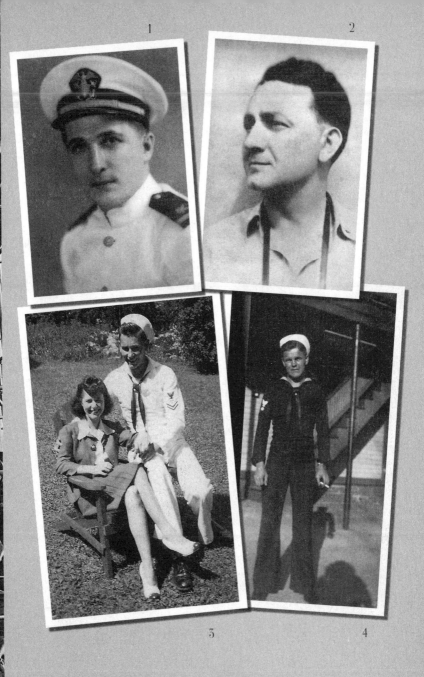

Opposite page: *Spence* sliding down the ways at Bath, Maine. U.S. NAVY

1. Al Krauchuanas, of Michigan, *Spence* supply officer. U.S. NAVY   2. James
Andrea, Annapolis class of 1937, the last commanding officer of *Spence*. U.S.
NAVY   3. Robert Strand, of Ridgway, Pennsylvania, *Spence* machinist's mate,
with his sweetheart, Jane. FAMILY PHOTOGRAPH   4. Edward Traceski, of Turner
Falls, Massachusetts, *Spence* quartermaster. FAMILY PHOTOGRAPH

USS *Tabberer*

1

2

3

1. The widow and the mother of hero naval aviator Charles A. Tabberer, for whom the hero ship was named, on christening day in Houston ship harbor. U.S. NAVY 2. Paul "Cookie" Phillips, of Texarkana, Texas, *Tabberer* cook, and former state amateur boxing champion. FAMILY PHOTOGRAPH 3. Frank Burbage, of Newark, New Jersey, *Tabberer* sonarman. FAMILY PHOTOGRAPH 4. *Tabberer* officers and men who were awarded individual medals for their rescue efforts. Seated from left, executive officer Robert Surdam, commanding officer Henry L. Plage, and gunnery officer Howard Korth. Standing from left, Torpedoman's Mate 1st Class Robert L. Cotton and Boatswain's Mate 1st Class Louis A. Purvis. U.S. NAVY

The destroyer escort *Tabberer* (DE-418), commissioned in 1944. U.S. NAVY

Third Fleet

Admiral William F. Halsey, Jr., commander of Third Fleet. U.S. NAVY

Commander George F. Kosco, Third Fleet aerologist. U.S. NAVY

Under the guns of the battleship *New Jersey*, Halsey with his weatherman Kosco. U.S. NAVY

Army Air Corps meterologist
Bryson Reid who first tracked
the typhoon and tried to warn
the fleet. U.S. ARMY

The typhoon that sunk three
destroyers as photographed on
the radar screen of the aircraft
carrier *Wasp* (CV-18) at 2:10 P.M.
on December 18, 1944. U.S. NAVY

Survivors of *Hull* and
Spence aboard *Tabberer*
following their rescue.
U.S. NAVY

not see Martin again until the chief was picked up by *Tabberer* the day after it found the storekeeper. After finding himself alone again, Drummond began hallucinating, seeing a vision of a Model A Ford that was "black with chrome around the headlights and grille." It seemed to be only about 10 feet away, and he thought if he could get on the hood, he might be able to rest. He swam toward the Ford, but it remained the same distance away. The next thing he saw was a "very bright light." He started blowing the whistle attached to his life jacket "as hard as I could possibly blow." He heard a voice say, "There he is!" A line was thrown to him. When Drummond became entangled, Purvis leaped off *Tabberer*'s deck into the water and supported the struggling sailor as he was hauled aboard.

The next man picked up by *Tabberer* had been *Hull* Seaman 1st Class Carl Webb, nineteen, born in Oklahoma and raised on a cattle ranch, where he "rode bucking horses" and helped his father farm 600 acres of cotton. Webb had the 4:00-to-8:00-A.M. watch at a 40 mm gun director, and when he was relieved he decided the sea was too rough to go back to the crew's quarters, so he stayed topside—a decision that saved his life. When the ship went over, he was tossed quickly into the sea, which was fine by Webb because he could "swim like hell" and knew it beat being "trapped below." As for the 50 percent of *Hull*'s crew of 258 men that Webb estimated were below in the crew's quarters—where they "dogged themselves in from the inside" via a secured hatch that "never was broken open" after the ship went over—it was his impression that "none got out and all went down with the ship." When about 10 minutes later the boiler blew up, it felt to Webb "like a big earthquake." After being rescued, Webb was given a shower and shown to the executive officer's empty cabin, where he quickly went to sleep in Surdam's bunk.

The growing contingent of *Hull* survivors aboard *Tabberer* that first night included radarman Michael "Frenchy" Franchak, who had escaped from the chart room when it was nearly underwater only to enter the sea amid the dead bodies of shipmates still being battered against the side of the ship. After losing hold of a life raft, Franchak found himself drifting alone in his life jacket, certain that he must be the only survivor. He soon realized he was "bleeding like a stuck pig" under one arm from something sharp protruding from the small flashlight attached to the life jacket. Several hours later, still bleeding and with his "bloodied dungaree shirttail floating free," Franchak saw a single fin. When it disappeared under the surface, "shivers went up and down" Franchak's spine. In the next instant, he took "quite a wallop" on his right side, enough to "jerk my head like a whiplash." Knowing he had been bumped by the shark, he figured "this was it." Remembering a *Reader's Digest* article about striking two objects together underwater to "scare sharks away," Franchak started frantically hitting the heels of his shoes together and didn't stop until "the shoelaces wore out and the shoes fell off." He never had "any more trouble from the shark." A couple of hours after dark, he heard voices and swam toward them. When he got close enough, he asked if they had a raft. The answer was, "No, have you?" Franchak replied, "Never mind. I'll keep going." The three *Hull* sailors—sharing one life jacket between them— "pleaded and begged" Franchak to come with them. He agreed, and soon had "these guys hanging on me." When Franchak went under and came up coughing, he had to knock one of the guys free of him with his elbow. They drifted along with things looking "absolutely hopeless" until suddenly a spotlight appeared on the horizon. Beseeching the three men to swim toward the light, Franchak pretty much had to carry them along, as they kept hanging on to the powerfully built radarman. When *Tabberer* came alongside and lines were thrown to the men, the other three were too weak to get them around themselves, and Franchak had to tie them so they could be pulled in. With the others aboard, a line was dropped in the water for Franchak, but at that moment the ship rolled away from him, after which they had "a tough time" rescuing him. He was finally pulled aboard at 2:30 A.M., the thirteenth and last man rescued by *Tabberer* that night.

At sunrise, Plage, still on the bridge, decided that rather than heading belatedly for the fleet's fueling rendezvous, *Tabberer*, still limping along without radar and limited radio communications, would keep searching for survivors in seas that had settled to a height of 8 to 10 feet with winds not exceeding 20 miles per hour. His decision was soon rewarded with positive results:

0605 Recovered fourteenth and fifteenth survivors.
0630 Picked up sixteenth survivor.
0708 Picked up seventeenth survivor.
0723 Picked up eighteenth survivor.

One of the survivors picked up that morning—all from *Hull*—was Chief Quartermaster Archie DeRyckere, the former youth boxer from Montana who, when the ship rolled over the last time, had assisted the injured Greil Gerstley to the highest point atop a searchlight platform next to the bridge. When the destroyer sank beneath them, DeRyckere had come struggling to the surface just as the boiler exploded. Gerstley, the executive officer who might have relieved Marks of command in an effort to save the ship except for his dread of being "tried for mutiny and hanged," was never seen again. DeRyckere did see Marks in the water, however. That evening, as *Hull*'s captain floated past in his life jacket, he asked DeRyckere if he would like to know the time. "Appreciate it," said the chief, without grasping at all why the time of day would be useful information. Checking his watch, Marks said, "Five after six," then drifted away. Some hours later, DeRyckere spotted a ship, which the quartermaster signaled in Morse code with blinking dots and dashes on his small flashlight: "SOS. Send help." The ship, apparently signaling another ship in the area, flashed a Morse code message that DeRyckere was able to read: "We are departing area." DeRyckere's angry thought weighed heavily on him throughout the long night: *They're leaving us out here.*[*]

At 8:30 A.M. on December 19, *Tabberer* rescued a solo survivor in a life jacket who turned out to be James Marks. When Plage greeted *Hull*'s commanding officer, he found him with "eyes as black and blue as could be," although the doctor could offer no reason other than "salt spray driving so hard into his face that it bruised him." Marks didn't have much to say

---

[*]In all likelihood, this ship-to-ship Morse code message was the departing *Dewey* signaling *Tabberer*.

other than that he had gotten so hungry in the water that he tried to chew on his whistle, and when that "didn't taste very good" he had ripped a piece of leather off a shoe and chewed on that.

0852 Rescued twenty first and twenty second survivors.
0854 Rescued twenty third survivor.
0908 Rescued twenty fourth survivor.
0920 Rescued twenty fifth survivor.
0930 Rescued twenty sixth survivor.
0950 Rescued twenty seventh survivor.

Picked up by *Tabberer* that morning alone in the water in a life jacket was Lieutenant (j.g.) Arthur L. Fabrick, twenty-four, of Gainesville, Florida. A graduate of the University of Florida, Fabrick—shy, slender, and soft-spoken—served as *Hull*'s radar officer. On the bridge when the ship went over, he experienced the event "in slow motion." After exiting the bridge and climbing to the side of the ship, he jumped quickly, thinking that the faster he got into the sea the better. It was "not a good idea," however, as he was "sucked down," and when he surfaced he was "bashed against the ship." But he was a "pretty good swimmer with a Red Cross lifesaving badge," and being in the ocean was nothing new to him as he had grown up swimming in the Atlantic and the Gulf of Mexico. He swam rapidly away from the ship and soon cleared it. When he saw a "bunch of guys banding together around life rafts and all trying to climb in at the same time," Fabrick decided "the hell with that" and kept swimming. At one point, he turned back—about half an hour after the ship went over—in time to see *Hull* "just slide down into the water" and sink from sight.

After the twenty-seventh survivor was picked up at 9:50 A.M.—in the vicinity of where several others were rescued—no one else was observed in the area. With the seas "flattening out," Plage decided it would be possible to expand the visual search to cover an area of 1,500 yards on each side of the ship. Ordering a speed of 10 knots from the engine room, Plage had dye markers dropped in the water at each turning point on a new leg in order to have a "visual check point."

At 10:05 A.M., two Curtiss SB2C Helldiver planes circled *Tabberer.* Plage quickly flashed a message asking them for assistance in searching for men in the water. The aircraft replied they did not have time to assist because they had their own patrol to complete, and flew away.

A radio message was received at 1:10 P.M. from Commander Third Fleet, directing *Tabberer* to "proceed to rendezvous with other damaged ships" 90 miles south of their current position, to arrive there by sunset, and then to proceed together to Ulithi for temporary repairs.

Plage "believed other men were in the water" but possibly were so scattered that it would be difficult to find them. He continued searching for "as long as possible." At 2:00 P.M., he ordered a new course in order to make the rendezvous. As soon as *Tabberer* turned to the new course, another man was spotted in the water, and he was brought aboard at 2:06 P.M.

Shortly after the course was "resumed for rendezvous," a drifting man in a kapok life jacket was spotted. As the ship came closer, an 8-foot shark was observed circling beneath the man, who appeared to be unconscious. Plage ordered "our tommy guns, .30 caliber rifles, .45 pistols and everything else" quickly handed out to sailors topside. Hurrying to bring the ship alongside, Plage snapped commands on the bridge: "Right full rudder," "All engines back full." The sound of gunfire filled the air as marksmen fired "all around trying to keep the shark away."

Awakening to the sound of machine guns and bullets landing nearby, Lieutenant (j.g.) Don Watkins thought "they were Japs" and that "this was the end." He soon deduced that the gunmen were either terrible shots or aiming at something else in the water. He then saw a line being thrown his way. Although it fell short by some distance and he was too weak to reach it, he knew an attempt was being made to rescue him. Since waving goodbye to Mackenzie some twenty-four hours earlier, Watkins had been drifting alone. During the night, he had seen lights of a ship in the distance and tried to swim toward them. It proved to be too much effort, however. Exhausted, he soon discovered that he could sleep in the life jacket, and did. The next thing he recalled was the rude awakening.

Tabberer was still 50 to 60 feet away. On deck, executive officer Surdam saw the situation and thought the man might be "attacked by the shark before we could get to him." Without any hesitation, Surdam, who did not take the time to don a life jacket, "dove in the water and swam out to the man." Grabbing him by the back of his life jacket, Surdam hauled Watkins toward the ship as the shark—the target of a "steady stream of bullets"—circled menacingly 20 feet away. Stationed halfway down the cargo net, waiting to help board the survivor, Torpedoman 1st Class Robert L. Cotton, a wiry-framed cowboy from Cheyenne, Wyoming, dove headfirst into the water and swam out to assist Surdam, who by then was "pretty tuckered out." Between them, Surdam and Cotton brought Watkins safely in.

More searching continued in the area, but Plage realized that unless they again turned for the rendezvous they would be late, and in violation of Halsey's latest directive. At 2:55 P.M., *Tabberer* was "again taken toward rendezvous," although everyone aboard wore a "hangdog look" because they "hated to give up the hunt." Dozens of men topside kept searching the sea for more survivors. At 3:15 P.M., a lookout yelled excitedly. A group of men had been sighted two miles away.

When the survivors—seven in all—were boarded fifteen minutes later, they turned out to be a group of *Hull* sailors under the command of the ship's engineering officer, George Sharp, the admiral's son, who had been "instrumental in saving the men." After abandoning a broken raft, five men wearing life jackets had come across a sixth man, Fireman Roderick Mackenzie, floating on a soggy mattress. Initially assisted by Don Watkins shortly after they both went in the water, Mackenzie was still without a life jacket. The others took turns keeping Mackenzie from drowning by holding him up or having him hang on their backs. Sharp was the last to join the group. After evaluating the situation and seeing that Mackenzie was a "heavy burden to seaworn men," the officer insisted on the men lashing themselves together in a circle. Mackenzie was then placed in the center so that everyone could help keep him afloat.

Tied up "like a bunch of asparagus," Sharp, a "scientist by training and nature" who had not followed in his father's footsteps to Annapolis only due to defective eyesight, knew they would be easier to spot in a group than as individuals in the water. The seven men all made it.

Plage decided at that point to continue the search. He figured that without a mast and electronics *Tabberer* "couldn't do the fleet any good," and although they were under orders to make the rendezvous, he decided to "hang with that, we're going to stay here" and look for more survivors. While he was "a little shaky inside because of [his] highhanded disregard for orders," he felt sure that staying and "hunting for these men" was the right thing to do, even if he ended up with a reprimand.

Among those rescued that afternoon was Boatswain's Mate Ray Schultz, who since being washed off the bridge of the sinking *Hull* had drifted mostly solo until just an hour before being rescued, when he'd joined a cluster of his shipmates floating along in the kapok life jackets he had ordered some months earlier. After being hauled aboard *Tabberer,* Schultz learned that among those rescued that day was "Jack Ass" Marks. Of all the great guys who had not made it off the ship, Schultz was dismayed that this bastard had gotten lucky. Had the two men come across each other in the water, Schultz had no doubt what he would have done: "held underwater the head" of his captain and nemesis until the life went out of him.

It came as no surprise to the *Hull* survivors that Marks failed to come below to greet them and check on their well-being. In fact, *Tabberer's* Cookie Phillips, hearing complaints about Marks from weakened men who still "could hardly walk," believed that Plage was "hiding Marks so he wouldn't go back in the water." One of Phillips' shipmates, mail clerk William McClain, decided after speaking with numerous *Hull* sailors that Marks "wasn't much of a man," which made him "the opposite of our skipper."

At 4:30 P.M., Plage received a message from Commander Third Fleet ordering *Tabberer* to "stay in the area and search until dawn" the following morning, when they would be relieved by other ships.

Plage went to the ship's public address system, announced their new orders, and heard his crew, who had been through a typhoon and not slept in more than two days, cheering wildly on deck and throughout the ship.

The search for survivors went on.

Eighteen

Tabberer was relieved of her independent search at 8:40 A.M. on December 20 by other ships assigned by Halsey to conduct what he would later term "the most exhaustive search in Navy history." Unlike *Tabberer,* many of them found nothing, for which Halsey would later have an explanation: "a man in a life jacket is almost impossible to spot in a rough sea."

Ordered to proceed to Ulithi, Plage set an easterly course at a speed of 16 knots while keeping extra lookouts posted. At 10:50 A.M., an object in the water was sighted. It turned out to be what was left of a life raft holding ten *Spence* sailors, all of whom were quickly boarded.

Senior man on the raft was Chief Water Tender James Felty, who had been transferring fuel in preparation for taking on water ballast before the destroyer went over. Felty, a seven-year Navy veteran who had slipped into his life jacket minutes before *Spence* rolled, ended up in the sea next to the ship just before she turned over. The raft had started out with some twenty men crammed inside and hanging on the outside, but their numbers steadily dwindled. Some men died quietly; others "hallucinated and swam away," never to be seen again.

The youngest man on the raft was Seaman 2nd Class Floyd Balliett, eighteen, of Plant City, Florida, who had joined the Navy shortly after graduating from high school. It seemed to Balliett that *Spence* was "underwater like a submarine" most of that last morning. After a "couple of steep rolls," he had gone to the galley and found a life jacket, which he put on. Going back to the hatch that opened onto the main deck, he was standing there when the ship "made a 75 degree roll and didn't come back." There were several men in front of him "screaming like babies," and Balliett "climbed out over the top of them." He grabbed a pole and "swung off into the water." He was "caught in the undertow" and held underwater for so long he didn't think he would ever come up. Just as he concluded his life was ending, he saw his mother "reach in the mailbox and take out that little note that said her son is missing in action." Popping back to the surface, he found himself swimming "in a sea of dead sailors." Balliett had made it 40 or 50 yards from the ship when the "boilers blew" and *Spence* "broke in half and went down" in two pieces. After spending the rest of the day alone in the sea—where the wind spray hit with such force that it felt like someone "throwing a handful of gravel" in his face—he came to the broken-up raft, already overcrowded. He saw a rope dragging behind it, and held on all night and into the next day until enough "guys disappeared" that there was room for him inside the raft.

Within twenty minutes of bringing aboard the raft survivors, *Tabberer* picked up three more *Spence* crewmen, including Seaman Ramon Zasadil, who had been assigned to the ship after radio school. When *Spence* went over, he was on watch in the radio shack, from which he "crawled out on my hands and knees." Not wearing a life jacket, Zasadil had on "nothing but skivvies" because his dungarees had been soaked when he had gone from the crew quarters to the radio shack, and he had taken them off and hung them up to dry. Unsure at first whether he should leap into the water, he decided to do so just before the ship "turned keel up." He met up with some guys in life jackets and hung on to them, treading water, through the night until meeting up in the morning with the raft party.

Plage sent out a radio message to other ships that they were still finding men in the water. *Tabberer* was soon joined by the escort carrier *Rudyerd Bay* (CVE-81) and other vessels. At 12:25 P.M., Plage was again ordered by Commander Third Fleet to set his ship's course for Ulithi, thereby ending the destroyer escort's role as hero rescue ship.

Tabberer steamed into Ulithi lagoon late on the morning of December 22 carrying fifty-five "storm-tossed and exhausted" survivors. Before their guests could be unloaded, Plage was directed alongside the moored oiler *Sebec* (AO-87). *Tabberer's* half-empty bunkers were filled to capacity, receiving some 160 tons of fuel oil in less than an hour.

Moving slowly through the crowded harbor, *Tabberer* anchored on the west side of berth eight at 1:30 P.M. Three hours later, the stream of guests—most wearing borrowed dungarees and shirts—began to depart amid handshakes, backslaps, and fond farewells. All were healthy enough for thirty days' survivor's leave—sent to the transport *General S. D. Sturgis* (AP-137), serving as a station receiving ship—except for one man who went to the hospital ship *Samaritan* (AH-10) for treatment.

Tabberer was so badly damaged that extensive repairs would be required at Pearl Harbor. The officers and enlisted men alike were more than willing to give credit where they knew it was most due: to their steady and quick-thinking captain, who had stayed on the bridge and at the conn during the typhoon and rescues for "three days and two nights"—at times, Robert Surdam thought, Plage "looked worse" than some of the survivors.

While the enlisted crew had long liked and respected their fair-minded, even-tempered captain, "we didn't really know how good he was," Cookie Phillips realized. "But we found out. So did those guys we picked up in the water." Phillips would hear Plage say that "the good Lord saved *Tabby* and all those men." Phillips agreed, at least in part: "The good Lord had some help from the captain." Sonarman Frank Burbage wrote his parents about "the wonderful way the captain handled the ship, and saving the lives of 55 men." Still, Burbage wondered: "What would have happened to all those men if that first guy's flashlight hadn't been working?"

Tabberer's crew—with much to be grateful for—celebrated Christmas at Ulithi. The supply officer and galley gang went all out with the menu: turkey soup and crackers, roast young tom turkey, giblet gravy, cranberry sauce, mashed potatoes, buttered peas and carrots, stuffed olives, fruit salad, cherry pie, ice cream, hot rolls, butter, nuts, and candy.

At 11:30 P.M. on December 29, Halsey came aboard *Tabberer* with members of his staff including Rear Admiral Robert Carney and four "four-striper captains tagging along." The admiral, who since *New Jersey*'s arrival at Ulithi on Christmas Eve had been dealing with questions from the highest echelons of the Navy in the aftermath of the tragic loss of ships and men in the typhoon, seemed to relish taking this time to visit the officers and sailors of one of his smallest ships.

Learning of *Tabberer*'s lifesaving work while she was still at sea, Halsey had wired: "Well done for a sturdy performance." *Tabberer*'s captain, he was told, "while ships around them were barely keeping afloat," maneuvered alongside and hauled men out of the water. Inquiring about the captain's background before he pinned a medal on him, Halsey admitted he "expected to learn that he had cut his teeth on a marlinespike," and was shocked when Henry Plage turned out to be not yet thirty years old or an Annapolis graduate but a reservist who had earned his commission at Georgia Tech ROTC. "How could any enemy," Halsey later wrote, "ever defeat a country that can pull boys like that out of its hat?"

Stepping up to a microphone that faced the assembled officers and crew, Halsey said: "Captain Plage, officers and men of *Tabberer*. I am greatly honored and privileged, and it is with great pleasure that I come over here this morning to tell you what I think of you, as the Commander of the Third Fleet. Your seamanship, endurance and courage and the plain guts that you exhibited during the typhoon that we went through is an epic of naval history and will long be remembered by your children and their children's children. It is this plain guts displayed throughout the world by the American forces of all branches that is winning the war for us. How those yellow bastards ever thought that they could lick American men is beyond my comprehension. Keep going until the final thing, and the final thing should be the complete destruction of the Japanese empire. If I had my way there would be not a Jap yellow bastard alive, but I guess I won't have it my way."

With that, Halsey nodded to one of his staff officers and began pinning medals on the five men lined up next to him at attention. The Navy–Marine Corps Medal for Heroism went to Lieutenants Robert Surdam and Howard Korth, Torpedoman Robert Cotton, and Boatswain's Mate Louis Purvis. Then it was Plage's turn. After Halsey read a short citation that made the young officer "blush from hairline to Adam's apple," the admiral pinned the coveted Legion of Merit Cross on Plage's chest. Plage was so "flabbergasted" that he didn't know whether to "salute or shake hands," and Surdam, overcome by thoughts of the hundreds of sailors who had perished at sea in this typhoon as well as the storm three months earlier that had sunk his former ship, *Warrington,* at such a great loss of life, didn't help matters by bursting into tears.

In the wardroom a few minutes later, Halsey was served coffee with pineapple upside-down cake, apple pie, and Commissary Steward Baker Chief Alan Lumb's acclaimed hot cinnamon buns. Halsey took a small bite of everything, then looked up, raising his bushy eyebrows. He stuffed in his mouth what was left of the bun, wiped his lips with a napkin, and asked to adjourn to Plage's quarters.

Soon after, Lumb was summoned to the captain's cabin, where Halsey and Plage were waiting, the latter with a much different look on his face than when he and his men were receiving accolades and medals.

Halsey complimented the chief on his cinnamon buns. Then he said, "Chief, pack your seabag. You've been transferred to my ship."

The only man pleased to see the chief baker go was Cookie Phillips, who with the departure of his boss was now "in charge of everything in the galley." As for Plage, he admitted he "hated to lose Lumb," although he allowed it would be "good for him to be with Admiral Halsey."

Tabberer left the next day for Pearl Harbor via Eniwetok.

On January 11, 1945, as the dismasted and beat-up destroyer escort entered the channel at Pearl Harbor, *Tabberer* presented an unusual profile. As the ship moved slowly toward her anchorage, a nearby battleship asked by signal light: "What type ship are you?"

As far as Plage was concerned, "that hurt." He believed "our type of fighting ship" should be recognizable with or without a mast. So he signaled back to one of the largest and most majestic ships in the fleet: "Destroyer escort. What type are you?" There was no reply.

No sooner had *Tabberer* tied up to the dock when a "sedan with a Marine driver" arrived with orders for Plage to report immediately to the headquarters of the Commander Destroyers Pacific Fleet, Vice Admiral Walden L. Ainsworth. After earlier winning a Navy Cross for "outstanding leadership, brilliant tactics and courageous conduct," Ainsworth was now in administrative command of all cruisers, destroyers, destroyer escorts, and patrol frigates of the Pacific Fleet.

Reporting as ordered, Plage was ushered into the admiral's office, where Ainsworth, "looking very stern," sat behind a gunmetal desk.

"Plage, did you ask that battleship what type of ship she was?"

"Yes, sir, I did."

The admiral stared at the young officer for a moment, then "put his head back and roared laughing." He stood up and came around the desk to shake Plage's hand, congratulating him for rescuing the scores of men in the typhoon as well as for putting destroyer escorts in the fleet "one up" on the battleships. An Annapolis classmate, Ainsworth explained, was a battleship division commander, and his flagship was the one Plage had signaled back. He had called Ainsworth to report the cheeky response received by the battlewagon from one of Ainsworth's smaller ships. "My battleship buddy thinks you must be a proud ship with a great crew."

The next day Plage was shown by his own division commander a copy of a letter the commander had sent to headquarters that included remarks about *Tabberer* being the "best ship in the division" and how during maneuvers Plage always had his ship "in correct position." Typically, Plage was quick to credit others, writing home to his wife that such praise was due to "the alertness of the OODs, the crew in the engine room, in CIC and on the bridge." He also added, "A part of the team must include the firemen on watch. We have such a swell bunch on board I can't help bragging."

On December 18, 1944, coincidentally the same the day the typhoon hit the Third Fleet, the Navy Unit Commendation had been established by order of the secretary of the Navy, James Forrestal, to be awarded to any ship, aircraft, detachment, or other unit of the U.S. Navy or Marine Corps that distinguished itself in action again the enemy or for non-combat service in support of military operations that was "outstanding when compared to other units performing similar service." *Tabberer*, within weeks, was awarded the first Navy Unit Commendation. The citation read, in part:

> *For extremely meritorious service in the rescue of the survivors in the Western Pacific typhoon of December 18, 1944. Unmaneuverable in the wind-lashed seas, fighting to maintain her course while repeatedly falling back into the trough, with her mast lost and all communication gone,* Tabberer *rode out the tropical typhoon and, with no opportunity to repair the damage, gallantly started her search for survivors. Steaming at ten knots, she stopped at short intervals and darkened her decks where the entire crew topside, without sleep or rest for 36 hours, stood watch to listen for the whistles and shouts of survivors and to scan the turbulent waters for the small lights attached to kapok jackets which appeared and then became obscured in troughs blocked off by heavy seas. Locating one survivor or a group,* Tabberer *stoutly maneuvered to windward, drifting down to her*

*objective and effecting rescues in safety despite the terrific rolling which plunged
her main deck under water, again and again conducted an expanding box search,
persevering in her hazardous mission for another day and night until she had
rescued 55 storm tossed and exhausted survivors and had brought them aboard to
be examined, treated and clothed. Brave and seaworthy in her ready service,*
Tabberer, *in this heroic achievement, has implemented the daring seamanship
and courage of her officers and men. All personnel attached to and serving on
board* Tabberer *during the above mentioned operation are hereby authorized to
wear the* NAVY UNIT COMMENDATION RIBBON.

James Forrestal
Secretary of the Navy

Nineteen

In a harbinger of the days that lay ahead, three hours after Halsey's flagship, *New Jersey*, dropped anchor at Ulithi on December 24, 1944, the battleship "received a direct hit on the main deck" from a 5-inch shell fired by a U.S. ship conducting antiaircraft practice outside the lagoon. The shell pierced the main deck, passed through the second deck, and "lay unexploded in a washroom." Three sailors were injured.

Arriving the next day by seaplane was a "very concerned, very upset" Fleet Admiral Chester Nimitz, the recipient a week earlier of his fifth star, a new rank created that month by Congress. Both Halsey and members of Congress "had expected" that Halsey too would receive a fifth star, but his promotion had not yet been approved by a military hierarchy that had not forgotten the admiral's strategic blunders at Leyte Gulf. Now any Halsey promotion would be "typhoon-delayed" as well.*

Nimitz came to Ulithi to hear from Halsey why he had taken his fleet into "the dangerous semicircle of a typhoon," which Nimitz and his Pacific Fleet staff believed to be a reflection on Halsey's "seamanship." Nimitz carried a top-secret cable from Chief of Naval Operations Ernest King, who according to a Nimitz aide "practically tore the Navy Department apart" over the typhoon losses. Addressing himself to both Nimitz and Halsey in the cable, King said he wanted "to know the circumstances which caused operating units of Third Fleet to encounter typhoon which resulted in the loss and crippling of so many combatant ships."

In a private meeting aboard *New Jersey*, Nimitz informed Halsey that he had already appointed a court of inquiry to investigate and hear testimony from various parties involved. Word had not yet been made public about the three lost destroyers, and Nimitz would hold off another two weeks before issuing a press communiqué about the disaster— by which time most of the next of kin would have been contacted. The human losses were staggering even to Navy commanders accustomed to weighty matters: with only some 90 survivors accounted for from the lost destroyers, more than 700 men were missing and presumed dead. When he did finally speak publicly about Halsey's run-in with the typhoon, the soft-spoken Nimitz pulled no punches, describing it as "the greatest loss that we have taken in the Pacific without compensatory return since the First Battle of Savo," the first major naval engagement of the Guadalcanal campaign.†

---

* Nimitz received his fifth star on December 19, one day after the typhoon struck the Third Fleet. Promoted days ahead of him were Admirals William D. Leahy (December 15), President Roosevelt's senior military advisor and in effect the first chairman of the Joint Chiefs of Staff, and Ernest King (December 17), the chief of naval operations.

† The battle near Savo Island was fought on August 8 and 9, 1942. A Japanese task force surprised and routed a larger Allied naval force, sinking three U.S. cruisers and one Australian cruiser while incurring only moderate damage to three ships. A total of 1,077 Allied seamen were killed, with Japanese losses reported as 58 killed.

Nimitz had brought with him from Hawaii the officer he had selected to lead the inquiry: Vice Admiral John H. Hoover. The compact Hoover, a fifty-eight-year-old native Ohioan, had graduated from Annapolis in 1907, three years after Halsey. Designated a naval aviator upon completion of flight training in 1929, Hoover was captain of the aircraft carrier *Lexington* at the time of the attack on Pearl Harbor. He now served as commander, forward area, Central Pacific, in which capacity he reported directly to Nimitz and had no dealings with Halsey. Years earlier, Hoover had been nicknamed "Genial John" because of his "dour personality." He was considered a "capable enforcer" of naval regulations and policies and a man who could "not be browbeaten," even by the famously irascible Ernie King, under whom Hoover had served in a number of commands.

In a formal letter to Hoover dated Christmas Day, Nimitz directed him to serve as president of the court of inquiry, which would consist of two other admirals. Nimitz wanted them to begin the next day, "or as soon thereafter as practicable," for the purpose of "inquiring into all the circumstances connected with the loss" of *Hull, Monaghan,* and *Spence* and the damage sustained by other ships of the Third Fleet "as a result of adverse weather on or about 18 December 1944." Nimitz directed Hoover to include in the court's findings a statement of facts and give its opinion as to whether "any offenses have been committed or serious blame incurred." In the event of the latter, Nimitz expected the court to "specifically recommend what further proceedings should be had."

By naval statute, the court of inquiry would be "more inquisitorial than adversarial"—an investigative and fact-finding proceeding. Yet courts of inquiry were "serious affairs" that could "cripple or wreck careers." There were more ominous possibilities, too. Based on the findings and recommendations of the court of inquiry, Nimitz could decide to hold a court-martial, in which defendants would be charged with infractions of military law; if they were convicted, a range of punishments could be levied, from losing rank to being drummed out of the service or imprisoned.

The court convened the morning after Christmas.

Nimitz had appointed as judge advocate Captain Herbert K. Gates, although he was untrained in the law. Rather, his specialty was marine engineering. A 1924 graduate of Annapolis, Gates, forty-three, a native of Michigan, was commanding officer of the destroyer tender *Cascade* (AD-16), equipped to provide destroyers with spare parts and repairs on everything from typewriters to guns and boilers. As the court of inquiry would be fact-finding, with few rules of procedure and evidence to follow, Nimitz had decided that Gates' expertise in naval engineering and ship repair would be invaluable to the members of the court as they strove to understand how the ships of the Third Fleet "reacted to storm conditions and storm damage." The technical-minded Gates would have much leeway to bring before the court any of a vast array of information he saw fit.

Gathered in the wardroom of *Cascade* were Gates, Hoover, and the other two members of the court: Vice Admiral George D. Murray, fifty-five (Annapolis 1911), a former commander of aircraft carriers (*Enterprise* and *Hornet*) who had fought under Halsey earlier in the war and now worked for Nimitz as commander air force, Pacific Fleet, and Rear Admiral Glenn B. Davis, fifty-two (Annapolis 1913), who had won the Navy Cross in the battle for Guadalcanal in November 1942 while commanding the battleship *Washington* (BB-56) and engaging and sinking the Japanese battleship *Kirishima* in the first head-to-head confrontation of battleships in the Pacific theater. Davis was now a battleship division commander who had been "in the thick of the action in the Pacific" when Nimitz summoned him to Ulithi to serve on the court.

After identifying four enlisted men who would alternate as court reporters in the preparation of the official transcript, the court's first order of business was the decision to "sit with closed doors," meaning that other naval personnel as well as war correspondents, "casual observers and the curious" would be excluded from all sessions.

After some other preliminaries, such as administering oaths to one another and reading aloud Nimitz' letter authorizing the formation and makeup of the court, Gates asked about the order in which witnesses were to be called, pointing out that the Third Fleet was scheduled to sortie from Ulithi in four days for a return to the Philippines. The court decided to call the "fleet witnesses first," followed by the survivors.

There was also what amounted to another formality: officially advising James Marks, the only commanding officer from the three sunken destroyers to survive, that "in view of the loss of *Hull*" he would be considered a "defendant." As officers who lost ships routinely faced such inquiries, it came as no surprise to Marks, who entered the wardroom with his appointed counsel, Captain Ira H. Nunn, an Annapolis (1924) and Harvard Law School (1934) graduate and former destroyer skipper who had won a Navy Cross two months earlier at Leyte Gulf for "pressing home numerous successful attacks against hostile aircraft and surface ships" while in command of a screening unit during amphibious landings.

Preston Mercer was also called before the court and advised that given the loss of *Hull* and *Monaghan,* both attached to his destroyer squadron, he had "an interest in the subject matter" and would be allowed to be present during the course of the inquiry, examine witnesses, and introduce information pertinent to the inquiry. As an interested party rather than a defendant, Mercer did not have counsel.

The first witness to be sworn in was Rear Admiral Robert Carney, who had commanded the cruiser *Denver* (CL-58) before being assigned as chief of staff to Halsey. Carney was asked by Gates to acquaint the court with the operations of the Third Fleet and "events leading up to the matter . . . which took place on December 17th and 18th."

Speaking uninterrupted for some minutes, the chief of staff explained that the fleet had been conducting air strikes on Luzon from December 14 through 16 in support of the landings on Mindoro, and that the fleet was "further obligated" to deliver additional strikes on Luzon beginning on the nineteenth. He explained that the plan called for "a retirement" toward the fueling rendezvous set for the morning of December 17. While the rendezvous was kept, Carney said, the wind and sea conditions made fueling difficult. Although the fleet was "very expert at fueling at sea" and had previously done so "in the wake of typhoons," such "expertness and experience testifies to the unusual conditions that existed that day."

According to Carney, Halsey concluded "about midday on the 17th that there was a tropical disturbance to the eastward of the fleet's position." The fuel situation in the destroyers was "somewhat critical by reason of three successive days and nights of operations" off Luzon, with "many destroyers reporting in the neighborhood of 15 percent oil." Carney said "the problem" was to get the fleet fueled in order "to resume offensive operations on the 19th." He said their best estimate at that time was that the storm was moving in a northwesterly direction and that it would recurve to the northeast. This, Carney explained, was "based on a study of the current weather map" and "the history of December typhoons." Using this information, the fleet set a new fueling rendezvous for the next morning.

That night, with the weather deteriorating, the fleet was "unable from available data to accurately determine the existence of a typhoon system nor to accurately estimate the storm's character or movement." In the early hours of the eighteenth, Carney said, a decision was made to "head in a generally southerly direction in order to be south of the storm track," where calmer conditions might be expected. Efforts were made at daylight "again to do some fueling but that had to be abandoned." It wasn't until "mid forenoon of the 18th before we had accurate information that a genuine typhoon did in fact exist, where it was, and what it was doing." Courses were "adjusted" at that point, Carney said, for "comfort and safety."

Still speaking without being asked questions, Carney gave—perhaps inadvertently—his most revealing insight into Halsey's mind-set after he first learned that ships had been lost in the typhoon.

"The fleet was refueled on the 19th at a rendezvous sufficiently south of the storm track to give good weather, and efforts were made to go over the fleet track to pick up survivors from the ships then definitely known to be lost. . . . Air searches and by surface vessels were initiated to the extent possible with the concurrent necessity for getting back to the Luzon area for the resumption of the offensive operations."

To the extent possible? Was Carney saying that continuing offensive operations had a higher priority than rescuing U.S. sailors in the water?

"Word had been received that survivors from *Hull* and *Spence* had been picked up and had stated that their ships had foundered," Carney went on. "The fact of *Monaghan* was not disclosed until the destroyer *Brown* reported picking up six survivors from *Monaghan* on the 21st."

Notwithstanding the reports concerning the loss of *Hull* and *Spence* and survivors being picked up in the water late on the eighteenth and throughout the nineteenth, the fleet began "on the night of the 20th a high speed run toward Luzon." Only when the "residual seas became increasingly heavy" as they "overtook the storm again" were the air strikes cancelled. "Upon retirement toward the fleet's base [Ulithi]," Carney said, "a final check by searchers, air and surface, was made in the hope of finding further survivors."

Again it sounded as if rescue operations—for Halsey and the Third Fleet—were a decidedly lower priority than was getting on with the war.

Returning Carney to his testimony about the fuel status of destroyers on the seventeenth—"15 percent fuel oil on board"—Gates asked whether that was a "normal amount" for destroyers to have before replenishing.

"No," Carney said. "The operations were extensive by the necessity for high speed run-ins initially on the night of the 13th and by three consecutive days of carrier strikes. The obligation [of the fleet] was a heavy one and those operations did not permit interim fueling or topping off."

Neither Gates nor the court had any further questions for Carney.

Not surprisingly, Nunn, the Harvard-trained lawyer representing Marks, did have questions—more than a dozen. One area of interest had to do with what orders were given by Halsey to the fueling groups—one of which *Hull* belonged to—"after the fueling was abandoned on December 17th due to bad weather."

"That would not be a matter for Commander Third Fleet to concern himself with," Carney said testily. "The rendezvous was given to the force and group commanders and they would issue their own orders."

The battle-hardened Nunn did not back down as he faced the higher-ranking officer. "In other words, a rendezvous with the fueling units was given by Commander Third Fleet and he expected them to meet that rendezvous?"

"Yes."

"Can you state the date and time when the Commander Third Fleet was first advised of the presence of dangerous weather?"

"About half past two in the afternoon of the 17th," said Carney, who added that "prior to that time" Halsey had "suspected the existence of such a disturbance."

Asked by Nunn when the fleet "first maneuvered for the sole purpose of avoiding bad weather," Carney answered that it had been at "midday the 17th when fueling was found to be impractical and a new fueling rendezvous was set to the westward. That being based on the first estimate that the storm was moving to the northwest and would recurve to the northeast."

"Admiral, did local observations prove helpful in determining the path of the storm?" asked Nunn.

"As I previously stated it was not until the forenoon of the 18th that we were able to determine the position, course and speed of the storm."

"Was that determination by local observation?"

"Yes."

When no one had further questions for Carney, he was told that he was "privileged to make any further statement covering anything relating to the subject matter of the inquiry" that had not been brought out by "the previous questioning." Like the vast majority of the more than fifty witnesses who were to testify over the next week, Carney, when asked, had nothing to volunteer.*

Also testifying the first day of the court of inquiry was Commander George Kosco. The Third Fleet's senior aerologist began by reviewing his qualifications and experience as a weather forecaster. As his background included a master's degree from MIT (1940), extensive duty as an aerology officer aboard a number of aircraft carriers (*Saratoga, Yorktown, Ranger*), and three months of "special hurricane research" in the West Indies, Kosco seemed impressively qualified on aerological matters, although he had been in the Pacific and on Halsey's staff for only two weeks prior to the typhoon.

In the first question posed by the judge advocate, Kosco was asked to take the court through the weather conditions "determined by you" on December 17 and 18, as well as indicating which factors influenced him to make "any recommendations to Commander Third Fleet."

Kosco's uninterrupted answer eventually covered five pages of single-spaced typed transcript—ranging from the technical to the vague to the incomprehensible. Knowing that his testimony would be key, he had prepared himself with notes and charts—providing copies of the latter to the court—but at times they seemed to confuse him more than they helped.

The first thing that became apparent was how little weight Kosco placed on his own observations or those of other aerological officers assigned to aircraft carriers in the Third Fleet—which Nunn had characterized during his cross-examination of Carney as "local observations"—and how much he relied on reports from other stations and forecasts from weather centrals, which, Kosco would admit, "fell short" of being adequate.

---

* Carney, a future four-star admiral and chief of naval operations (1953–55), had more to say about the typhoon years later in his oral history. "MacArthur was counting on Halsey's carrier air support. . . . Halsey could have deserted MacArthur and we could have headed south for a hundred miles or so. Then we'd have bypassed [the typhoon]. But Halsey felt that we had to stay until the last minute. This was his decision and nobody was disposed to argue with it. So the enemy in this case was the typhoon, and it inflicted serious damage on us. It was a terrible tragedy."

In addition, Kosco explained how heavily he based his own weather forecasts on historical data. "There is only one typhoon normally for December," he explained. "According to the record for the last 50 years, 75 percent of these storms pass off to the northwest and about 25 percent pass off into the Philippines. The fact that this storm was somewhere around Ulithi and west of Guam, gave it a free range to move off to the northeast, and I so indicated it on my map." Kosco was acknowledging not only that he had placed great faith in the historical record as well as in the work of others but also that these sources seemed to him at least as important, if not more so, than wind and sea conditions he could observe for himself.

Still without any prompting from the court, Kosco attempted in his long narration to pinpoint exactly when he knew he was dealing with a typhoon. "About 8 or 9 o'clock [on December 18] it became rather apparent it was not an ordinary tropical storm, but was starting to get into typhoon conditions, and the best estimate was that it was a tropical storm that was developing into a typhoon." Yet that morning—still hours before any ships were lost—no typhoon warning was sent out by Third Fleet to its more than one hundred vessels.

As if he wasn't paying attention to his own testimony, Kosco went on, "About 1300 on the 18th we sent out a typhoon warning. You will find that in the TBS log. Until that time we had thought we were dealing with a tropical storm. The first mention of typhoon, which is about the worst weather report that can be sent out, was dispatched about 1300, and this was the first mention from any source of a typhoon or the possibility of a typhoon. In other words, we didn't think that we were dealing with a storm as severe as a typhoon until we were within 100 miles of it."

As if he heard a critical voice in his own head about the substandard quality of his weather forecasting until a storm was on top of the fleet, Kosco blurted out, "By this report, I don't mean that we didn't know about it before 1300, but by 1300 it became apparent that the outside world and weather centrals should know that this intense storm was in this location."

Then, as if to get himself off the hook, Kosco added: "I have not made an exhaustive study of this typhoon, because I would have to get a lot of records from places like Guam, Leyte and Ulithi. They are not available on board the flagship [New Jersey]."

Not only had the fleet aerologist given direct testimony, but he had seemingly cross-examined himself as well, in the process tripping himself up as surely as any skilled opposing lawyer might have. Perhaps realizing it was time to shut his mouth, Kosco abruptly said, "That is all I have."

The judge advocate asked:

Q: "Did you as the Aerological Officer on the staff of Commander Third Fleet receive information from all the other fleet units having aerological equipment or personnel aboard?"

A: "No, I did not."

Q: "Did the operation plan require various units to furnish you with that information?"

A: "Not unless the situation warranted breaking radio silence, and the storm was the tantamount thing at the time. The breaking of radio silence was then up to the task group commanders or the individual ships."

The court was understandably confused as to the precise time Kosco had realized the storm was a typhoon. One of the admirals asked: "You first diagnosed this as a typhoon at about 1300 on the 18th?"

"No, about 8 o'clock on the morning of the 18th. But we were fighting the typhoon and before I sent a message to the other weather centrals, it was about 1300, although I had sent a message at 8:30 A.M., saying that it was increasing in intensity."

No one—none of the admirals, not Gates as judge advocate, not Nunn for Marks, not Preston Mercer—attempted to pin down Kosco about why he had waited some five hours before sending out a typhoon warning, an hour *after* the destroyers were already sunk and hundreds of men dead.

Declining the opportunity to volunteer anything further, Kosco stepped down, no doubt thinking he was finished with his ordeal.

The court's first day ended with a succession of ship captains, squadron commanders, and flag officers in charge of various task units and groups on the witness stand. While the skippers testified to the damage the typhoon had caused their ships, the higher-ups were asked about reports and forecasts by their own aerologists, most of whom were not on the witness list. While the quality of the weather forecasting by others generally seemed to be on a par with Kosco's own confused meanderings, there were exceptions. The aerologist on the carrier *Monterey* (CVL-26), for example, provided a more timely forecast for the approaching typhoon. "We knew that a typhoon was somewhere around our area on the 17th," said *Monterey's* commanding officer, Captain Stuart H. Ingersoll. That was one full day before Kosco and the Third Fleet came to the same conclusion. Yet *Monterey,* "the only large ship near the position of the three destroyers when they capsized," had not broadcast a typhoon warning. Ingersoll was not asked why, although he might have answered that doing so was not his responsibility.

The last to testify that first day was Captain George H. DeBaun, commanding officer of the light carrier *Cowpens* (CVL-25), which had suffered damage similar to *Monterey*—tied-down aircraft breaking loose on the flight and hangar decks and careening into each other, exploding and causing fires—as well as personnel casualties.* If the fleet's aerological officers—including his own aboard *Cowpens*—had been unsure about the storm's severity and what it portended, DeBaun had not been.

"The weather followed the book description of a typhoon," said the straight-talking DeBaun, an Annapolis graduate (1921) and designated naval aviator. "We had swells, increasing winds, barometer dropping, all that. A very good example of what is written up in Knight's 'Seamanship.' There was no trouble realizing there was a typhoon. This was all evident to me from 8 o'clock in the morning [of December 18]."

More ship and group commanders testified the next day. Soon a topic of inquiry besides aerological reports and forecasts emerged: the issue of ballasting, first addressed by Captain Jasper T. Acuff, the com-

---

* A total of 146 Third Fleet aircraft were lost or damaged beyond repair by the typhoon.

mander of the oiler task group that included *Hull* and *Monaghan,* along
with *Spence,* which had been so low on fuel that she was left behind the
night of the 17th to remain close to the oilers. Acuff testified that he
had gone on the TBS to recommend to any ship low on fuel that it "par-
tially ballast" to 50 percent in order to get through the night. His idea
was for such ships to fuel "in two parts"—filling to capacity with oil in
the morning, then moving away, deballasting the seawater, and returning for more oil. Of course, fueling never took place the next morning.
While Acuff pointed out that the ships assigned to his group—including
Hull and *Monaghan*—"were not low on fuel," the question remained un-
answered for the court whether *Spence,* down to 10 percent capacity by
December 18, had ballasted during the night, as suggested by Acuff.

"Have you any idea as to why these destroyers sunk?" asked one of
the admirals.

"No, sir," Acuff answered. "Unless they failed to ballast."

Punctuating the importance of this issue, Captain William T. Kenny,
commander of a destroyer squadron, testified that his flagship, the
Fletcher-class destroyer *Hickox* (DD-673), a sister ship of *Spence,* had received
Acuff's message to ballast. Kenny testified that for the morning fuel report on December 17 *Spence* had reported 15 percent fuel and *Hickox* 14
percent. Rather than ballasting to 50 percent during the night, however,
Hickox was "fully ballast[ed]" as of 9:30 A.M. on December 18. And still,
Kenny said, *Hickox* rolled up to 70 degrees and nearly capsized on two occasions "when the issue was in doubt." Asked if *Hickox* might have gone
over if the ship had had less ballast, Kenny said, "It is quite possible."

The focus of the court returned to aerological questions with the
testimony of Captain Michael H. Kernodle, commanding officer of the
light carrier *San Jacinto* (CVL-30) and Annapolis classmate (1921) of
Acuff's. After confirming that an aerologist was assigned to his ship,
Kernodle was asked by Gates: "Was he able to predict the storm that you
encountered?"

"Yes."

"Did you receive any warnings from any outside sources as to the
approach of the storm?"

"Yes."

"Please tell the court approximately when and from whom you received that information."

"I received warnings continuously for 24 hours before I got into the storm, from my aerographer, from the action of the ship, and condition of the sea. I was fully aware of the storm, and that it was going to be severe. In addition to that, I also heard reports from other vessels who were in desperate trouble. . . . I had all the warnings any one could possibly have."

Whatever standing Kosco still had with the court had plummeted. The Third Fleet's senior weatherman—indeed, Halsey himself—certainly had had available the same reports that Kernodle had seen, as well as local weather reports and—if one looked out any porthole on any ship—the "condition of the sea." Why, then, was the approaching typhoon not as evident to them as it had been to one skipper of an aircraft carrier?

Captain Preston Mercer, switching hats from interested party to witness, was now called to the stand by the judge advocate. Allowed to make an opening narrative statement, Mercer described at length the movements and storm experiences of his destroyer squadron—and especially his flagship, *Dewey*—on December 17 and 18. He described making several suggestions to *Dewey*'s captain, Charles R. Calhoun, about ballasting and speed during the worst of the storm, rather immodestly suggesting that his sage advice may have helped save the ship from sinking.

He then said he would "like to say a little about the stability characteristics" of the *Farragut*-class destroyers that made up his destroyer squadron. "When in the Navy Yard recently, there was the usual effort by various people to add lockers and various items topside, which I resisted vigorously," Mercer said. He described pursuing the "new type 20 mm mounts" and having them installed on his squadron destroyers, "which reduced topside weight by 3,250 pounds" per ship. In spite of such efforts, Mercer stated, "all commanding officers and most of the officers and men of the squadron who have been in the ships any length of time were very much aware of the lack of stability" of the *Farraguts*.

As for his two lost destroyers, Mercer said that on the morning of December 18, "when *Monaghan* reported inability to come to the southerly [fleet] course," he considered "offering her some advice" but decided not to "in view of the communications situation and my inability to visualize what was happening in the ship." He did not suggest what nature of advice he might have given. As for *Hull*, "I never knew she was in difficulties." Mercer testified that *Dewey* had received the message by light signal from *Tabberer* on the night of the eighteenth that she was "picking up some survivors," although Mercer said it was unclear to him at that time which ship had gone down. In defending his decision to withdraw *Dewey* from the search for men in the water, Mercer said: "*Dewey* was on a course about 220, speed 3 knots, and although we would have attempted to rescue survivors had we seen any, I did not consider it advisable to make a search."*

Once again, the search for survivors seemed to be a lower priority.

One of the first questions asked of Mercer was whether *Spence* had been under his tactical command. He confirmed that she had been until he turned over command of the task unit about 10:00 A.M. on December 18.

"At any time during that period did you receive any indication that *Spence* might be in trouble?"

"No, none other than she was unable to fuel," said Mercer.

"Have you anything to say regarding the seamanship of *Spence* during the time that she was under your tactical command?"

"I noted nothing unusual, except that she dropped behind when my task unit headed south at about 8:20 A.M. on the morning of the 18th.

---

* The storm damage to *Dewey*, which Mercer had been concerned about aggravating in pounding seas when he directed that the destroyer not assist *Tabberer* in search operations, took six weeks to repair. With no major structural or hull damage, *Dewey*'s repairs were completed alongside a tender at Ulithi. The "most critical" deficiency was the construction of a new forward stack. Repairs were also made to the 5-inch gun director, internal communications circuits, steering control, and radar. Ironically, the damage to *Tabberer* proved more extensive, requiring the destroyer escort's return to the fleet repair facilities at Pearl Harbor.

With the communication difficulties being encountered at that time, I didn't think it remarkable that a ship had failed to receive a change of course signal, and I believe that she reported she was making the fleet speed on the fleet course after she fell out of position."

Mercer was asked to "compare the experience and capabilities" of the commanding officers of *Hull* and *Monaghan*, Jim Marks and Bruce Garrett, respectively.

"The commanding officer of *Monaghan* was in the squadron for such a very short time that I had practically no opportunity to make a sound estimate. He handled his ship well in formation, kept her on station in the screen, but that was the limit of my opportunity to observe him. The commanding officer of *Hull* has been separated from my squadron a great deal [while *Hull* was assigned to other units]. Likewise, when she was with us for a very short time, his ship was handled well, and I have no criticism whatever of his ability. [Marks] has served in the North Atlantic and experienced very heavy weather, but perhaps he did not appreciate that *Hull* was not as stable as previous destroyers in which he was embarked. I believe the commanding officers of *Monaghan* and *Hull* have at least average ability and judgment compared with their contemporaries."

This line of questioning continued for several exchanges.

Q: "How does the service experience of the commanding officers of your squadron compare with that of the commanding officers of other squadrons?"

A: "The commanding officers of the ships of my squadron are the most junior in destroyers, being of the Naval Academy class of 1938. The commanding officer of *Spence* was in the class of 1937, and now there are a few commanding officers of the class of 1938 in other destroyers."

Q: "I gather from your answer that these destroyer commanders of the class of 1938 are the junior ones in the fleet. Is that correct?"

A: "That's right, sir. They are just beginning to send a few of them to other ships."

Q: "Would you venture an opinion, or could you state, which destroyer squadron of the fleet has [ships with] the least stability?"

A: "There is no question in my mind but that it is Destroyer Squadron One."

Inexplicably, the judge advocate had just provided Mercer with cover in the event anyone in authority questioned how and why two of his squadron's destroyers—the oldest destroyers in the fleet, being commanded by the youngest skippers—and a third, newer destroyer temporarily assigned to him had been lost on December 18 while so many other ships, some also low on fuel, managed to survive the typhoon.

Mercer was not questioned about failing to join *Tabberer,* seemingly the one ship in the fleet that put a top priority on rescue, in the search for survivors in the water—men who happened to be from his own squadron.*

THE THIRD DAY of the court of inquiry began with admirals.

Rear Admiral Frederick C. Sherman, Annapolis class of 1910 and a former submarine commander during World War I who later became a naval aviator, was one of the Third Fleet's most experienced seamen and now served as one of Halsey's carrier division commanders. Like other admirals called to the stand, his questioning was conducted by the admirals on the court rather than handed over to the judge advocate, a mere captain.

Q: "Did you have timely warning or know that a severe storm was approaching?"

A: "I wouldn't say that I did, no. The aerologist on my staff kept reporting a typhoon 500 miles to the northeast. That was on the 17th.

---

* *Tabberer* officer Howard Korth, winner of the Navy–Marine Corps Medal for Heroism for his own participation in the rescue operations, wrote in his diary about a December 29 press conference with war correspondents at which Preston Mercer "held the limelight describing the activities of *Dewey* and how he saved her. She is the ship that ran off and left us all alone searching for survivors—she doesn't rate very high in our book nor does Capt. Mercer after yesterday."

The wind was about 060 and increasing, the barometer was falling, which according to my experience, indicated a storm. I put it to the southeast and much closer than 500 miles."

Q: "You were the northern task group of [Third Fleet], therefore probably the storm went as close to you as any of the others."

A: "I think it went closer except for some of the tankers and destroyer escorts that were northeast of us right in the path of the storm."

Q: "There were three fueling rendezvous set for the morning of the 18th. In view of the fueling rendezvous set for the morning of the 18th, did you feel that the storm would strike your task group?"

A: "I was not particularly happy over the last rendezvous."

Q: "Did you make your ideas on this matter known to any higher authority?"

A: "No, sir."

Sherman described his group's unsuccessful attempts to fuel two destroyers low on fuel the morning of the 18th. The refuelings were "called off by higher authority" after an hour due to the "line of weather getting worse."

Asked if he had any further observations he would like to make about the storm, Sherman did not hesitate to speak a piece of his mind. "Without meaning any particular criticism of our present day aerologists, I'm inclined to think that they have been brought up to depend on a lot of readings they get from other stations. I think they are much weaker than older officers in judging the weather by what they actually see. Whether anything can be done along these lines to either encourage or instruct them to watch weather that is then existing without waiting for reports from Pearl Harbor or other stations, I don't know. I think they should be taught to judge the weather by what they actually see."

Another of Halsey's carrier group commanders, Rear Admiral Gerald Bogan, testified to being told early on the afternoon of the 17th by the aerology officer of the carrier *Lexington,* his flagship, about a "cyclonic storm forming to the northeast of us."

Bogan gave testimony similar to Sherman's regarding his uneasi-

ness over the setting of the final fueling rendezvous on the morning of
the eighteenth. This time when the question was asked as to whether
he had so informed "high authority," the response was quite different:
"Yes, sir."

"Please tell the court what it was."

"I sent a signal to [Halsey] stating that the Lexington weather estimate
indicated that improved conditions would be found further to the
southward."

Instead, the fleet had plowed northward into the path of the typhoon that morning in an unsuccessful attempt to fuel.

Vice Admiral John S. McCain (Annapolis 1906), Halsey's task force
commander, came to the witness stand next. Even though he was
Halsey's highest-ranking subordinate, McCain claimed not to have been
involved in any of the fleet's decisions to set fueling rendezvous. He
could "only venture a guess" as to why certain decisions were made by
Halsey, but no one on the court was much interested in having McCain
provide such speculation.

Q: "At what time did storm considerations begin to govern the disposition and movement of your task force, if at all?"

A: "The morning of the 18th, I believe."

Q: "You had no cause for alarm until the weather markedly deteriorated the morning of the 18th?"

A: "That's true as far as I was individually concerned, yes."

An admiral on the court now told McCain something he might not
have known, given that all testimony had been behind closed doors and
after the first witness had been "classified as secret."

"There has been testimony that indications were plain to certain
commanders that the storm was approaching and increasing in violence during the 17th and that perhaps aerographers in the fleet did not
estimate on local conditions sufficiently, but relied mostly on reports
from outside stations. What is your opinion of this?"

McCain sounded surprised. "I have no opinion of value on that."
When it came to providing illumination to the subject of the inquiry,
McCain had been of no value, either.

The next witness was the one man who would be unable to claim that he was not part of the decision-making process involving the Third Fleet or otherwise not responsible for any of its actions or movements. When the name was called by the judge advocate, an orderly hurried from the room to summon the witness, who was waiting in the passageway. Everyone watched the door. Soon it opened, and a barrel-chested admiral, tieless and in pressed khakis like the other officers in the wardroom, strode forward. Once seated, the witness was asked by the judge advocate to state his name, rank, and present station. It would be the last question asked by Gates, as all further questioning would be handled by the court's own admirals: "William F. Halsey, admiral, U.S. Navy, commander Third Fleet, U.S. Pacific Fleet."

Q: "Admiral, did you consider that you had timely warning or did you know that a severe storm was approaching around the 16th and 17th of December?"

A: "I did not have timely warning. I'll put it another way. I had no warning."

Q: "There has been testimony from other commanders that the local conditions indicated the approach of the storm. Was that evident to you?"

A: "The local conditions commencing on the 17th were very bad. So bad that I ordered the destroyers that were alongside tankers and heavy ships [for fueling] to clear. A disturbance was indicated, but whether it was a severe storm or merely a local disturbance, there was no way of determining. We still thought it was a storm that had curved away to the northward and eastward and we determined to get away from it."

Q: "When fueling had to be stopped on the 17th of December due to increasing bad weather, what were your considerations?"

A: "The general picture was sour. I had numerous destroyers that were very short of fuel. I was under obligation to make a strike on Luzon, but of course a strike could not be made until the fleet was fueled. I was also obligated to avoid by that time what I considered a storm the magnitude of which I did not know. . . . Up to the forenoon of the 18th

December, when an unsuccessful attempt had been made to fuel, I was still under the impression that the tropical disturbance would curve to the northward and the eastward and its severity was not indicated."

Q: "At what time did the storm considerations begin to govern the disposition and movement of the fleet, if at all?"

A: "On the forenoon of the 18th it was very definitely apparent that we were very close to a violent disturbance which I believed was a typhoon. We were completely cornered and in the dangerous semicircle. The consideration then was [to find] the fastest way to get out of the dangerous semicircle and get to a position where our destroyers could be fueled."

When asked what was "wrong with the weather service," Halsey characterized it as "nonexistent" and went on to describe late and missing weather reports from outside sources. "As I recollect, there was only one report of a disturbance that came in. . . . It is the first time in the four months that I've been operating in this area that I haven't had reports to enable me to track a storm."

"Had you any idea there were any vessels in your force that were very low in stability when low on fuel?"

"Having spent a great many years in destroyers and having been in some very severe weather in ships ranging from 160 tons to 1200 tons, I knew there had been grave doubts as to their stability from time to time, particularly when in a light condition," Halsey answered. "I believe that some time before we got into the worst of this storm we sent out a general signal advising everybody to ballast down."

Halsey had stated something not mentioned by any other witness during the course of the proceedings. The suggestion that the Third Fleet sent out a "general signal" recommending ships take on seawater ballast "before we got into the worst" of the typhoon was offered by Halsey alone. No copy of such a communiqué from Halsey or his Third Fleet staff was ever produced.

Q: "Comparing the conditions of the 17th fueling with those of the early morning of the 18th, what is your estimate of the weather conditions?"

A: "On the morning of the 17th I was under the impression that we were on the fringes of a disturbance. On the morning of the 18th there was no doubt in my mind that we were approaching a storm of major proportions and that it was almost too late to do anything."

One question now begged to be asked: if Halsey had thought his fleet was approaching a "storm of major proportions" and that it was "almost too late to do anything" on the morning of the 18th, why had he ordered one final, disastrous attempt to fuel, which turned his fleet into the wind and in the direct path of the typhoon? Any veteran mariner would know that a safer course at that point would have been *away* from the storm, to the south. A southerly run of several hours by his ships might well have taken them all—including *Hull, Spence,* and *Monaghan*—out of harm's way.

Unfortunately, that question was not asked.

Rather, Halsey was dismissed as a witness. Before leaving, he told the court he had "an interest in the subject matter of the inquiry" and wished to be named as an interested person. He regretted, however, that he would be unable to appear at future sessions due to an upcoming fleet movement. Therefore, he wished to waive his right to be present at the inquiry, and asked that he be represented by counsel. Halsey's request was granted.

On its fourth day in session, the court of inquiry met briefly in *Cascade's* wardroom, then adjourned to the larger pilothouse, where there was room to set up chairs for the survivors.

As he had walked along the pier that morning, Don Watkins was surprised to see a fellow officer who looked familiar heading up the gangway to *Cascade's* quarterdeck. *My God, that looks like Ed Brooks,* thought Watkins, who then dismissed the notion, believing that his fellow officer and friend—the proper "southern gentleman" from Virginia— had not survived. Watkins knew only about his *Hull* shipmates who, like him, had been picked up by *Tabberer:* a total of forty-one, five officers and thirty-six enlisted men. He did not yet know about his twenty-one shipmates picked up by other ships, including Brooks, who had been rescued off the raft with a dozen other men by the destroyer *Brown*.* When they finally faced each other in the crowded pilothouse, a smiling Watkins shook his friend's hand and said, "Glad you made it, Ed." Other *Hull* men had similar reunions that morning with shipmates.

The session in the wheelhouse had no sooner begun when the court directed *Hull's* commanding officer, James Marks, to read his official "narrative statement concerning the capsizing and sinking" of his ship. For those who had seen Marks on the bridge the morning of the typhoon, it was a remarkable transformation. What had been a stooped figure in soaked khakis and life jacket clinging to the navigation equipment, his face pale and contorted with fear, had turned into an erect, confident, and even handsome young officer looking like one of the Navy's finest, even though the uniform in which Marks stood before the court was borrowed—including skivvies—from the taller Henry Plage.

Marks had been in the Navy long enough to know what it meant when a captain loses his ship. Had they lived, Bruce Garrett and Jim Andrea would have been standing here, too. They would, in fact, now be there in absentia, with their actions and decisions documented and judged to the best of the court's ability. But because they hadn't survived, Marks alone would face what he called "the question and answer business." With the lid on any news of the typhoon losses as next of kin were still being notified, Marks had not yet been able to tell even his brother, Arthur, an Annapolis graduate (1927) and Navy officer serving in Washington, D.C. He had, however, come up with a way to send word home. Prior to his taking command of *Hull*, his fiancée, Virginia Fritchman, a Connecticut coed ten years his junior to whom he had become engaged on his last leave home, had convinced him to take his clarinet with him, as "he might have a chance to play it." The former Naval Academy swing band leader had not had time to do so, of course, and bringing the instrument along probably had been a mistake.

---

* Sixty-two *Hull* crew members survived the typhoon. In addition to the forty-one men picked up by *Tabberer* and the thirteen men rescued by *Brown*, four other *Hull* survivors were pulled from the water by the destroyer escort *Robert E. Keller* (DE-419), three by the destroyer *Knapp* (DD-653), and one by the destroyer *Cogswell* (DD-651).

But a few days earlier, he had written to Virginia from Ulithi: "I lost my clarinet." Upon reading those words, Virginia "knew right away Jim had lost his ship," and the first thing she did was call her brother-in-law, Arthur.

Sitting in the same gray metal chair used by the previous witnesses, Marks read the statement that came strictly from his own memory of events, as all of *Hull's* "records, papers [and] publications" had gone down with the ship. Everyone from the highest-ranking admiral to the lowliest seaman listened, with the only sound other than Marks' courteous and firm voice coming from the soft hum of *Cascade's* generators. Marks began with routine matters, such as the nature of the screening duties assigned to *Hull* on December 17 and the ship's movements that day. Then he came to the day all hands were waiting to hear about.

"The next morning the sea remained quite rough. The sky was heavily overcast. From time to time the course of the fueling group was changed and the screening units maneuvered to attain new screening stations each time. During the forenoon the sea increased steadily in roughness and the barometer readings were dropping."

Marks said that to his "best recollection" the fueling unit's course was changed to a southeasterly course about 11:00 A.M. Marks told how the heavy rain and sea spray caused difficulties with electrical equipment, including the radio and radar. "It was during [this] period," he went on, "while the ship was proceeding to her new screening station that her capsizing and sinking occurred."

Marks claimed he had the ship inspected that morning for "security of stowages and watertightness." He also explained that he had received a report from the engineering officer that morning that the ship was "well above the required ballasting point," having between "125 and 120 thousand gallons of fuel aboard," which represented "a little over 70%" capacity. "In view of the fact that the ship was riding the seas satisfactorily at the time and that I estimated that we would be fueled on short notice as soon as the heavy weather abated I did not consider ballasting advisable."

So far, Marks had said nothing that might serve as a red flag.

"Roughly about 1130 the seas became mountainous and the wind increased to hurricane proportions. At this point I wish to state that there had at no time been any storm warnings received from any source whatsoever, although we had been keeping careful watch for same. In endeavoring to alleviate the heavy rolling of the ship, I tried every possible combination of rudder and engines, with little avail. An attempt was made to bring the ship's head into the sea but she would not respond. Then an attempt was made to turn away from the wind and bring it as far on the port quarter as possible, but again the ship would not answer. It was apparent that no matter what was done with the rudder and engines, the ship was being blown bodily before the wind and sea.

"Shortly before twelve o'clock the ship withstood what I estimated to be the worst punishment any storm could offer. She had rolled about 70 degrees and righted herself just as soon as the wind gust reduced a bit. I have served in destroyers in some of the worst storms in the North Atlantic and believed that no wind could be worse than that I had just witnessed. Just at this point the wind velocity increased to an unbelievable high point which I estimated at 110 knots. The force of this wind laid the ship steadily over on her starboard side and held her down in the water until the seas came flowing into the pilothouse itself. The ship remained over on her starboard side at an angle of 80 degrees or more as the water flooded into her upper structures. I remained on the port wing of the bridge until the water flooded up to me, and I stepped off into the water as the ship rolled over on her way down. The suction effect was felt but it was not very strong. Shortly after I felt the concussion of the boilers exploding under water. The effect was not very strong and caused me no ill effects. I concentrated my efforts thereafter to trying to keep alive in the mountainous seas which pounded us. I could only see a few feet while in the water as the sea was whipped to a froth and the air full of spray."

When he was finished, Marks was asked by one of the admirals if the statement he had just read was a "true statement of the loss" of *Hull*.

"It is, sir."

"Have you any complaint to make against any of the surviving officers and crew of the said ship on that occasion?"

Among the survivors, Chief Quartermaster Archie DeRyckere wondered what would come next. Certainly there were men Marks might
now name: those who had been insolent on the bridge, those who had
spoken of mutiny that morning. DeRyckere had seen and heard much.

"I do not, sir."

Marks had waived any possibility of his pointing a finger at anyone.

Now it was time to ask the same question of the crew.

Addressing the surviving officers and enlisted men seated in the pilothouse, the admiral asked, "Have you any objection to make in regard
to the narrative just read to the court, or anything to lay to the charge
of any officers or man with regard to the loss of the United States Ship
Hull?"

Five survivors spoke up, all briefly and none critical of Marks.

One officer clarified that the power in CIC went out "about 30 minutes before we went over," and engineering officer George Sharp explained that "at all times" the number two boiler forward was in use,
with number four aft "cutting in on the main steam line 20 minutes
before the ship capsized."

Gunner's Mate John Valverde said that although they had started
with 70 percent fuel that morning, every time "we took a roll to starboard fuel would pour out of the tanks." He theorized that the forward
fuel tanks were "all empty and that was what caused the loss of the ship.
She was top heavy."

Chief Radioman Francis Martin gave a rather rambling discourse
about a decoded message he had given to Marks about 7:30 A.M. on the
morning *Hull* went down, speculating it may have been a storm warning. But then he added that he hadn't read it in its entirety and couldn't
be sure whether it had to do with "our fueling that day" or if it was "a
storm warning or not."

Then there was silence.

DeRyckere, like his shipmates Ray Schultz and others, felt there was
no alternative but to keep quiet. Marks, the Annapolis man being

judged by other Annapolis men, had not pointed the finger when given the opportunity to do so. Who were they—"lowly enlisted men"—to accuse an officer, the captain of their lost ship, before a court of admirals during a time of war? The young chief, who had turned twenty-five a week earlier, had it in mind to stay in the Navy after the war. Calling out one's superior officer—commanding officer, at that—for incompetence would not be a great start to his career. All any of the *Hull* men wanted now was to "get home as quickly as possible" for the thirty days of survivors' leave that the Navy gave all who lived through the loss of their ship. It would have been different had Greil Gerstley taken over command of *Hull* and lived. He, and they, might be defending themselves at a court-martial, charged with mutiny. Even had Gerstley not taken over the ship that morning, had he lived, details from the respected executive officer about what had taken place on the bridge that morning might have surfaced. Then again, maybe not. No matter; Gerstley was gone.*

From where he sat observing the proceedings, Watkins was also thinking about Marks. Although he hadn't been on the bridge that morning, he certainly believed Marks to be "incompetent." He could accept a scenario in which Marks, a bad ship handler under the best of conditions, had mishandled the conn in the confused seas and otherwise "not assessed the situation properly." There was no doubt that Gerstley or any number of *Hull* officers could have handled the ship better that morning, perhaps even saving her. Yet had Marks been so incompetent "at that particular moment to sink the ship"? Since Watkins had not observed the captain on the bridge that morning, he could not

---

* Archival records searched for the period between World War I and World War II, and in the years thereafter, show no instance of a court-martial for mutiny resulting from the relief of a Navy captain at sea. After the publication of the novel *The Caine Mutiny* (1951), concerning the court-martial of an executive officer who relieved an incompetent destroyer captain during a typhoon in order to save the ship, Halsey was asked by a correspondent how he would have handled such an incident had it happened under his command. "I would have sent a senior officer with long destroyer experience over to investigate. I would have hoped, and feel quite confident, I could have settled the question very quickly." Presumably, Halsey was suggesting handling such an instance without filing charges.

say. He would hear snippets of what others were saying and claiming; some of it he believed and some of it he did not. In any case, Watkins, like the other *Hull* officers, had for days been preparing himself to testify by writing down events as best he could recall them. His notes focused on what he had done and seen. He was "a little nervous" about the prospect of being a witness, but ready to do so.

When no further replies were forthcoming from the group of survivors, the court adjourned in order to return to the wardroom, where they would again meet behind closed doors. When the court members had reassembled, Marks was recalled to the stand and warned that the oath he had taken to tell the truth "was still binding." In all, the court asked thirty-five questions of him, with the queries and his answers covering approximately five pages of transcript. Many questions simply solicited further details about issues Marks had covered in his narrative. Also, the judge advocate went over two points raised by *Hull* crewmen in the pilothouse.

Asked about the chief radioman's suggestion that a storm warning might possibly have been delivered to him that morning, Marks said, "I know of no such report having been received."

"At the same time another member of your crew indicated that he thought a large amount of fuel oil had been lost from the forward tanks. Have you anything to say in regard to this?"

"Yes, the oil of which the man spoke I believe came from the fuel tank vents when are very small pipes and could not possibly pass the quantity of oil in those tanks over a short period of time. I believe since the man is a gunner's mate and not greatly acquainted with the fuel system of the ship that it was his impression that large amounts of oil had been lost."

On the subject of stability, Gates asked, "On any previous occasions had your ship acted in a manner which caused you to doubt its stability?"

"Having spent five months in a 2200-ton destroyer shortly before taking command of *Hull,* it was quite evident to me on taking command and maneuvering the ship that her stability was very poor compared to that of the destroyer on which I had just served. However, in full power trials in which rudder tests were conducted throwing the rudder hard over in one direction and quickly reversing it hard over in the opposite direction and making complete circles at maximum speeds, the stability was within satisfactory limits in the relatively calm water in which the tests were conducted."

An admiral now had a question. "How long before you rolled over had you decided that the ship was in such a hazardous position that you felt free to maneuver independently?"

"I would say about a half hour before the ship went down."

"Up to that time," the admiral went on, "you had continued your efforts to maintain your station. Is that correct?"

"That is correct."

Asked if he had any idea of the number of men trapped inside the ship, Marks estimated "about 100 men were unable to make their way clear of the ship" due either to their location in the ship or to the difficulty of getting through closed hatches from spaces below.

Revealingly, the last question asked of Marks by the court was obviously aimed not at the destroyer captain but at a certain fleet admiral.

"If you had decided to maneuver independently at let us say 8 o'clock in the morning would your ship have behaved better and gotten along better?"

"Yes. If I could have steamed clear to the southward to get clear I am sure I might have avoided the storm center completely."

That was, of course, based on information about the storm Marks only learned after the fact. More important, steaming clear to the south at 8:00 A.M. was not a decision Marks could have made on his own. He knew, as did everyone in the wardroom, that the only way *Hull* or any other ship might have escaped to the south would have been had Halsey taken the Third Fleet southward much earlier that morning than he did.

Marks seemed to have scored well with the court. Neither the admirals nor the judge advocate had been particularly harsh on him. They had asked direct questions that lacked a sharp edge, and he had appeared to answer them fully and competently.

Only a few *Hull* witnesses were called by the court, and Watkins, notwithstanding his days of preparation, was not among them. In fact, the only officer to testify was engineer George Sharp, who gave further details about the situation with the boilers, and also the amount of flooding in the forward engine room—"2 or 3 feet of water at 11:30 when I told them to pump bilges." Asked why he thought *Hull* was unable to recover from her final roll, Sharp said it was possible that the longitudinal bulkheads between three fuel tanks had been "carried away" when the ship lurched to starboard about 11:30 A.M. He thought that later rolls were worse because the "oil would flow over" in the direction of the roll.

Machinist's Mate Roy Lester, the throttle man in *Hull*'s engine room on the 8:00-to-noon watch on the day the ship went down, was asked the maximum speed made by the ship on his watch.

"Flank speed, 22 knots."

"Was this on both engines?"

"Yes, it was."

"Can you tell the court approximately what time this was?"

"Just before we went over."

It was the first hint of possible confusion on the bridge. Even though the admirals would know that flank speed would not be appropriate for a ship caught in a typhoon, they did not pursue the matter.

Fireman 2nd Class Roy Morgan was called to verify that there were countless men caught in "flooded spaces" below when the ship went over.

"In your opinion did everybody get out of the surrounding spaces?"

"No, sir, I wouldn't say they all got out."

By the time the court was finished with the *Hull* witnesses, it was apparent that the admirals were not going to blame *Hull*'s captain or crew for the loss of their ship. In most instances of ships being lost in non-combat situations, the actions of the captain and crew were carefully scrutinized. But in this case involving a typhoon that had hit a fleet—

sinking and damaging other ships as well as their own—the court's sights seemed to be aimed higher.

After a two-hour lunch break, court reconvened in the wardroom. The judge advocate recalled George Kosco to the stand. This time, the questions all came from the admirals, and their tone was adversarial.

"The court asked Admiral Halsey whether he considered that he had timely warning or did he know that a severe storm was approaching around the 16th and 17th of December. He answered, 'I did not have timely warning. I will put it another way—I had no warning.' In view of the fact that you are the aerology officer for Admiral Halsey, how do you account for this answer?"

"I take it that the admiral means he had no warning that a severe storm was approaching, although he did have a warning that there was a light, moderate storm in the area," Kosco said. "On the morning of the 16th, I showed him the weather map and told him that it looked like a small storm was developing between Ulithi and Guam. I continued telling him that this storm wasn't indicated to be very much of a storm."

While he might not have been helping his own cause, Kosco was at least showing a willingness to fall on his sword for his fleet commander.

"On the 17th, when we decided to knock off fueling operations, I framed a dispatch on the storm and also indicated to the admiral that a storm was somewhere to the east of us. No typhoon warning was given at any time. That is about all I can answer to that question, sir."

The admiral confronted Kosco with more of Halsey's testimony.

"The court asked Admiral Halsey what seemed to be wrong with the weather service in this case, and his answer was in part: 'It was nonexistent. That is the only way I can express it. . . . It is the first time in four months that I have been operating in this area that I haven't had reports that enabled me to track a storm.' What have you to say in regard to this answer?"

"The reason I think the admiral said that is that in the case of the other storms, they were so far away from his location that he had warnings of them two or three days before they came to his operating area, so that he could have notice to get out of the way. In this case the storm formed almost on top of him, and he was the first one to report it, so that he didn't have the advance information that he had in other storms. That is the only plausible answer that I could give to that."

With that, Kosco was gone, leaving the court with two seemingly contradictory positions. While he had spotted the storm days before it hit the fleet and admittedly underestimated its size in briefings to Halsey, Kosco also described the storm having formed on top of them with no advance warning. Clearly, both situations could not be true.

Following a brief recess, the court reconvened at 3:45 P.M. to hear from the *Monaghan* survivors. There were so few of them that there was no need to meet in the larger pilothouse. Heartbreakingly, the six sailors were all that remained of a crew of 262 officers and enlisted men.*

The senior survivor, Water Tender Joseph McCrane, read his narrative, which, curiously, was a page longer than Marks'. McCrane's view of events was narrower, too, because he had not been on the bridge the morning of the sinking or in communication with other sections of the ship.

McCrane told of *Monaghan*'s unsuccessful efforts to fuel on December 17th. That night "the weather was so bad and the ship was rocking and rolling so much that it was impossible for any of us to sleep. We all lay in our sacks and held on to keep from rolling out."

McCrane described receiving "word from the bridge" about 10:30 A.M.

---

* Recently, Keith N. Abbott, of Whittier, California, has represented himself as a seventh *Monaghan* survivor, claiming in a 2007 nonfiction book, *Halsey's Typhoon*, that he had "never told a soul, not even his spouse," until 2004, when he and his wife were on a South Pacific cruise with other World War II veterans. Abbott, a radar repair expert in the Navy, claimed to have been assigned to *Monaghan* for temporary duty and to have been on the bridge of the destroyer when it sank during the typhoon. Abbott's military personnel records, however, tell a different story. The Utah native served temporary duty on *Monaghan* from May 19 to August 8, 1944, at which time he was transferred off *Monaghan* at Pearl Harbor to attend Radar Material School. Following completion of the specialty school, Abbott returned to his original assignment on the destroyer escort *Emery* (DE-28), boarding her on September 18, 1944, at Purvis Bay. He remained on *Emery* through December 1944. On December 18, 1944, *Emery*, with Keith Abbott aboard, according to crew muster rolls, was anchored at Saipan, some 1,200 miles east of the typhoon that sank *Monaghan*. "There's only ever been six of us," says Evan Fenn, eighty-four, of Saint David, Arizona, today the sole living survivor of *Monaghan*.

on December 18 to pump seawater ballast into two empty fuel tanks aft, and how he began the process until "the lights went out" at 11:30 A.M. Minutes later, the ship "went over on her side." He spoke about the courage of Gunner's Mate Joe Guio in assisting men from the sinking ship, and how he ended up with "thirteen of us hanging on the outside of a raft." McCrane went on for several pages about their ordeal in the water, and the rescue of the six survivors by the destroyer *Brown* on December 21.

When McCrane was finished, the *Monaghan* sailors were given their chance to "lay to the charge of any officer or man," and all declined.

McCrane, however, did wish to register a complaint that would add weight to the possibility that *Farragut*-class stability problems had contributed to the sinking of *Monaghan* and, by extension, perhaps *Hull*. "The only thing I could complain about is ever since we left [Seattle] the ship seemed top heavy. I was on there for two years. Ever since we left [the shipyard] in October 1944, she seemed to roll worse than she ever did. Even in the calmest weather and even when anchored, she seemed to roll lots more than she used to."

While McCrane and four other survivors were called to testify, few new details emerged as to how or why *Monaghan* sank on December 18. All any of them could recount was that ballasting had begun (no doubt belatedly), lights and communications failed, and then the ship rolled over, unexpectedly and seemingly without much of a struggle.

Asked how many men had made it off *Monaghan,* Water Tender 2nd Class James T. Story estimated "at least 40 fellows" who had come out of the same hatch on the port side as him were "washed over the side." The rest, more than 200 men, were never seen in the water. Presumably they had gone down with the quickly sinking ship, trapped in spaces below.

On December 30, the same morning the Third Fleet weighed anchor in Ulithi lagoon and headed westward—during the first week of 1945, Halsey's carriers would conduct air strikes on Formosa, reporting 111 enemy planes destroyed, and Luzon, covering MacArthur's landings at Lingayen Gulf—the court of inquiry convened for the fifth day.

Before beginning testimony that morning, the court announced that Preston Mercer would no longer be considered "an interested party." Although the court gave no reason, it could be assumed that the admirals believed Mercer had had no complicity in the sinking of the three destroyers. At that point, Marks requested that Mercer be appointed to replace his counsel, Ira Nunn, who had informed the court the day before that his ship would be leaving with the Third Fleet. The court agreed to have Mercer act as stand-in counsel, and he took a seat next to Marks.

It now came down to *Spence,* the newest of the three destroyers lost in the typhoon, and also the one with the least amount of fuel aboard.

The reading of the official narrative fell to supply officer Al Krauchunas, the only officer to live and therefore the senior survivor. Krauchunas and five other "shivering figures still clinging to the floater net" had been picked up by the destroyer escort *Swearer* (DE-186) in the wee hours of December 20. They had already had one near miss that night: they thought they had been spotted in the dark by a passing aircraft carrier, but it had not slowed, and "our hopes dashed away with the disappearing carrier." When a rescue swimmer from *Swearer* appeared with a lifeline, one of the survivors asked: "What took you so long?" Among those rescued with Krauchunas after forty hours in the sea were Water Tender Charles Wohlleb, Quartermaster Edward Traceski, Torpedoman Al Rosley, and the Boy Scout from New Jersey, Edward Miller, who had been advised by an uncle to join the Navy so he would have a "dry bunk and three hot meals a day."

There were 24 survivors from *Spence*'s crew of 339 men. A short list of 23 other names had been compiled by the survivors identifying shipmates seen to have made it off the ship into the water, suggesting the strong likelihood that most of the 315 missing men had been trapped inside the capsized and rapidly sinking ship.*

Among the dead was Machinist's Mate Robert Strand, the amateur

---

* Of the twenty-four *Spence* survivors, fourteen men were picked up by *Tabberer*, nine men by *Swearer*, and one man by the destroyer *Gatlin* (DD-671).

bowler whose duty station was in the after engine room, and who had hoped to return home to Pennsylvania and marry his girlfriend, Jane, and own the local bowling alley one day. Like other families, the Strands would not receive the telegram advising them that Bob was "missing while in the service of his country" until January. On February 9, they received a second telegram from the chief of naval personnel stating that "there is no hope for his survival" and explaining that Bob had "lost his life as a result of typhoon on 18 December." It was only then Jane recalled a strange incident that had happened a week before Christmas. When Bob came home on leave he would sometimes arrive on the train after midnight and go directly to her family's house and throw pebbles at her second-story bedroom window so he could say hello before walking home. When she learned the date of *Spence*'s loss and Bob's death, Jane remembered being awakened that night by "pebbles striking the window." Thinking it might be Bob surprising everyone by making it home for the holidays, she had jumped up and looked out her window. Not seeing anyone, she had returned disappointed to bed but had a difficult time getting back to sleep. Only now—and for as long as she lived—was Jane certain it *had* been Bob at her window that night, saying goodbye.

With the pilothouse of *Cascade* again crowded with survivors, Krauchunas read his narrative. He told of *Spence* being down to 10 percent fuel capacity as of 4:30 P.M. on December 17, but at that time "no attempt was made to ballast the ship because the captain, James Andrea, thought it possible we could fuel that evening"; indeed, *Spence* had been left with the fueling group for that purpose. An attempt to fuel alongside an oiler was made at about 5:30 P.M., but after "10 or 15 minutes the lines parted." Still another unsuccessful attempt was made "before darkness set in," said Krauchunas. After a "lengthy informal discussion over the TBS" between various ship commanders, it was decided to fuel the destroyers in the morning. "From this decision," Krauchunas went on, "I would say that our captain did not attempt to take on any ballast. The ship was not in danger of turning over, nor was there any thought in that line. Normal routine was carried on through the night."

On the morning of December 18, "the weather became worse and no attempt was made to fuel." Krauchunas described the scene over breakfast in the wardroom, where officers conversed with the captain, and the engineering officer told Andrea that the ship was operating on one boiler and had enough fuel for another twenty-four hours. Krauchunas said he understood the captain as wanting to take on ballast if fueling was not imminent. He told of going to his stateroom, then learning of water flooding the engine room and electrical circuits, and "all the lights going out" about 10:50 A.M. About twenty minutes later, "we took a hard list to port and stayed there momentarily, and then came back. Our next roll to port capsized the ship."

The court heard testimony from six *Spence* survivors. Some of the most detailed information emerged from Quartermaster Edward Traceski due to his position on the bridge the morning of the typhoon. Traceski, who would never forget "how many men called out for God" as the ship went over, confirmed Andrea's order to begin ballasting about 10:30 A.M., not quite an hour before *Spence* sank. He also described in detail the sea and wind conditions. He remembered the swells coming from "directly aft," which would tend to eliminate them as the cause of *Spence*'s fatal roll. However, Beaufort force 12 winds (over 75 miles per hour) were blowing from *Spence*'s starboard side, suggesting they could have been a culprit in causing—or at least worsening—the last, steep roll to port. When asked about which damage control condition was set, Traceski said, "Baker was set. The normal wartime condition." This allowed transit between spaces through open hatches, as opposed to condition Affirm, in which more hatches to watertight spaces were closed, restricting traffic and limiting any flooding.

The next witness called was Chief Water Tender James Felty, one of the ten raft survivors rescued by *Tabberer*. The judge advocate returned to the topic of ballasting, as the loss of stability from being light of fuel and riding so high in the water—in conjunction with the typhoon's powerful winds—was looming as a likely cause for the capsizing. Felty testified to *Spence* receiving approximately 12,000 gallons of seawater ballast, a relatively small amount given the ship's low fuel level.

Another water tender, Chief George Johnson, twenty-five, of Fresno, California, also among the raft survivors saved by *Tabberer,* thought *Spence* might have received about 16,000 gallons of ballast, and estimated it would have required another hour and a half to take on an additional 30,000 gallons of seawater.

A picture could be drawn from the testimony of Traceski, Felty, and Johnson of a concerned Andrea on the bridge, aware of his ship's precarious condition and realizing that fueling would not happen anytime soon. In the last hour of his and *Spence*'s life, Jim Andrea, desperate to ballast for stability, simply started too late and ran out of time.

Johnson was asked if he saw his ship after he went into the water.

"Yes, sir."

"In what condition was the ship then?" asked Gates.

"She stayed up, I'd say, about seven or eight minutes. She broke in half and went down."

"Did you observe any explosion after the ship capsized?"

"Yes, sir."

"Was this before or after the ship broke in half?"

"After."

With the end of testimony from selected survivors, the court of inquiry moved toward concluding the investigative process. It met again on New Year's Eve and New Year's Day—the court's last day of proceedings—to query a number of experts in various fields, including weather forecasting and ship engineering, on largely technical issues.

One of the final witnesses was Captain Wilbur M. Lockhart (Annapolis 1918), the Pacific Fleet's senior aerology officer, based at Pearl Harbor Weather Central, which was responsible for sending out regular weather advisories and maps to U.S. ships and stations. Initially without current weather observations or reports "within a radius of 600 miles of the Third Fleet," Pearl Harbor was "lacking" in early knowledge about the developing typhoon, Lockhart stated. He was neither asked nor volunteered information about Pearl Harbor receiving early typhoon warnings from the naval weather station at Saipan

or elsewhere.* As "indications" came in on the seventeenth that dif-
fered from what the Third Fleet had broadcast in terms of the location,
intensity, and track of the storm, Pearl Harbor decided not to place
"another forecast on the air entirely different from what Third Fleet
had sent [because] it was decided that it would rile the waters too much
to change this forecast." Not until the afternoon of December 18, as the
latest reports left "little doubt" about the storm's position and track,
did Pearl Harbor broadcast warnings of a "deep typhoon" at 15°N,
128°E—the approximate area where *Hull, Monaghan,* and *Spence* were already down.

Asked about Kosco's inaccuracies in forecasting the approaching typhoon, Lockhart said, "With the report as sent out [by Third Fleet on
December 17], I believe the evaluation was considerably wrong and that
the storm no doubt was much closer." Lockhart's indictment did not
stop with Kosco, however. "I do not believe that most forecasters in the
fleet broke out a compass and laid down the possible positions that the
typhoon might be in. As indicated by the averages over a great number
of years, there is a possibility of about three [out of four] storms forming
where this one formed and moving to the northwest or north, but there
always remains one possibility that it could move to the west. This possibility was apparently lost sight of completely."

On January 3, 1945, having "thoroughly inquired into all the facts
and circumstances" asked of them by Nimitz in his authorizing letter,

----

*U.S. Army Air Corps Captain Reid Bryson, the meteorologist stationed on Saipan who sent the
teletyped message on the morning of December 17 to the local Navy weather office warning of
a typhoon on a track toward the Third Fleet and was dismissed with the reply, "We don't believe
you," heard "six or seven years after the war" from the Navy radioman who received his message. By then teaching meteorology at the University of Wisconsin, Bryson was in his office
when the phone rang. The voice on the other end asked, "Are you the Bryson who was on Saipan
in 1944?" Yes, said Bryson. "Did you send a message to the Fleet Weather Central [on Saipan]
about a typhoon?" Yes again. "Well, I was the Navy radio operator who received your message,
and I passed it on. I just wanted you to know that it didn't stop with me." Reports Bryson, today
emeritus professor of meteorology, geology, and environmental studies at Wisconsin: "All those
years later, and it was still on his conscience that something could have been done about those
dead seamen and was not. And it is still on mine, too. Somehow we could not get across that the
presence of the typhoon was not a guess. It was not some vague anticipation of the future. It was
a scientific observation on the spot. Ignoring science cost a lot of American lives."

and after considering "the evidence adduced," the court of inquiry announced its finding of facts, opinions, and recommendations.

Halsey's fleet movements—including his attempts to fuel his ships—were "logical" in view of his war commitments. At the first sign of the growing storm, however, Halsey should have ordered special weather flights and reports from vessels, and the "large errors made in predicting the location and path of this storm" were ultimately his responsibility. Halsey was also at fault for not broadcasting definite danger warnings to all his vessels early on the morning of December 18 "in order that preparations might be made and that inexperienced commanding officers might have sooner realized the seriousness of their situations." Another finding was that in the group containing *Hull, Monaghan,* and *Spence,* "vessels were maneuvering to maintain formation up to the times they were disabled or lost," suggesting criticism of Halsey for not issuing orders "to the fleet as a whole to disregard formation keeping and take best courses and speeds" to ride out the storm before it was too late.

Although the court found the "aerological talent assisting Halsey was inadequate in practical experience and service background" and would recommend "that older and more experienced aerological officers be assigned" to the Third Fleet, Kosco got off lighter than some of the testimony—including his own—indicated that he might. The court judged that on the morning of December 18 the Third Fleet aerology officer had given his "best information" to Halsey on the position of the storm and advised him "correctly as to the best course to steer to get in the safe [navigable] semi-circle of the storm which was then in progress." The northeasterly courses steered by the fleet from about 7:00 A.M. on December 18 in an attempt to fuel certain destroyers contributed to the disaster; "this maneuver held the fleet in or near the path of the storm center and was an error in judgment" on the part of Halsey.

The court offered opinions as to why each ship sank. In the cases of the *Farragut*-class destroyers, the losses of *Hull* and *Monaghan* were "due primarily to insufficient dynamic stability to absorb the combined effects of wind and sea." The court did not believe that *Hull* and *Monaghan* "would have survived" that morning even had they ballasted, given the basic instability of the *Farragut*-class ships, which "is materially less than other destroyers." With the high winds and mountainous seas, there was "no safety margin left for stability." While the quantity of free surface water that may have been present in *Hull* and *Monaghan* "is not known," there was little doubt in the mind of the admirals that "under the existing weather some amount of free water was present in the ships" and that such a condition could "contribute to their instability." In conclusion, the court decided, "it is believed that the sea and wind then existing, together with the fact that both ships had lost steering and were rolling heavily in the trough of the sea, could have combined [with their instability] so as to cause capsizing."

The newer *Spence*, which had "greater stability" and could be expected to "withstand heavier seas and winds than the *Farragut* destroyers," was another matter. In addition to effects of the typhoon, the factors contributing to her loss included the small amount of fuel in the ship, little or no water ballast, and the presence of free water below deck.

The commanding officers of all three ships "failed to realize sufficiently in advance" the necessity for them to "give up the attempt to maintain position" and to give "all their attention to saving their ships." The court acknowledged that such "good judgment" could "require more experience than had the commanding officers" of the three destroyers.

The court concluded that although there was "extremely limited . . . reliable and direct evidence" about *Spence*, the "available evidence indicates the commanding officer, Lieutenant Commander James P. Andrea, probably failed to set the highest material condition (Affirm); that he delayed ballasting until too late; that topside ammunition, provisions and other portable topside weights were not struck below." Inasmuch as Andrea was "lost with his ship," however, the admirals pointed out that "he is not available to the court" for further questioning or proceedings.

As for the captains of *Hull* and *Monaghan*, the court decided that "no offenses have been committed or serious blame incurred" by James Marks or Bruce Garrett. As a result, "no further proceedings" were recommended for the only defendant named in the inquiry: Marks, who two days later would write a letter to Henry Plage arranging for the return of the "gear" and $150 Marks had borrowed from him, and revealing also that he "expects to emerge from this 'deal' with a clean record as everyone seems to realize the situation we went through was a *most* unusual one!"

The court of inquiry formally listed three reasons for the "storm damage and losses incurred" by the Third Fleet:

1. Maneuvering the Fleet unknowingly into or near the path of a typhoon under a false sense of security and a belief that danger did not exist until too late. The lack of appreciation of storm danger was in turn due to unsound aerological advice based on insufficient data, coupled with wishful reasoning in connection with desired operations.
2. A certain amount of delay and maneuvering in the face of the storm and near its track, in an effort to fuel destroyers, after the storm had struck the Fleet.
3. In some cases a lack of appreciation by subordinate commanders and commanding officers that really dangerous weather conditions existed until the storm had taken charge of the situation, thus delaying until too late the preparation for security which would be expected.

"The preponderance of responsibility for the above falls on Commander Third Fleet, Admiral William F. Halsey, U.S. Navy. In analyzing the mistakes, errors and faults included therein, the court classifies them as errors in judgment under stress of war operations and not as offenses. Those pertaining to subordinates are attributed to errors in judgment, or in experience, or both. The court fully realizes that a certain degree of blame attaches to those in command in all disasters, unless they are manifestly 'Acts of God.' The extent of blame as it applies to Commander Third Fleet or others, is impractical to assess."

A FEW DAYS LATER, an old wooden minesweeper chugged out of Ulithi lagoon, headed in a "column of ships" for Eniwetok. Aboard were three

Hull officers who were to pick up other transportation at Eniwetok to Pearl Harbor, and from there to fly home on a PBY amphibious plane. Don Watkins and Ed Brooks, who had held their surprise reunion at the court of inquiry, were on the minesweeper along with Lloyd Rust, the CIC officer and new Texas lawyer who, tumbling endlessly in the churning sea upon his escape from the sinking ship, had resigned himself to possibly dying.

After making it through his first night in the water, Rust in the morning had kicked off his shoes and removed his pants, "figuring it would be easier to swim." At one point he pulled his white undershorts off, too, and waved them at a plane crossing overhead that kept going. Doubting there would be many more chances for rescue, he again "made my peace." As it started to get dark, "a ship came up" heading right for him. Close enough to spot "men walking back and forth on the deck," he could see "she was one of our destroyers." Rust hollered, but the ship turned away. That was "another low point." As the ship steamed from sight, Rust got "mad at God." Was he being teased? Tortured? As it got dark and the sky cleared, twinkling stars appeared. Rust got to thinking that *God doesn't torture people.* He decided the only reason a ship in the middle of the ocean would make a "hard right turn" would be as part of a search pattern. If that were the case, the ship would be turning at right angles, heading back in his direction. Figuring he would swim closer to the ship's next leg, he swam until the moon came up and he became "completely worn out." Saying to himself, *You've done all you can do,* he passed out. The next thing he knew, a rescue swimmer was in the water with him, pulling him by his life jacket toward the destroyer *Knapp* (DD-654). It was 8:28 A.M. on December 20. Rust had been in the water for forty-four hours. Resting on the ship, he began to think he might practice law after all. That morning *Knapp* picked up two more survivors and a third man who was dead—only "half a body" left by sharks.

The minesweeper bound for Eniwetok had a small complement of officers, so the *Hull* officers stood watches on the bridge. When it was Watkins' turn, he got the lowdown from one of the watch officers. "Oh, yeah, the fire bell is out there under the bridge," the officer said before departing. *Guess I won't need that,* Watkins thought. A short time later, the bridge filled with smoke, and Watkins ran for the alarm. When the minesweeper went dead in the water, a ship behind them collided with a ship from up ahead that had turned around to assist. By then the other *Hull* officers had rushed to the bridge. "You guys want to fight the fire or go back in the water?" one of them asked. The three agreed it would be best to fight the fire.

THE RECORD OF THE COURT of inquiry, as well as its findings and recommendations, was reviewed and accepted by all higher authorities— Admirals Nimitz and King, followed by Secretary of Navy James Forrestal. King believed the characterization of Halsey's mistakes— which Nimitz called "a commendable desire to meet military commitments"—needed to be tougher. The Navy's top admiral added a line to the official record: "The mistakes made were errors in judgment resulting from insufficient information, committed under stress of war operations, and stemmed from the firm determination to meet military commitments."

In February 1945 Nimitz signed a four-page typed letter—classified as confidential but copied to every ship commander in the Pacific Fleet—with the heading "Lessons of Damage in Typhoon." In what amounted to a lecture on the history and lessons of marine navigation and safety, Nimitz suggested further readings, such as the Bowditch and Knight classics, as well as books on ballasting instructions, damage control, and stability. On the latter subject, Nimitz said, "steps must be taken to insure that commanding officers of all vessels, particularly destroyers and smaller craft, are fully aware of the stability characteristics of their ships."

Nimitz concluded that at times in bad weather "a ship's safety must take precedence over further efforts to keep up with the formation or to execute the assigned task. This time will always be a matter of personal judgment. Naturally no commander is going to cut thin the margin between staying afloat and foundering, but he may nevertheless unwittingly pass the danger point even though his ship is yet in extremis. The time for taking all measures for a ship's safety is while still able to do so. Nothing is more dangerous than for a seaman to be grudging in taking precautions lest they turn out to have been unnecessary. Safety at sea for a thousand years has depended on exactly the opposite philosophy."

ACTING ON A RECOMMENDATION of the court of inquiry, the Bureau of Ships conducted stability testing on two *Farragut*-class ships. The decade-old destroyers, however, were deemed "stable," with "no major alterations to improve stability" necessary. The five remaining ships of that class—*Farragut, Dewey, Macdonough, Dale,* and *Aylwin*—served out the war in the Pacific, although one of them, *Aylwin,* caught in a second typhoon in June 1945, "again rolled excessively and came dangerously close to capsizing." They all earned ten or more battle stars. Eight weeks after the war ended, the Navy declared those five "unfit for other than limited service" and concluded that "the poor stability characteristics of the vessels preclude [their] being altered and equipped for other than limited service."

The *Farragut*-class ships, once the "goldplaters" of the U.S. Navy, were sold for scrap. The appraised value of their steel was set at $4,000 per ship.

Postscript

Operating off Okinawa on the night of June 2, 1945, Halsey's Third Fleet received typhoon warnings from a number of sources, including one of "Japanese origin which predicted that the typhoon would pass over Okinawa."

Commander George Kosco, still serving as Halsey's aerologist, "digested all the forecasts" but did not believe that the fleet was in danger of being hit by a typhoon for the second time in six months. Kosco estimated Okinawa "would not be in the path of any typhoon that could develop from the present weather situation." He recommended to Halsey that the "best course of action" was to "maintain [our] present position and await future developments."

"You are probably right," Halsey told

Kosco, "but I can't take a chance. If possible, I would like to be south of the typhoon."

On the afternoon of June 3, with weather improving on their southerly course, Halsey decided to redirect the fleet to the east in order to resume fueling and replenishment operations. Continuing to head south took him farther away from enemy targets, and he was not comfortable going westward, where he would be "in shallow waters with no room to maneuver and in range of Japanese aircraft from China." Although "not sure at that time where the typhoon was located," Halsey steamed his ships easterly through the night, "assuming that the storm" was on a "northerly course" heading "safely" away from them. In fact, Halsey's fleet was heading directly into its path.

The typhoon, with a maximum diameter of "less than 100 miles," was smaller but faster and even "surpassed in intensity" the December typhoon off Luzon. It struck Okinawa and the surrounding waters on the afternoon of June 4, when it was "much too late" for ships to maneuver to avoid its fury.

This time the worst damage was to Halsey's larger ships, with four aircraft carriers sustaining collapsed flight decks. Two battleships and three cruisers received major damage, including *Pittsburgh* (CA-72), which lost a major portion of her bow "with almost surgical neatness." In all, 33 ships were damaged and 146 planes destroyed. A total of six men were killed or lost overboard, and four men were seriously injured. By a combination of good luck and seamanship, as well as captains who took on ballast as needed, the destroyers and destroyer escorts "fared well as a class."

Another court of inquiry was held, again headed by Admiral John Hoover, who privately thought Halsey "deserved a general court-martial." Officially finding that Halsey "in the face of increasing heavy weather and storm conditions continued to maneuver large forces as a Fleet on set courses and speeds" in violation of the "spirit and letter" of Nimitz' fleet directive, the court recommended "serious consideration" be given to assigning Halsey and his task group commander, John McCain, to "other duty." Also, the court strongly recommended that a "more experienced and expertly qualified office of mature judgment be assigned as aerology officer on the staff of Third Fleet."

Nimitz, however, "disapproved" the court's recommendations regarding Halsey and McCain. Adding notes to the record, Nimitz said that Halsey had demonstrated his "skill and determination time and again in combat with the enemy," and credited McCain with rendering "services of great value in prosecuting the war against our enemies."

When the court of inquiry's official record reached Washington D.C., King agreed with the court that Halsey had been "inept in acting upon the weather warnings" and should have avoided the second typhoon. But with the war nearly over and Halsey a national hero, King had "no stomach for publicly reprimanding Halsey," which, with victory in the Pacific on the horizon, could have "ruined the Navy's finest hour." When Navy Secretary Forrestal, less charitable, was "on the point of retiring Halsey," he was argued out of his intention by King and Nimitz due to Halsey's status as a "national hero whose removal would impair American morale and boost that of the enemy."

On August 29, 1945, Halsey—ever wary of a last surprise attack by an enemy he so hated—took the Third Fleet into Tokyo Bay for the signing of the instrument of Japan's surrender, which took place two days later on the deck of his flagship.

Eight months after the typhoon of December 1944, and 1,364 days after December 7, 1941, the war ended.

Dramatis Personae

Andrea, Jean and Judy. Wife and daughter of James P. Andrea, commanding officer, *Spence*. In January 1945, Jean heard on the radio that three unnamed destroyers were missing in a Pacific typhoon and hoped against hope that it was not her husband's ship. She received a telegram a short time later. Before *Spence* departed San Francisco for the last time, Jean had visited the ship with two-year-old Judy, who today says, "I'm told I was the apple of my father's eye." With the assistance and lifelong friendship of Arleigh Burke, *Spence*'s former squadron commander who later became chief of naval operations, and his wife, Bobbie, Jean had a successful career working in naval intelligence at the Pentagon. She did not remarry for nineteen years. "She always said Dad was such a romantic that it was difficult for another man to measure up." Mother and daughter attended a *Spence* reunion in 1984, and were seated at the head table with Arleigh and Bobbie Burke. "According to Mom," says Judy (Andrea) Mahood-Cochran, "Admiral Burke quietly blamed Halsey for the loss of the three destroyers." Jean (Bailey), a widow for the second time, died in 2005 at age eighty-eight.

Bryson, Reid A. U.S. Army Air Corps meteorologist, Saipan. Shortly after the "Halsey disaster" in the December 1944 typhoon, Bryson discussed with his boss, Colonel William Stone, later a general and commandant of the Air Force Academy, whether he should contact authorities investigating the matter and turn over his chart of the typhoon's estimated track—which proved accurate—and reveal his ignored warning sent to the naval weather service. "The colonel told me to drop it because we couldn't prove when I had made my chart." Bryson would never forget how the Third Fleet, with the eight hours warning his message might have provided, "could have turned onto another course and avoided the typhoon." Those witnesses at the court of inquiry who described the typhoon that hit the Third Fleet as "sudden and undetectable were flat out wrong," Bryson says. "I think Halsey—the reason his nickname was 'Bull' was because he was bull-headed—killed more than 700 American sailors." Bryson, eighty-seven, senior scientist at the Center for Climatic Research and emeritus professor at the University of Wisconsin, is the author of five science books and more than 230 academic articles. He resides in Madison, Wisconsin.

Burke, Arleigh A. Commander, Destroyer Squadron 23. After leading "The Little Beavers"—the only destroyer squadron to win a Presidential Unit Citation—Burke transferred in 1944 to a fast carrier task force, where he served as chief of staff to Admiral Marc Mitscher for the rest of the war. Burke, who served an unprecedented six years (1955–61) as chief of naval operations, never forgot *Spence* and her crew—"a wonderful ship that fought mightily and always did more than her share." The Navy named a class of modern guided-missile destroyers the *Arleigh Burke*-class. The four-star admiral died in 1996 at age ninety-four.

Consolvo, Charles W. Commanding officer, *Hull*. After teaching at Annapolis for the rest of the war, Consolvo spent two years in China teaching at the naval academy in Shanghai. When a wealthy uncle made him an offer to go into the hotel business, Consolvo resigned his regular Navy commission, although he stayed in the reserves. Soon after, his uncle died and Consolvo was out of work. He volunteered for active duty, and served as executive officer of a destroyer tender. Following promotion to captain, he was assigned to NATO in Naples. Not long after his return to a Pentagon assignment, he was "kicked out of the Navy" with nineteen years of service and no pension. He tried a number of businesses, all of which failed. "My father became an alcoholic," says Charles Consolvo Jr. "I suspect he was depressed by two things: he felt guilty that had he stayed on *Hull* she would not have been lost and all those men killed—and a string of bad business decisions." Largely cut off from family and friends, Consolvo, *Hull*'s most popular wartime commanding officer, took his own life in 1959, at age forty-nine.

DeRyckere, Archie G. Chief quartermaster, *Hull*. Making a career of the Navy, DeRyckere received a commission and retired after twenty-eight years with the rank of Lieutenant Commander. Employed by Rohr Aircraft Corporation, he worked on the F-14 fighter and later for the City of El Cajon as an engineer, retiring in 1983. Throughout the years, DeRyckere has thought a lot about what happened on *Hull's* bridge the morning of the sinking. Also weighing heavily on his mind: neither he nor anyone else spoke up at the court of inquiry. "We were just a bunch of enlisted men in front of the most prestigious court of inquiry the Navy had ever had. We felt overwhelmed by the occasion. We all had our opinions, but no one wanted to hear them. I have a lot of regrets that I didn't stand up and tell the admirals what I knew. I would have pointed the finger directly at Marks. *Hull* had stability problems, sure, but with a little help she could have weathered that storm. With Marks at the conn, we didn't stand a chance." DeRyckere, eighty-seven, resides in San Diego, California, with his wife, Jacky.

Halsey, William F., Jr. Admiral, 3rd Fleet. Halsey received his delayed fifth star after the war ended. He retired in December 1946 following forty-two years of service. Halsey's memoir, *Admiral Halsey's Story*, was published in 1947. He wrote of the "tragic experience" of going through both typhoons, but not about either court of inquiry. As for the December 1944 storm, Halsey could "only imagine what it was like on a destroyer, one-twentieth the *New Jersey*'s size." The typhoon "tossed our enormous ship as if she were a canoe." He claimed that after a turn to the south on the morning of December 18, "most of our ships cleared the center [of the typhoon] but a few stragglers didn't. Some of them managed to ride it out. The rest we never saw again." Halsey spent much time in his latter years defending his actions at Leyte Gulf, confusion that he blamed on faulty communications in a split command structure between 3rd and 7th Fleets, under the overall command of Nimitz and MacArthur, respectively. He died in 1959 at age seventy-seven, and is buried at Arlington National Cemetery.

Kosco, George F. Senior aerologist, U.S. Third Fleet. Kosco remained on Halsey's staff until the end of the war. Thereafter, he served in both the North and South Poles. He retired in 1960 as a captain. In 1946, Kosco took his wife, Bernadette, to the New York Waldorf-Astoria Hotel, where Halsey was hosting a "big wingding." Bernadette found that the admiral "didn't take himself seriously. He was a great, fun-loving guy, and very personable. George loved working for him." Kosco's book, *Halsey's Typhoons*, was published in 1967. In it, Kosco does not discuss the court of inquiry following the December 1944 typhoon, and only briefly mentions the second board of inquiry; however, he does name both typhoons—previously unnamed. The December 1944 typhoon Kosco belatedly tabbed Cobra, and the June 1945 typhoon, Viper. Kosco died in 1985 at age seventy-seven.

Krauchunas, Alphonso S. "Al." Supply officer, *Spence*. As the senior surviving officer, Krauchunas was flown to Washington, D.C., for temporary duty—assigned to write letters of condolence to each parent, wife, or next of kin for his lost shipmates. "The greatest task," he wrote a former shipmate in March 1945, was "seeing [the] many parents who came to Washington, D.C.," hoping to get more information about their loved ones. "When a mother cries her heart out in front of you, it is unbearable. I have written over 500 letters and more come in [to answer] each day." Krauchunas was still "finding it hard to believe how so many died as they did in their compartments, without any light and utter confusion and hysteria going on. All of this happened so suddenly that even the captain was not able to get off the bridge, or any other officers." Returning to his native Grand Rapids, Michigan, he worked as director of parks and recreation, then spent many years in private industry as a personnel administrator. The former Chicago White Sox prospect continued his interest in sports, volunteering as a high school, college, and semipro referee for thirty years. He also loved golf, and shot a hole-in-one a month before suffering a stroke. He died in 1994 at age seventy-four.

Kreidler, Portia. Wife of sonarman John Kreidler, *Hull*. One of the thirteen wives who were pregnant when *Hull* departed Seattle for the Pacific in October 1944, Portia celebrated her 22nd birthday on December 17—the day before the typhoon. Happy with her pregnancy, she was certain that her husband—"the great love of my life"—would survive the war "right up until I saw the man with a telegram." Instead, they had "only 77 days together as husband and wife." She delivered on July 14, 1945, her first child, John David Kreidler, one of ten babies fathered by *Hull* men killed in the typhoon (including Greil Gerstley's son and namesake, Greil Gerstley Marcus, born June 19, 1945). In 1987, mother and son attended a *Hull* reunion, where they were welcomed by typhoon survivors—including Ray Schultz and Pat Douhan—many of whom wore baseball caps wih the slogan HALSEY'S SWIM TEAM. Portia, eighty-five, lives in Alameda, California.

Marks, James A. Commanding officer, *Hull*. After temporary duty in Washington, D.C., writing letters of condolence to families of his lost crew, Marks was given command of the destroyer *Clarence K. Bronson*, just finishing an overhaul at Mare Island, California. Bronson arrived at Pearl Harbor in July and took part in the bombardment of Wake Island; after the war, *Bronson* participated in the occupation by patrolling Japanese waters. After other sea and shore assignments, Marks attended the Imperial Defense College in London in 1956. Thereafter he commanded a Florida-based amphibious squadron placed on alert during the Cuban Missile Crisis. After a thirty-four-year career, Marks retired a captain in

1968. His retirement was "not a good one," according to his wife, Virginia, who said her husband found little to occupy him other than golf, which he eventually had to stop playing due to a bad back, first injured during the typhoon. "He should have taught or done something with himself, but he wasn't interested. I wonder if he didn't suffer a delayed depression from what happened in the typhoon. He talked so little about it, I just don't know." One day in 1986, Virginia came home; as she closed the kitchen door she heard a loud bang. In the living room, she found her husband dead of a self-inflicted gunshot wound to the head. Marks, *Hull*'s most unpopular commanding officer, was seventy-one.

Plage, Henry L. Commanding officer, *Tabberer*. The hero of the rescue operation, Plage was among a limited field of reserve officers to receive regular Navy commissions after the war. While serving as commanding officer of the minelayer *Terror* (CM-5) in 1947, Plage dove off the bridge into the water to rescue a crew member who had fallen between the ship and pier. Plage pulled him to safety, and was awarded another medal. Suffering hearing loss, Plage was discharged in 1954. He became a pharmaceutical distributor. He died in 2003 at age eighty-eight.

Rust, Lloyd G. Jr. CIC officer, *Hull*. After the war, Rust returned home to Wharton, Texas, where he practiced law until elected to the state house of representatives. After spending eighteen years as a county attorney, he was elected to a judgeship in 1974, serving on the bench until retiring in 1983. Believing that "any other officer" aboard *Hull* on the morning of December 18, 1944 "could have saved the ship" had they had been at the conn, Rust felt that "one hundred percent of the reason we sank was because Marks was not a good ship handler." Rust's wife died after twelve years of marriage, leaving him with four daughters, ages six to ten, to raise. After initially struggling with "survivor's guilt" for having made it when so many of his shipmates did not, Rust came to believe that he was "spared from dying in the typhoon so that he could live up to the obligation of being the only parent to the children he brought into the world." He died in 2006 at age eighty-six.

Source Notes

Complete book publication details are supplied in the bibliography. World War II U.S. Navy records such as deck logs, action reports, and war diaries are available at the National Archives II, College Park, Maryland. Other naval documents, such as oral and command histories, and communications, are collected at the Naval Historical Center, Washington Navy Yard, Washington, D.C. Military personnel records are available at the National Personnel Records Center, St. Louis, Missouri. Information about destroyer (DD) and destroyer escort (DE) veterans is available from the National Association of Destroyer Veterans (www.destroyers.org).

Prologue

pages xi–xiv

xii "a Nebraska farm kid": Charles Wohlleb interview.

xiii "Jesus . . . After fire room" and further quotations attributed to Franklin Horkey: Wohlleb interview.

Chapter One

pages 1–19

2 "in readiness": Log of *Monaghan,* Dec. 6, 1941.

3–4 "an old World . . . 'We have attacked' ": Roscoe, *Destroyer Operations,* p. 45.

4 "Proceed immediately . . . sound general quarters" and further quotations attributed to William P. Burford and the ship's lookout, Action Report, *Monaghan,* Dec. 30, 1941.

6–7 "wanting to get . . . Well, Curtiss . . . an over and . . . had to depth": Prange, *December 7, 1941,* pp. 235–36.

7 "Okay, Captain": Lord, *Day of Infamy,* p. 122.

7n. "unexplained and almost": Morison, *The Rising Sun in the Pacific,* p. 97.

8 "could scarcely": *New York Times,* Dec. 7, 2006.

8 "smoldering lust": *Commercial Appeal* (Memphis, Tenn.), Jan. 26, 1945.

10 "which will live in infamy": Franklin D. Roosevelt speech, Dec. 8, 1941.

10 "inspected as to quality": Log of *Hull,* Dec. 7, 1941.

11–12 "quite a few . . . only man aboard . . . goofy guy . . . The Japs are . . . wasn't off . . . one solid blast . . . something was always": Ray Schultz interview.

11n. "on a scale": Cant, *America's Navy in World War II,* p. 41.

11n. "still a couple of": Morison, *The Struggle for Guadalcanal,* p. 286.

12 "take and no give": Morison, *The Rising Sun in the Pacific,* p. 118.

12 "out of . . . directed toward": DuCharme, *Recollections of 7 December 1941,* p. 7.

13 "squirting a garden hose": Schultz interview.

13–14 "no drill . . . the real McCoy . . . just a mess": Tom Stealey interview.

15 "I can swim . . . all over the place": Stealey interview.

16 "so smashing a victory": Morison, *The Rising Sun in the Pacific,* p. 125.

16n. 1,465 U.S. servicemen: Ryan, *The Longest Day,* p. 303.

17 "tin cans . . . too thin": Halsey, *Admiral Halsey's Story,* p. 43.

17 "never mastered": ibid., p. 59.

17n. "guessed that the blame": ibid., p. 82.

17 "seafarers and adventurers": ibid., p. 2.

18 "There are exceptions": ibid., p. xiii.

18 "fighting-cock stance . . . beetle-brows": Potter, *Bull Halsey,* p. 1.

18 "Admiral": Halsey, *Admiral Halsey's Story,* p. 77.

18 "My God": Halsey dictated memoirs; Thomas, *Sea of Thunder,* p. 37.

18 "scene after scene": Prange, *December 7, 1941,* p. 372.

18–19 "see enough . . . the worst . . . Before we're": Halsey, *Admiral Halsey's Story,* p. 81.

Chapter Two pages 20–29

21 "Our Navy has . . . are grimly": Bath *Daily Times,* Oct. 27, 1942.

21 "the Stradivarius of destroyers": Jones, *Destroyer Squadron 23,* p. 39.

22 "throughout . . . could be felt": Bath *Daily Times,* Oct. 27, 1942.

22 "fast, roomy": Friedman, *U.S. Destroyers,* p. 111.

22n. "the heart and soul": Holland, *The Navy,* p. 115.

23 "We will fight . . . the enlisted men": Bath *Daily Times,* Oct. 27, 1942.

24 "perpetual mist": Morison, *Aleutians, Gilberts and Marshalls,* p. 3.

25 "Roman holiday": ibid., p. 24.

25–26 "just thrilled . . . Cut 'em in . . . all be over": Joseph Candelaria interview.

25 "revved up": DuCharme, *Recollections of 7 December 1941*, p. 24.

27 "sitting duck . . . little chance . . . magnificent and": Morison, *Aleutians, Gilberts and Marshalls*, p. 31.

27–28 "expendable . . . Torpedoes! . . . I wouldn't": Candelaria interview.

28 "I should have . . . in the states": Ernest Stahlberg interview.

28 "We were goners": Candelaria interview.

28–29 "smothering them . . . seemed impossible . . . outstandingly valiant": Morison, *Aleutians, Gilberts and Marshalls*, p. 32.

29 "boy, what a sight": Joseph Guio letter, Dec. 21, 1943.

29 "what seemed almost": Morison, *Aleutians, Gilberts and Marshalls*, p. 31.

29 "no fights . . . Just having": Candelaria interview.

Chapter Three   *pages 30–39*

30 "the first . . . surprised and": Morison, *The Struggle for Guadalcanal*, p. 15.

32 "various holes . . . traumatic amputation": Action Report, *Hull*, Aug. 11, 1942.

32–33 "in for . . . What makes . . . There's smoke . . . sunk a truck . . . ran Japs . . . short chain . . . no match . . . the trees . . . nothing to eat . . . artillery duty": Schultz interview.

34 "British and warm": Michael Franchak interview.

35 "headed up north . . . You're going": Patrick Douhan interview.

36 "thawed into . . . in a fog": Roscoe, *Destroyer Operations*, pp. 249, 250.

36 "as unlike those": Morison, *Aleutians, Gilberts and Marshalls*, p. 38.

37 "frozen stiff . . . about twenty-one . . . wiped out . . . the whole ship": Franchak interview.

37–39 "fog hung over . . . thicker than . . . glowing mushroom . . . like a dead . . . remarkable exploit": Roscoe, *Destroyer Operations*, pp. 252, 253.

38 "by three feet . . . But I . . . I trembled": Guio letter, Dec. 21, 1943.

39 "long, tedious": Candelaria interview.

39 "possibly the world's": Roscoe, *Destroyer Operations*, p. 246a (caption).

Chapter Four   *pages 40–52*

40 "green as grass": *Time*, "King of the Cans," July 17, 1944.

41 "5 percent . . . screwed up . . . tough but": Al Bunin interview.

41 "some of us": Jones, *Destroyer Squadron 23*, p. 37.

41 "You have over": ibid., p. 219.

41 "refresher course": Parkin, *Blood on the Sea*, p. 264.

42 "hard-driving": Potter, *Admiral Arleigh Burke*, p. 51.

42 "a thousand miles . . . work like hell": Potter, *Admiral Arleigh Burke*, pp. 3, 15.

42 "trademark that . . . the little beavers": Potter, *Admiral Arleigh Burke*, p. 93.

42 "how he loved": *Converse* newsletter, Mar. 1983.

43 "I'm heading": Potter, *Admiral Arleigh Burke*, p. 93.

43 "administered to": Jones, *Destroyer Squadron 23*, p. 5.

43 "superior cruiser": Morison, *Breaking the Bismarcks Barrier*, p. 313.

44 "to attack upon": Jones, *Destroyer Squadron 23*, p. 22.

44 "the first to": Sigismund L. Koperniak diary, Nov. 3, 1943.

44 "wracked by a murderous": Parkin, *Blood on the Sea*, p. 264.

45 "Execute turn . . . roughly parallel . . . sparks flew . . . handsome silver . . . the good saint": Jones, *Destroyer Squadron 23*, pp. 230, 231.

45–46 "right standard . . . Ship approaching . . . Full right . . . in night actions": Action Report, *Spence*, Nov. 8, 1943.

47 "feisty Spence . . . columns of fire": Parkin, *Blood on the Sea*, p. 265.

47 "We have . . . Cease firing!": Potter, *Admiral Arleigh Burke*, pp. 97, 98.

47 "Sorry, but": Jones, *Destroyer Squadron 23*, p. 234.

48 "scared plenty . . . overcame fear": Koperniak diary, Nov. 3, 1943.

48 "sprawled over . . . suddenly came": Jones, *Destroyer Squadron 23*, p. 240.

48n. "swift continuous . . . pouring out . . . masterpieces": Morison, *Breaking the Bismarcks Barrier*, p. 322.

49 "fanatical speech . . . swarming": Parkin, *Blood on the Sea*, p. 265.

49 "the most weird": Koperniak diary, Nov. 10, 1943.

49 "very good intelligence": Jones, *Destroyer Squadron 23*, p. 240.

49 "Please, Arleigh": Jones and Kelley, *Admiral Arleigh (31-Knot) Burke*, p. 133.

50 "31-Knot Burke": Potter, *Admiral Arleigh Burke*, p. 102.

50 "gentle reproach": Jones, *Destroyer Squadron 23*, p. 40.

50 'sardonic rib': Jones and Kelley, *Admiral Arleigh (31-Knot) Burke*, p. 133.

50 "with true instinct": Morison, *Breaking the Bismarcks Barrier*, p. 354.

50 "any Allied": Potter, *Admiral Arleigh Burke*, p. 103.

50 "a foe . . . a wait . . . detonations boomed . . . orange flame": Roscoe, *Destroyer Operations*, p. 264.

51 "in hot pursuit . . . One more": Morison, *Breaking the Bismarcks Barrier*, p. 356.

51 "fuel line . . . King of": *Time*, July 17, 1944.

51 "Never had the": Potter, *Admiral Arleigh Burke*, p. 106.

52 "not even . . . an almost . . . fortune of . . . we reached": Morison, *Breaking the Bismarcks Barrier*, pp. 358–59.

52 "hoisted into": Jones, *Destroyer Squadron 23*, p. 264.

52 "Yesterday was": Robert Strand letter, Nov. 26, 1943.

Chapter Five

pages 53–66

54 "viciously intercepted . . . gallantly pressed . . . courageous fighting . . . played a major": Distinguished Flying Cross citation.

54 "dogfighting a Zero": Robert J. Cressman, Naval Historical Center, "Tabberer," revised Nov. 2006.

56 "immediate active . . . assistant athletic": Henry L. Plage letter, Jan. 7, 1941.

56 "cruiser, battleship": Plage letter, Mar. 6, 1942.

57 "engulfed himself . . . one step": Arthur H. Plage interview.

57 "above average": Report of Fitness, Henry Lee Plage, Jan. 31, 1944.

57n. "I believe": Navy News Service, Aug. 1, 2005.

58 "90 Day . . . couldn't get": Henry L. Plage letter, circa 2001.

58 "run the ship": Paul Phillips interview.

59 "engines going": Log of *Tabberer,* May 29, 1944.

59 "very capable": Frank Burbage interview.

59 "great ship handler": Tom Bellino interview.

60 "Cruising in the": Log of *Tabberer,* June 11, 1944.

60 "hardly stay up": Howard Korth interview.

60 "youthful exuberance": Henry L. Plage statement, Aug. 30, 1986.

61 "very high . . . treated the enlisted": Burbage interview.

61 "attached to . . . from time . . . everybody loved": Bellino interview.

61–62 "best chow . . . couldn't get fat . . . best cinnamon . . . not too tough . . . cutting back . . . not many dogs . . . they didn't look": Phillips interview.

62–63 "establishment that . . . made a few . . . took on . . . bloody, with . . . a real crew . . . not supposed . . . the first time . . . developed real . . . so very": Plage statement, Aug. 30, 1986.

63 "pretty good system": Adamson and Kosco, *Halsey's Typhoons,* p. 130.

64 "there always . . . Sir, you": Phillips interview.

65 "under war . . . Ensign Surdam": Report of Fitness, Robert McClellan Surdam, Dec. 8, 1941.

65 "As officer of ": Report of Fitness, Surdam, Dec. 12, 1943.

66 "the fitting": Navy Dept. Bureau of Personnel letter, May 8, 1944.

Chapter Six

pages 67–75

68 "goners": Candelaria interview.

68 "more valuable . . . mother would . . . thinking of ": Candelaria interview.

69 "murderous fire . . . the Marines": Candelaria interview.

69–70 "the President . . . we are going . . . The Monaghan is . . . lots and lots . . . Bill, I've changed . . . putting a fake": Guio letter, Dec. 21, 1943.

70 "never-ending cheerfulness": Joseph C. McCrane letter, Mar. 15, 1945.

71 "real laid-back . . . like a father": Russell Friesen interview.

71 "exceptionally well": Calhoun, *Typhoon: The Other Enemy*, p. 16.

71 "GQ Wendt . . . every five . . . everyone wanted . . . all wrapped up": Joseph C. McCrane interview.

71 "a very religious . . . Mother McCrane": Candelaria interview.

72–73 "knows where . . . Now I'm going . . . Is she dancing . . . I thought he . . . wouldn't have . . . Louie walk": Candelaria interview.

73 "falling over . . . over a mile . . . closer than": Statement Concerning Finding of Death, May 1, 1945.

73 "they took the . . . *looked* like": McCrane interview.

73 "pounded the": Candelaria, *Tin Cans*, p. 198.

74 "all night . . . monotonous . . . back and forth . . . not doing much": Candelaria interview.

74 "when the Japs . . . filled their . . . started at . . . good ship . . . pretty busy": Evan Fenn interview.

Chapter Seven   pages 76–91

76–77 "quite a few . . . murderous journey": Franchak interview.

77–78 "dim, flashing . . . a float . . . a hole . . . presenting a . . . column of . . . mighty roar . . . men, planes . . . fragments of . . . flared up . . . spreading pool . . . frightful condition . . . thickly covered": Morison, *Aleutians, Gilberts and Marshalls*, pp. 140–41.

77 "an uneasy . . . blew up . . . but never": Douhan interview.

78 "entered the oil": Action Report, *Monaghan*, Dec. 10, 1943.

79 "ability to . . . ship handling . . . Remain in . . . splendid contribution . . . task force": Report of Fitness, Charles Willard Consolvo, Jan. 18, 1944.

79 "holding a broomstick . . . didactic": Charles W. Consolvo Jr. interview.

79 "officers of the": C. Donald Watkins interview.

79 "high-stress": James D. Torres interview.

80 "a car traveling . . . cleanly, without": McCain and Salter, *Faith of My Fathers*, p. 138.

81–82 "his whole body . . . beloved . . . so important . . . make it home . . . his duty . . . help the . . . the war . . . a privilege to . . . never treated": Rosemary Rust interview.

81–83 "tremendous seaman . . . 100 percent . . . handle the . . . a whole lot": Lloyd G. Rust Jr. interview.

82 "Jewish movie-star . . . privileged environment": Greil Marcus interview.

82 "natty ZBTs . . . urbane": *Cornell Magazine,* May-June 2004.

82 "services to the": Report of Interview with Applicant, Ltjg. Howard Baldwin, Feb. 24, 1942.

83 "a young man": Frederick George Marcham letter, January 15, 1942.

83 "member of ": Louis C. Boochever letter, January 15, 1942.

83 "unimpeachable integrity": Horace Stern letter, January 15, 1942.

83 "I am getting . . . I wouldn't swap": Greil I. Gerstley letters, Aug. 27, 1942, and Sept. 28, 1942.

83 "good judgment . . . cheerful disposition": James A. Marks letter, Feb. 1, 1944.

83 "had to get": Kenneth Drummond letter, Jan. 7, 1944.

83–84 "get a look . . . looked her . . . greatly resented . . . fell in love": Greil Gerstley Marcus interview.

84 "dull business": History of *Hull,* Ships' Histories Branch, Navy Department.

84 "either to land": Schultz interview.

85–86 "regular guy . . . best ship . . . make the . . . take the . . . 40-degree . . . terrific . . . too top-heavy . . . never make": Schultz interview.

86 "rather relaxed . . . get acquainted . . . developing teamwork": Dawes, *The Dragon's Breath,* p. 112.

86 "fashionable broken-deck": ibid., p. 94.

86–87 "were not . . . dependable performance . . . services were . . . well over . . . greatly in": Calhoun, *Typhoon,* pp. 3–5.

87 "between our . . . probably 65 . . . climbing the . . . perfectly calm": Douhan interview.

87 "I'm leaving the ship" and other dialogue between Schultz and E. M. Toland, Schultz interview.

89 "pretty well . . . wore out": Schultz interview.

90 "the only enlisted . . . Gabby . . . well spoken . . . I understand": Robert Coyne interview.

90–91 "never fired . . . without enough . . . didn't want . . . stepped right . . . *Oh, what* . . . beautiful destroyer . . . an old . . . messed-up . . . never touched . . . Sheet metal" and dialog between Stealey and rated petty officer, Stealey interview.

Chapter Eight   pages 92–106

92 "the nearest": Jones, *Destroyer Squadron 23,* p. 263.

92 "utterly one-sided": Potter, *Admiral Arleigh Burke,* p. 107.

92 "we were into": Albert Rosley interview.

93 "young ship . . . We have proved . . . We have a": *Ye Olde Dis-Spence-er* newsletter, Apr. 9, 1944.

93 "plastering . . . uncomfortably close . . . By radical": Action Report, *Spence*, Feb. 27, 1944.

93–94 "engaged in . . . not dared . . . their full . . . target listed . . . act of respect . . . - gallant enemy . . . opened ineffective": Potter, *Admiral Arleigh Burke*, pp. 108–9.

94 "salvo chasing . . . thanks to . . . straddling salvo . . . keep clear . . . quite possibly": Action Report, *Spence*, Feb. 27, 1944.

94 "remarkable record": Potter, *Admiral Arleigh Burke*, p. 108.

94 "conquered its": ibid., p. viii.

94 "devastated . . . Somebody's trying": Jones and Kelley, *Admiral Arleigh (31-Knot) Burke*, p. 142.

94–95 "conceal the . . . always keep . . . Tell the boys": Potter, *Admiral Arleigh Burke*, p. 111.

95–96 "pride and . . . athletic training . . . dove off": Alphonso Krauchunas, "USS *Spence:* The Typhoon and the Senior Survivor," 1992.

95 "a strong swimmer": Jim Krauchunas interview.

96 "executed colors . . . mustered at . . . pursuant to": *Spence* deck log, July 8, 1944.

96 "impressive ceremony": Krauchunas, "USS *Spence*."

96 "taut in all": Jones, *Destroyer Squadron 23*, p. 40.

96 "I relieve you": Regulations for the Government of the United States Navy, Article 3, Section 1, #72.

96 "very good . . . not afraid": Wohlebb interview.

96 "all business . . . I trained": Rosley interview.

97 "wasn't a natural . . . keen but gentle": Rene (Jack) Hoyle interview.

97 "distinguished himself": *Ye Olde Dis-Spence-er* newsletter, Oct. 1944.

97 "An actor . . . the Dorothy Dix . . . listen to . . . almost didn't . . . couldn't swim . . . The Navigator": Judith (Andrea) Mahood-Cochrane interview.

97 "going into the": Helen Andrea interview.

97 "Jimmie has": *The Lucky Bag 1937.*

98 "all Academy . . . you couldn't . . . named one": *Sumner* reunion newsletter, vol. 15, no. 4, Dec. 1994.

98 "a direct hit . . . making an . . . disintegrated in . . . sunk without": Action Report, *Sumner*, Dec. 11, 1941.

98 "seen firing": Hoyle interview.

98–99 "much younger . . . not much . . . ordinary fellow": Rosley interview.

99 "An efficient . . . It would only": *Ye Olde Dis-Spence-er* newsletter, Oct. 1944.

99 "a friendly visit . . . to talk . . . an enlisted": U.S. Navy press release, Feb. 15, 1945.

99 "5 percent . . . never wanted": Bunin interview.

100 "they needed": Ramon Zasadil interview.

100 "Bob got . . . all of the . . . going to catch": Richard Strand interview.

101   "Guess I wasn't": Robert Strand letter, Sept. 6, 1944.

101   "all the big": Robert Strand letter, Sept. 21, 1944.

101   "very, very sore . . . pulling for . . . it is no": Robert Strand letter, Sept. 25, 1944.

101   "waterborne": *Spence* deck log, Sept. 14, 1944.

101   "in love . . . lots of": Mahood-Cochrane interview.

102   "backing into . . . standing down": *Spence* deck log, Sept. 30, 1944.

102   "everyone was": Wohlleb interview.

102   "such a romantic": Mahood-Cochran interview.

103   "secured from drill": *Spence* deck log, Nov. 3, 1944.

103   "over a sand . . . Condition of Readiness": *Spence* deck log, Nov. 5, 1944.

103–104   "high degree . . . no longer . . . impressed upon . . . to reacquaint . . . of great . . . thumb rules . . . no conception . . . bad luck . . . only if we": Calhoun, *Typhoon,* pp. 16–18.

104   "At the present": Knight, *Modern Seamanship,* p. 686.

104   "Cyclonic Storms . . . Fixing the . . . Handling the . . . thumb rules": Bowditch, *American Practical Navigator,* pp. 272, 288, 290.

104   "dangerous semicircle . . . navigable semicircle": Bowditch, *American Practical Navigator,* p. 290.

104   "seas within": Bowditch, *American Practical Navigator,* p. 283.

105–106   "pitch-black . . . All of a . . . full reverse . . . the whole . . . looked at . . . what the hell . . . huge shadow . . . almost hit . . . got to thinking . . . nice, regular . . . just didn't . . . first mistake . . . tough bastard . . . gotten us . . . never did": Wohlleb interview.

Chapter Nine   pages 107–124

107   "gasket blew": *Monaghan* deck log, Sept. 27, 1944.

108–109   "something was . . . more sluggish . . . lurched . . . snap back . . . a long time . . . radical turns . . . imprudent . . . serious stability . . . beyond the . . . stability might . . . basically stable . . . in light . . . feasible . . . went about": Calhoun, *Typhoon,* pp. 5–7.

108   "very aware . . . matter of constant": Calhoun, *Typhoon,* p. 133.

109   "mustered the . . . absentees": *Monaghan* deck log, Oct. 1, 1944.

109*n.*   "had been substantially": ibid., p. 7.

109*n.*   "theoretical computations": ibid., p. 200.

109*n.*   "idea seemed preposterous": "Typhoon!" *Reader's Digest,* Jan. 1959.

110   "The Navy will": Guio letter, Sept. 10, 1944.

110   "I didn't get": Guio letter, Oct. 27, 1944.

110–11   "miracle . . . no . . . go . . . never caught . . . over $200": Candelaria interview.

111 "Commander Consolvo . . . No change": Report of Fitness, Charles Willard Consolvo, Aug. 8, 1944.

112–14 "a very good . . . high morale . . . plenty of . . . aware of . . . the very . . . lay a ship . . . visibility dropped . . . almost upon . . . breathed a . . . How does . . . Second link . . . No, sit . . . going out . . . make history . . . We're going": Kenneth Drummond interview.

113 "private in the": Consolvo Jr. interview.

113 "took care . . . kept the . . . made a lot . . . strange looks": Drummond interview.

113–14 "changing his . . . I don't go . . . Douhan didn't . . . no profanity": Douhan interview.

114 "detached as": *Hull* deck log, Oct. 2, 1944.

114 "the most junior": Preston Mercer testimony, Court of Inquiry, Dec. 27, 1944.

114 "short and slight . . . very serious": Calhoun, *Typhoon*, p. 22.

114 "Valedictorian of": *The Lucky Bag 1938*, p. 311.

115 "pierced . . . never forgot": James A. Marks Jr. interview.

115–16 "stellar captain . . . hard act . . . remote . . . his problems . . . asocial . . . did not eat . . . in a hurry . . . afraid the . . . closing up": Watkins interview.

116 "improper uniform": *Hull* deck log, Oct. 13, 1944.

117 "use of foul": ibid., Nov. 2, 1944.

117 "20 days . . . full ration": ibid., Oct. 29, 1944.

117 "hard-fisted": Carl Webb interview.

117 "just couldn't . . . all for his . . . but our new . . . every unlegal": William Zentner, "The Last Cruise."

117 "three or . . . available at . . . could sleep . . . stay afloat . . . totally unrealistic": Watkins interview.

117 "on various courses": *Hull* deck log, Oct. 10, 1944.

118 "giving him . . . never been whipped . . . rolled over . . . sat back": Archie DeRyckere interview.

118–19 "relatively calm . . . thrown hard . . . quite evident . . . very poor . . . within satisfactory": James A. Marks testimony, Court of Inquiry, Dec. 29, 1944.

119 "more alterations . . . add slightly": Mercer testimony, Court of Inquiry.

119 "hang out . . . having the . . . going to . . . heads off . . . it stuck": Zentner, "The Last Cruise."

119 "a sure sign": Dawes, *The Dragon's Breath*, p. 27.

119 "tried to talk . . . lose the ship": Douhan interview.

120 "simulated depth": *Hull* deck log, Oct. 17, 1944.

120 "would not . . . a dangerous": Watkins interview.

120 "hunting cabin . . . not too sure": Portia Kreidler Albee interview.

120 "participated in operations": J. S. Monroe, Commanding, *PC-476*, June 17, 1944.

121 "attacking with . . . resupply mission": William C. Meyer, Commanding, *PC-476*, Jan. 4, 1943, and Dec. 12, 1942.

121 "not be back": Douhan interview.

121 "really embarrassing . . . no dock . . . take the . . . I don't": Schultz interview.

122 "getting into . . . get rid . . . Consolvo's crew . . . spoiled . . . everyone else . . . tough . . . destroyed by": Douhan interview.

122–23 "good ship . . . gentlemanly . . . never saying . . . a lot of . . . very big . . . main ambition . . . joked about . . . an awful": Watkins interview.

122 "after my . . . not amused . . . go back": DeRyckere interview.

123 "capable of . . . incapable . . . so bad in . . . bad driver": Rust interview.

123 "southern gentleman . . . calling Marks": Edwin B. Brooks III interview.

124 "My only hope": Kenneth Drummond letter, Dec. 2, 1944.

Chapter Ten

125 "a Pacific base": George Spangler, "Ulithi," Mar. 1998.

126 "first female . . . skunky drunk . . . short, slight": F. Bruce Garrett III interview.

126 "to study for": Floyd Bruce Garrett Jr. U.S. Navy discharge, Feb. 17, 1934.

126 "smallest and . . . quiet dignity": Calhoun, *Typhoon*, p. 24.

126 "small in size": *The Lucky Bag 1938*, p. 63.

126 "Ensign Garrett": Floyd B. Garrett Jr. fitness report, Apr. 1939.

127 "I now request": Floyd B. Garrett Jr. letter, Nov. 12, 1940.

127 "impractical to approve": Chief of Bureau of Navigation letter, Nov. 16, 1940.

127 "because I would": Floyd B. Garrett Jr. letter, Jan. 26, 1943.

127 "an outstanding piece": Henry E. Eccles letter, Mar. 4, 1942.

128 "Experience possessed": William J. Giles Jr. letter, Jan. 26, 1943.

128 "sudden and . . . I am going": F. Bruce Garrett III letter, Nov. 14, 2006.

128 "be retained . . . experience and . . . answered by": *Cowell* telegram, Nov. 27, 1944.

129 "intelligent, demanding . . . more like . . . thoroughly seasoned . . . destroyers in": Calhoun, *Typhoon*, p. 9.

129 "watertight integrity . . . up to . . . very much": Mercer testimony, Court of Inquiry.

129 "the confidence . . . kind and . . . our duties": Joseph C. McCrane letter, Apr. 18, 1945.

129 "southern charm": F. Bruce Garrett III letter, Nov. 14, 2006.

129n. "a man's man": *The Lucky Bag 1938*, p. 152.

129n. "life in general": ibid., p. 230.

129n. "good-natured": ibid., p. 319.

129n. "Long Island's": *The Lucky Bay 1937*, p. 222.

129n. "loyal, energetic": ibid., p. 146.

129   "great pride": Calhoun, *Typhoon*, p. 24.

129–30 "crimson, yellow . . . caps of": Adamson and Kosco, *Halsey's Typhoons*, p. 1.

130   "The greatest fleet . . . fresh from": Baldwin, *Sea Fights and Shipwrecks*, p. 19.

130   "neat as . . . shone like . . . As was . . . head up": Adamson and Kosco, *Halsey's Typhoons,* pp. 3, 4.

130   "Japan's capacity . . . terminated": Potter, *Bull Halsey*, p. 306.

130   "aware of . . . Pearl Harbor . . . no discernible . . . morale rose": ibid., p. 52.

130   "essentially pinpricks": Thomas, *Sea of Thunder*, p. 50.

130–31 "first American": Potter, *Bull Halsey*, p. 155.

131   "completely reversed . . . on the defensive": ibid., p. xiii.

131   "successful turning": ibid., p. 181.

131   "a legendary figure": Thomas, *Sea of Thunder*, p. 118.

131   "sixty-five": Morison, *Leyte*, p. 196.

131   "the last . . . glimpsed the . . . recklessly": McCain and Salter, *Faith of My Fathers*, p. 42.

131   "childish . . . to guard": Thomas, *Sea of Thunder*, p. 316.

131   "We will go": Harold Stassen letter, Jan. 13, 1959.

131n. "Kill Japs": Dawes, *The Dragon's Breath*, p. 11.

132   "one hell . . . heading right": Thomas, *Sea of Thunder*, p. 233.

132   "rather impatient . . . Yes, yes . . . alert mind . . . mass of . . . a decoy": Morison, *Leyte*, p. 195.

132   "formidable aggregation": ibid., p. 162.

132   "nothing but empty": ibid., p. 196.

132   "Where is": Potter, *Nimitz*, p. 340.

132n. "not bomb . . . bait . . . draw it . . . nothing worse . . . to bother . . . He's busy": Potter, *Bull Halsey*, p. 297.

133   "padding to": McCain and Salter, *Faith of My Fathers*, p. 40.

133   "taunting him . . . broke into sobs . . . Stop it . . . not until . . . Where is Task": ibid., p. 303.

133   "stunned as if . . . mad he": Potter, *Nimitz*, p. 340.

133   "pale with anger": Adamson and Kosco, *Halsey's Typhoons*, p. 26.

133   "What right does": Solberg, *Decision and Dissent*, p. 154; Thomas, *Sea of Thunder*, p. 301.

133   "furious . . . gratuitous insult": Halsey letter, Feb. 5, 1959.

133n. "strongly opposed . . . nowhere near . . . as a nudge . . . wisest use": Potter, *Halsey*, p. 403.

133n. "interfering with": Potter, *Nimitz*, p. 339.

133–34 "like a tiger . . . up and down . . . [Halsey] has left . . . golden opportunity": Clark and Reynolds, *Carrier Admiral*, p. 201.

134   "persistent grumbling . . . careless ways": Thomas, *Sea of Thunder*, p. 103.

134   "not sufficiently": Potter, *Bull Halsey*, p. xii.

134   "prone to . . . guilty of . . . often vague . . . complex offensive . . . inefficiency was": Reynolds, *The Fast Carriers*, pp. 281–82.

134   "shorten the . . . guarantee Japan's": ibid., p. 253.

134   "within forty": McCain and Salter, *Faith of My Fathers*, p. 41.

134   "Battle of Bull's": Potter, *Bull Halsey*, p. 304.

134   "the bugaboo": ibid., p. 216.

134–35   "failure   to . . . gushing . . . Everyone   here . . . verbal   castigation . . . charged . . . threatening the . . . repeatedly stated": Thomas, *Sea of Thunder*, p. 325.

135   "He's still a": ibid., p. 307.

135–36   "under the . . . hated . . . an adequate . . . tired in . . . MacArthur's obliging . . . rested and": Halsey, *Admiral Halsey's Story*, pp. 234–36.

135   "thousands of superb": Reynolds, *The Fast Carriers*, p. 253.

135–36   "break off . . . longed to . . . bitter disappointment . . . no wings": Morison, *Leyte*, pp. 354–55.

136   "unexpected opportunity": Report on Operations of the Third Fleet, Jan. 10, 1945.

136   "too long": Potter, *Bull Halsey*, p. 316.

136   "sitting in . . . Well, just . . . a bad . . . But lucky . . . Wish everyone": Robert Strand letter, Dec. 10, 1944.

137–38   "neutralization . . . through restricted . . . admittedly hazardous . . . knocked out . . . by conference . . . support the . . . high speed . . . tactical surprise . . . with its . . . Very few . . . paralyzing": Report on the Operations of the Third Fleet, Jan. 10, 1945.

137–38   "intercepting our . . . not a . . . the congestion": Halsey, *Admiral Halsey's Story*, p. 236.

137–38   "masterminded . . . early warning": Adamson and Kosco, *Halsey's Typhoons*, pp. 21–22.

138–39   "nearest spot . . . dangerously low": Morison, *Liberation of the Philippines*, p. 60.

138n.   "a huge . . . impossible to": ibid., p. 24.

139   "get back . . . imperative": Calhoun, *Typhoon*, p. 29.

139   "driven . . . his fullest": Adamson and Kosco, *Halsey's Typhoons*, p. 26.

139   "became standard . . . shortest in": Clark, *Carrier Admiral*, p. 153.

Chapter Eleven   *pages 140–164*

140–46   "potential typhoon": All quotations attributed to Reid Bryson and others: Reid A. Bryson interview; Reid A. Bryson, "Typhoon Forecasting, 1944, or,

The Making of a Cynic," *Bulletin of the American Meteorological Society,* Oct. 2000.

146 "storm indications . . . very weak . . . move off . . . didn't expect . . . very small": George F. Kosco testimony, Court of Inquiry, Dec. 26, 1944.

147 "found a way": Bernadette Kosco interview.

147 "Well, if I": *The Lucky Bag 1930.*

147 "one of the . . . broken steel . . . giving all": Associated Press, Nov. 6, 1931.

148 "prison ship": Potter, *Bull Halsey,* p. 123.

148 "not hurt": William G. Kosco interview.

148–49 "a little bad . . . quite delayed . . . sort of . . . wind with . . . all normal": Kosco testimony, Court of Inquiry, Dec. 26, 1944.

149 "exceptional cases": Morison, *Liberation of the Philippines,* p. 60.

149 "reassuring 29.84 . . . even existing . . . not given": Adamson and Kosco, *Halsey's Typhoons,* pp. 31–32.

149 "unsteady barometer . . . rules for establishing . . . 50 to 80 . . . surrounding the": Bowditch, *American Practical Navigator,* p. 286.

149 "disturbance . . . lay in the": Morison, *Liberation of the Philippines,* pp. 60–61.

149 "chased out . . . the same thing": Halsey letter, Oct. 3, 1944.

150 "steaming in close . . . congested area": Calhoun, *Typhoon,* p. 32.

151 "collisions were": Adamson and Kosco, *Halsey's Typhoon,* p. 31.

151 "noticeable change . . . latent power . . . rhythmically lifted . . . let to settle": Erwin S. Jackson, "Trapped in a Typhoon," *Naval History,* Dec. 2004.

151 "for dear . . . swaying, swirling . . . fighting air": Adamson and Kosco, *Halsey's Typhoons,* p. 31.

151 "too rough": Morison, *Liberation of the Philippines,* p. 61.

151–53 "maintaining station . . . lashed and . . . shocked . . . allowed to . . . incredible . . . never known . . . deemed inconsistent . . . unhappy impression . . . remiss in . . . I wouldn't . . . provide a": Calhoun, *Typhoon,* pp. 32–33.

152 "nice guy": Wohlleb interview.

152 "operated at": Halsey, *Admiral Halsey's Story,* p. 238.

153–54 "high-speed . . . paint on": Krauchunas, "USS *Spence:* The Typhoon and the Senior Survivor."

153 "on hand": *New Jersey* deck log, Dec. 17, 1944.

153–54 "almost landed . . . almost even . . . Get that . . . Mother Nature . . . too young": Wohlleb interview.

154 "6,000 gallons . . . 22,265 gallons": *New Jersey* deck log, Dec. 17, 1944.

154–56 "upon the . . . upper works . . . so fast . . . violence of . . . isolated instance . . . weather evaluation . . . his great . . . data had been . . . a weak . . . frontal zone . . . storm center . . . about 400 . . . all indications . . . only a tropical . . .

move off . . . at right angles . . . safe distance . . . cutting things . . . storm or no":
Adamson and Kosco, *Halsey's Typhoons*, pp. 34–37.

155   "knock off . . . right away . . . getting a . . . didn't think": Kosco testimony,
Court of Inquiry.

155–57   "falling steadily . . . 12 to 15 . . . mixing diesel": Action Report, Third Fleet, Dec.
25, 1944.

156n.   "hindsight . . . 10 knots . . . clear of . . . at least": Wilbur M. Lockhart testimony,
Court of Inquiry.

157   "not be expected": Gerald Bogan testimony, Court of Inquiry.

157–59   "valiant but . . . no typhoon . . . one of those . . . troublesome enough . . . capable of . . . pros and cons . . . every report . . . round-table . . . vainly for . . . predict its . . . kill time . . . seek better . . . reversing": Adamson and Kosco, *Halsey's
Typhoons*, pp. 38–39.

157–58   "lack of . . . inflated canvas . . . word was received": Krauchunas, "USS *Spence*:
The Typhoon and the Senior Survivor."

157   "high on the . . . sail": Adamson and Kosco, *Halsey's Typhoons*, p. 99.

157   "relatively simple": Torres interview.

158   "six hours": Potter, *Bull Halsey*, p. 321.

158   "severe cyclonic": Bogan testimony.

158   "less than 200": Halsey, *Admiral Halsey's Story*, p. 237.

158   "400 miles": Howard S. Smith, Memorandum to Commander in Chief, Pacific
Fleet.

158–61   "might be . . . at the time . . . to make a . . . something was . . . hit us . . . increasing in . . . hand-to . . . long range . . . out of the": Kosco testimony, Court of Inquiry.

159   "historical data": William G. Kosco interview.

159   "bit of shut-eye . . . safest . . . southerly course . . . What do . . . danger . . . nearer
the . . . Severe . . . in a weather": Adamson and Kosco, *Halsey's Typhoons*, pp. 41–42.

160   "not to be": Adamson and Kosco, *Halsey's Typhoons*, p. 38.

160   "counting on . . . live up . . . deserted . . . bypassed . . . stay until . . . disposed
to": Taussig, *A Warrior for Freedom*, pp. 114–15.

160–61   "hand-to-hand . . . long range . . . out of": Adamson and Kosco, *Halsey's Typhoons*, p. 42.

161   "short and . . . turbulent": ibid., p. 47.

161   "things . . . due to . . . almost due . . . What do . . . We turn": Kosco testimony,
Court of Inquiry.

161–62   "cyclonic storm . . . under existing": Bogan testimony.

161–62   "field of . . . snap judgment . . . mere hunch . . . reluctantly decided": Adamson
and Kosco, *Halsey's Typhoons*, p. 48.

162n.   "in about . . . not to": Bryson interview; Lt. Daniel F. Rex, "An Aerological Report of the Typhoon of 15–21 December 1944."

162   "Loath to . . . commence fueling . . . collision course": Morison, *Liberation of the Philippines*, pp. 66–67.

162   "clearly demonstrated . . . not feasible": Calhoun, *Typhoon*, p. 45.

162   "blacker than pitch": Adamson and Kosco, *Halsey's Typhoons*, p. 49.

162–63   "mountains of . . . a hell's chorus . . . just stay . . . riding as though": Charles R. Calhoun, "Typhoon!" *Reader's Digest,* Jan. 1959.

163   "sea and": Wouk, *Caine Mutiny*, p. 326.

163   "voiced an . . . probably take": Kosco testimony.

163   "rather than . . . stay in . . . carry out": Taussig, *A Warrior for Freedom*, p. 114.

163   "very rough": *New Jersey* deck log, Dec. 18, 1944.

163   "oilers [should] not": Jasper T. Acuff testimony, Court of Inquiry.

163   "didn't like": Bogan testimony.

163–64   "present conditions . . . Yarnall 20 percent;": Action Report, Third Fleet, Dec. 25, 1944.

164   "regretfully notified . . . could not . . . pounding heavily": Halsey, *Admiral Halsey's Story*, p. 238.

164   "rather apparent . . . ordinary tropical . . . typhoon conditions . . . falling very": Kosco testimony.

164   "typhoon": Frederick C. Sherman testimony, Court of Inquiry.

164   "ceiling rested": Adamson and Kosco, *Halsey's Typhoons*, p. 51.

164   "typical barometric . . . sure sign": Morison, *Liberation of the Philippines*, p. 68.

164   "wind went . . . all ships . . . improperly . . . much difference . . . making a race": Kosco testimony.

164   "last, vain": Morison, *Liberation of the Philippines*, p. 66.

Chapter Twelve   *pages 165–173*

165–66   "recording all . . . maintain station . . . no possible": Edward D. Traceski testimony, Court of Inquiry.

165   "Everyplace we": Greenfield (Mass.) *Recorder,* Oct. 1983.

166   "rough ride": James Traceski interview.

166   "not sure . . . done the": Louise Traceski interview.

166   "slammed into . . . simply out": Parkin, *Blood on the Sea,* pp. 267–68.

166–67   "dry bunk . . . loved the . . . were like . . . blowing the": Edward A. Miller interview.

167–68   "conversed with . . . problems they . . . present themselves . . . 24 hours . . . in danger . . . normal routine": A. S. Krauchunas narrative, Dec. 29, 1944.

167   "9,000 gallons": James Felty testimony, Court of Inquiry.

167   "300 to": Traceski testimony, Court of Inquiry.

168 "about 40 percent": Calhoun, *Typhoon*, p. 158.

168 "normal wartime": ibid., p. 79.

168–70 "took off . . . good ship . . . all four . . . right then . . . routine staff . . . go up . . . In complete . . . it was my . . . scared sailors": James Felty interview.

169 "30,000 gallons": George W. Johnson testimony, Court of Inquiry.

170–72 "pitched . . . hazardous . . . narrowly missing . . . slowly and . . . blinding mist . . . completely drenched . . . seek relief . . . hitting the . . . You better . . . concern with . . . concerned look . . . not at . . . struggling to . . . spouting a . . . words of . . . sudden roll . . . agonizing moments . . . almost completely . . . a hopeless . . . Is this . . . gurgling and": Krauchunas, "USS *Spence*: The Typhoon and the Senior Survivor."

172 "all the reports": Traceski testimony, Court of Inquiry.

172 "the safety": Krauchunas, "USS *Spence*: The Typhoon and the Senior Survivor."

172 "all personnel . . . should wear": Krauchunas testimony.

173 "43 degrees . . . a little left . . . supposed to . . . hard right . . . didn't have . . . gradually capsized . . . 11:23": Traceski testimony, Court of Inquiry.

173 "My last": Melton, *Sea Cobra*, p. 174.

173 "well over . . . huge . . . Caught in a": Parking, *Blood on the Sea*, pp. 268–69.

Chapter Thirteen   *pages 174–189*

174–76 "really tough . . . because it . . . trying to . . . a great . . . good ole . . . not become . . . no one . . . all thought . . . nothing anyone . . . really started": Drummond interview.

176 "those confused": Morison, *Liberation of the Philippines*, p. 68.

176–77 "two of the . . . this typhoon . . . small boys . . . Typhoon! . . . chance to . . . fooled by . . . maneuvering right": DeRyckere interview.

176 "so thick and": Morison, *Liberation of the Philippines*, p. 69.

176 "fixing the . . . approximately 10": Bowditch, *American Practical Navigator*, p. 288.

177–78 "Jack Ass . . . ready the . . . too top . . . button the . . . stow below . . . Stow the . . . dipping in . . . Permission denied . . . Get off": Schultz interview.

177 "report to . . . got his": John Valverde interview.

178 "expected to . . . about 124,000 . . . above the": Marks testimony.

179–80 "steady 70 . . . washed right . . . about twelve . . . frozen in . . . Sir . . . What are . . . All hands . . . Sure, sir . . . Why don't . . . dirty look": Schultz interview.

181 "fairly well . . . leaks in . . . troubles in": Action Report, *Hull*, Jan. 1, 1945.

181–82 "pandemonium on . . . huddle around . . . obeyed for . . . silly . . . ten feet . . . not to be . . . felt safer . . . did not . . . damage control . . . denied . . . made fools . . . swallow . . . going to . . . too much . . . reduce the . . . very salty": DeRyckere interview.

183 "We'll be": Schultz interview.

183 "Sir!": Drummond interview.

183 "in irons": DeRyckere interview.

183–84 "every combination . . . ship's head . . . turn away . . . blown bodily . . . the trough": Action Report, *Hull*, Jan. 1, 1945.

184–85 "let anything . . . ride the . . . incompetency . . . in a state . . . did not . . . He's sinking . . . If I take . . . The bastard . . . Even if": Schultz interview.

184 "functioning properly": Wouk, *Caine Mutiny*, p. 329.

184 "could be . . . action was": DeRyckere interview.

185 "two or three . . . basket for": George Sharp testimony, Court of Inquiry.

185 "many tons": DeRyckere interview.

185 "not throwing": Schultz interview.

186 "tremendous noise": Action Report, *Hull*, Jan. 1, 1945.

186–89 "Let's go . . . did not . . . Within 20 . . . 400 pounds . . . hanging on . . . kidding and . . . put us . . . to settle . . . the ceiling . . . more than . . . done for . . . - trembling . . . turned into . . . really whistling . . . breaking over . . . whole bunch . . . beat to death . . . We gotta . . . faster than": Stealey interview.

Chapter Fourteen   *pages 190–207*

190–91 "I am unable . . . heading of . . . caught in . . . given up": Calhoun, *Typhoon,* pp. 75–76.

191 "manned by . . . ": Adamson and Kosco, *Halsey's Typhoons,* p. 59.

191 "Use more . . . Cannot get . . . had the": Adamson and Kosco, *Halsey's Typhoons,* p. 88.

191 "Have tried": Calhoun, *Typhoon,* p. 75.

191 "out of control": ibid., p. 75.

191 "You are . . . racked and": Baldwin, *Sea Fights and Shipwrecks,* p. 28.

191 "slammed . . . grave jeopardy": Adamson and Kosco, *Halsey's Typhoons,* pp. 88–89.

191–94 "between 122,000 . . . putting on . . . come back . . . not without . . . flying around . . . coming too . . . luckily it . . . rolling so . . . fire and . . . pump seawater . . . just plain . . . feel the . . . the storm broke . . . holding on . . . praying as . . . all communication . . . couldn't find . . . volunteered without . . . deserved some": Joseph McCrane, Narrative Statement of Senior Surviving Petty Officer, Court of Inquiry.

193 "say for . . . stopped quite . . . pick up": Joseph McCrane testimony, Court of Inquiry.

194 "shudderingly . . . Thanks, Dear": Baldwin, *Sea Fights and Shipwrecks,* p. 28.

194–95 "about forty . . . Please bring . . . seven or . . . difficult job . . . beating up . . . the fellows . . . no confusion . . . everyone trying . . . with absolutely . . . pulling everyone . . . about the . . . so nervous . . . Waves were . . . carrying fellows . . . lost all . . . a whirl-pool . . . knocking up . . . beat my . . . right on . . . took me . . . thrown out . . . strength wouldn't . . . impossible to . . . almost zero . . . an eternity . . . and thanked": McCrane narrative statement.

194 "very few": McCrane interview.

195 "seventy feet": Baldwin, *Sea Fights and Shipwrecks*, p. 25.

195 "beaten off by": U.S. Navy press release, Feb. 11, 1945.

195–96 "water-gauge . . . everything on . . . waited to . . . tangled up . . . helluva time . . . big husky . . . spray was": Fenn interview.

196 "began to rip . . . bridge ripped": Doil Carpenter testimony, Court of Inquiry.

196 "never once": Fenn interview.

196 "one or two": McCrane testimony, Court of Inquiry.

197 "gagging on . . . relived an . . . regretted now . . . freshly-painted . . . bulwark of . . . screaming and . . . the chaos . . . to stay . . . shocked daze . . . suction of": Krauchunas, "USS *Spence*: The Typhoon and the Senior Survivor."

197–98 "essential records . . . no thought . . . push the . . . gusts of": Krauchunas narrative.

198–99 "rumbling blast . . . out from . . . pretty conscious . . . full of water . . . two new . . . funny feeling": Wohlleb interview.

199 "climbed the . . . just fell . . . just clung": Greenfield (Mass.) *Recorder*, Oct. 1983.

199 "rolled over . . . bobbing like . . . very long . . . rolled the . . . blown way . . . the pressure": Rosley interview.

200 "huge round . . . the most . . . sink slowly": Krauchunas narrative.

200 "Mr. Krauchunas . . . You'll have": Wohlleb interview.

200 "about 70 . . . increased to . . . laid the . . . held her . . . 80 degrees": James A. Marks Narrative Statement, Court of Inquiry.

200–201 "to be in . . . broken and . . . Would you . . . didn't come": Schultz interview.

201 "pleading with": Ernie Price, "The U.S.S. *Hull* DD 350," Owin, *Typhoon Cobra*, p. 40.

201 "When we . . . Yes, sir . . . guns and . . . kept hitting . . . just sunk": DeRyckere interview.

201–203 "sweating it . . . one of the best . . . one of the men . . . terrific pounding . . . wet from . . . annoyed for . . . plunking drops . . . the ship . . . Panic ensued . . . 60 or 80 . . . ship descended . . . kicked and . . . chin barely . . . halyards . . . making an . . . steaming hot . . . the most . . . angry enough . . . panic-stricken . . . all got . . . hammered . . . like breaking . . . loose timber . . . about thirty . . . spinning like . . . shot up . . . inhale as . . . just as . . . had a little . . . chances to": Franchak interview.

203–204   "not going . . . toughest guy . . . What are . . . sitting on . . . too much . . . very cold . . . assumed . . . goner . . . *This is really*": Drummond interview.

204–205   "couldn't see . . . worst thing . . . put the . . . the last . . . all kinds . . . quite upset . . . not doing . . . kept changing . . . stuck it . . . not been . . . incompetent ship handler . . . It hit . . . hitting the . . . came back . . . rushed in": Rust interview.

205–206   "not known . . . probably saved . . . didn't make . . . taking on . . . Where's your . . . Couldn't get . . . Hang on . . . weren't doing . . . swim over": Watkins interview.

205   "answering bells . . . 17 knots . . . lurched to . . . possibly carried": Sharp testimony, Court of Inquiry.

206–207 "hung up . . . a wave . . . almost drowned . . . it swallowed . . . so much . . . going to . . . resigned to": Rust interview.

Chapter Fifteen

208   "latticework": Fenn interview.

208–14   "let the . . . fish around . . . tired and . . . shaky condition . . . waves were . . . started to . . . who was . . . a large piece . . . a big hole . . . very cold . . . pray to . . . The stars . . . I can't see . . . went to . . . hold for . . . spent very . . . just absorbed . . . very good . . . plenty scared . . . fruitless . . . as fast as . . . seemed to . . . on edge . . . if we brushed . . . difficult time . . . slapped him . . . all in vain . . . some of . . . too water . . . unscrewed the . . . still so . . . impossible to . . . no one . . . push it . . . pretty bad . . . Farther out most of . . . put salt . . . whistle and . . . Go ahead . . . Where are . . . Don't worry . . . short swim . . . right back . . . calmer than . . . all in . . . keep up . . . more planes . . . prayed like . . . to let . . . catch ourselves . . . good drinking . . . right over . . . almost speechless . . . Thinking of . . . prayer of . . . the most wonderful . . . steaming at . . . dying day . . . never forget . . . the Mighty . . . brilliant role . . . fight for . . . trying to": McCrane narrative, Court of Inquiry.

209   "force-feedings": Adamson and Kosco, *Halsey's Typhoons*, p. 92.

209–10   "I don't . . . just circling": Fenn interview.

209   "I can't see": Navy Department press and radio release, Feb. 11, 1945.

213   "little fazed": Fenn interview.

213–14   "nosing over . . . deafening roar . . . befitted the": Adamson and Kosco, *Halsey's Typhoons*, p. 92.

Chapter Sixteen

215–19 "right into ... lungs would ... one big ... deep funnel ... started flapping ... right on ... sailing 30 ... stripped off ... beat the ... beaten to death ... clear up ... the big ... circles of ... swimming with ... thirteen guys ... one guy ... pulled him ... running on ... a mess ... nobody in ... started to ... dead, just ... exposure, exertion ... down the ... They're gonna ... egged him ... You know ... When you ... in torture ... bareassed ... little fish ... didn't even ... eaten by ... as much ... starting to ... in the back ... came back ... monkey's fist ... no muscle ... unbelievably good ... I'd sure ... Fix you": Stealey interview.

219–22 "to the point ... cussing the ... headed into ... at the mercy ... pure panic ... hanging on to ... our only ... in order ... kicking my ... never get ... foolish thought ... thrown out ... tied to a ... higher power ... praying mother ... swept up ... pinned against ... a great guy ... high winds ... Fight on ... kept being ... came together ... heads popped ... all alone ... a lot of ... deep enough ... ears would ... impossible to ... chopped up ... riding with ... depth was ... if any ... When will ... keep me ... never lost ... beautiful wife ... to be a widow ... never to ... thinking of ... finding an ... little light ... Who's there ... Pat Douhan ... Get over ... some company": Douhan interview.

222 "self-assured": Edwin B. Brooks III interview.

223 "started to settle ... pulled under ... it was all over ... compartment filling ... killed by ... their heads ... pulled me ... more than": Richmond (Va.) *News Leader,* March 30, 1945.

223 "not in very": Douhan interview.

223–24 "picking fruit ... get the machine ... wasn't qualified ... right over ... What do ... Don't give ... wasn't anything ... sucked down ... ears were bursting ... sucked back ... a whole bunch ... floating by ... protection from ... kill Marks ... keep away": John Valverde interview.

224–25 "looking for ... our legs ... made a big ... no one really ... started calling ... some kind ... sharks still ... go down ... Model T": Douhan interview.

225 "did not know ... didn't set ... had to be": Edward J. Price interview.

225–26 "pounced on ... strange delusions ... how deep ... I guess": *Time,* "Perils of the Sea," Dec. 29, 1944.

226 "Task force ... a little ... rubbed the ... sure enough ... little on ... all our waving ... make us cry ... lighting off ... our rescue ... knowing they ...": Douhan interview.

Chapter Seventeen

227   "poor little Tabby . . . The gang is pretty": Henry L. Plage letter, Oct. 27, 1944.

228   "rugged character": Henry L. Plage and Robert M. Surdam radio and press conference, Jan. 16, 1944.

228   "smart-aleck . . . his weight and": McClain interview.

228–29 "Waitin' on Purvis . . . Purvis eats . . . I run this . . . But this is . . . Purvis is out . . . My God, what . . . Don't worry . . . Look what Cookie . . . Stay away . . . good buddies . . . ": Phillips interview.

229–30 "pretty rough . . . rode it okay . . . saved money on . . . weren't very hungry . . . the bridge would call . . . damn phone rang . . . the cool water . . . rest of the coat . . . in all my glory . . . maintain discipline . . . everyone is laughing . . . Vienna sausage": Henry L. Plage letter, Nov. 14, 1944.

230   "everybody who eats . . . evenhanded fairness": William A. McClain interview.

230   "finally got some . . . meats, oranges . . . back up": Henry L. Plage letter, Dec. 5, 1944.

230–31 "really have taken . . . different person teaching . . . I am very . . . It means a lot . . . of course with no . . . we have some . . . choir or glee . . . nine or ten fellows . . . usually drift off . . . joins in . . . there is a dead . . . singing some song . . . happy hour . . . kids who . . . just plain longing": Henry L. Plage letter, Nov. 14, 1944.

231   "first return": Howard Korth diary, Nov. 17, 1944.

231   "fine ship . . . first-class guy . . . someone you could": Korth interview.

232   "18,582 gallons . . . 79,256 gallons": Log of *Tabberer*, Dec. 15 and 17, 1944.

232   "force 12 . . . above 75 . . . seaman's description . . . calm . . . light": Bowditch, *American Practical Navigator*, p. 52.

232–36 "close proximity . . . get clear of . . . put the vessel . . . ahead full on . . . riding quite well . . . rolling up to . . . over 100 knots . . . greatest ferocity . . . rapidly with no . . . this type of vessel . . . withstand rolls . . . broken loose . . . created a delicate situation . . . in the trough . . . swaying about eight . . . operated without serious": Commanding Officer, USS *Tabberer*, Report of Storm Damage, Dec. 24, 1944.

233–34 "high morale . . . those big waves . . . ripples compared to . . . handling the ship . . . nervous or taking . . . the captain jokingly . . . more than when . . . persistent pounding . . . would give out . . . whole ship would . . . crack the seams . . . watery hell of": Frank Burbage letter, Jan. 13, 1944.

234   "finding my sea . . . stopped throwing up . . . walking the bulkheads . . . wash off the . . . in full control . . . didn't try to . . . let the typhoon . . . would have been": George Pacanovsky interview.

234–36 "almost a certainty . . . just looking at . . . encourage exaggeration . . . they looked like . . . 93 feet above . . . nine seconds for . . . blew away at . . . life nets all ... quick 60-degree ... such precious devices ... no one could ... all in the laps ... without even denting ... freed of the irons": Adamson and Kosco, *Halsey's Typhoon,* pp. 118–20.

234–35 "close-knit crew ... loved the skipper": Bellino interview.

236–37 "As deaf as ... highly populated ... courses to steer ... pitched and pounded ... weird sounds ... left the scene": Adamson and Kosco, *Halsey's Typhoon,*p. 122.

237 "terrific rolling ... until two days": *Boston Globe,* Jan. 24, 1945.

237 "Man overboard": *Tabberer* Report of Storm Damage.

237 "terrifying word ... All hands": *Tabberer* anonymous officer's diary, Jan. 14, 1945.

237 "not one of ... another ship": Plage and Surdam radio and press conference.

237–38 "rose and sank ... waving man ... large seas and ... calculated and highly ... in their grip ... huge hunk of": Adamson and Kosco, *Halsey's Typhoons,* p. 123.

239 "demonstrated outstanding ability": Henry L. Plage letter to the Chief of Naval Personnel, Sept. 28, 1944.

239 "flashed through ... green water ... ": Adamson and Kosco, *Halsey's Typhoons,* pp. 123–24.

239–40 "There are probably ... We'll look for ... only 150 to ... hunting submarines ... start at the ... get the most": Plage and Surdam radio and press conference.

240 "2215": Log of *Tabberer,* Dec. 18, 1944.

240–41 "the loom ... must be *Tabberer* ... might have found ... headed over to ... *Dewey* could be ... more directly ... pitch and pound ... had come to ... understand the reason ... draped over the ... turn back to ... some risk that ... advice and abandoned": Calhoun, *Typhoon,* pp. 67–68, 99, 111.

241 "exhaustion from overexposure": F. W. Cleary, Report of Casualties and Rescue of Survivors from Sunken Vessels, Dec. 22, 1944.

241–42 "some 25 square ... so many minutes": Adamson and Kosco, *Halsey's Typhoons,* pp. 125–26.

241–42 "about 15 or 16 ... tremendous suction ... banged against it ... lost him ... pumped out": Korth interview.

242 "Dammit, I bet ... painful and often": Adamson and Kosco, *Halsey's Typhoons,* pp. 128–29.

242–43 "popping out of ... butt first ... You don't look ... Don't believe ... flying off ... black with chrome ... very bright light ... as hard as ... There he is": Drummond interview.

243 "rode bucking horses ... swim like hell ... trapped below ... dogged themselves ... never was broken ... none got out ... like a big earthquake": Carl Webb interview.

244 "bleeding like a ... bloodied dungaree ... shivers went up ... quite a wallop ... jerk my head ... this was it ... the shoelaces wore ... any more trouble ... No, have you ... Never mind ... pleaded and begged ... these guys hanging ... absolutely hopeless ... a tough time": Franchak interview.

245 "0605 Recovered": Log of *Tabberer*, Dec. 19, 1944.

245 "tried for mutiny": Schultz interview.

245 "Appreciate it ... Five after six ... SOS. Send help ... We are departing ... *They're leaving*": DeRyckere interview.

245–46 "eyes as black ... salt spray driving ... didn't taste very": Plage and Surdam radio and press conference.

246 "0852 Rescued": Log of *Tabberer*, Dec. 19, 1944.

246 "in slow motion ... not a good idea ... sucked down ... bashed against ... pretty good ... bunch of guys ... the hell with ... just slide": Arthur L. Fabrick interview.

246 "flattening out ... visual check point": Adamson and Kosco, *Halsey's Typhoons*, p. 132.

247–50 "proceed to rendezvous ... believed other men ... as long as ... resumed for rendezvous ... our tommy guns ... Right full ... all around ... again taken toward ... hangdog look ... hated to give ... to continue ... stay in the": Commanding Officer, USS *Tabberer*, Rescue of Survivors, Dec. 24, 1944.

247–49 "they were Japs ... this was the ... instrumental in saving ... scientist by training ... ": Watkins interview.

248 "attacked by the ... dove in the ... pretty tuckered ... " Plage and the Surdam radio and press conference.

248–49 "heavy burden ... like a bunch ... couldn't do the ... hang with that ... a little shaky": Adamson and Kosco, *Halsey's Typhoons*, p. 133.

249 "hunting for these": Plage and Surdam radio and press conference.

249 "held underwater": Schultz interview.

249 "could hardly walk ... hiding Marks": Phillips interview.

249 "wasn't much ... the opposite": McClain interview.

Chapter Eighteen   pages 251–258

251 "the most exhaustive ... a man in": Halsey, *Admiral Halsey's Story*, p. 240.

252–53 "hallucinated and ... crawled out on ... nothing but skivvies ... turned keel up": Zasadil interview.

252 "underwater like a ... couple of steep ... made a 75 ... screaming like babies ... climbed out over ... swung off into ... caught in the ... reach in

the . . . in a sea . . . boilers blew . . . broke in half . . . throwing a handful . . . guys disappeared": Floyd Balliett interview.

253 "storm-tossed and": James Forrestal, secretary of the Navy, Navy Unit Commendation Ribbon awarded to *Tabberer.*

253 "three days and . . . looked worse": Plage and Surdam radio and press conference.

253–54 "we didn't really . . . But we . . . the good Lord": Phillips interview.

254 "the wonderful way . . . What would have": Burbage letter, Jan. 13, 1944.

254 "four-striper Captains": Henry L. Plage letter, Dec. 29, 1944.

254 "Well done": Henry Plage memorandum to all hands, Dec. 20, 1944.

254 "while ships around . . . expected to learn . . . How could any": Halsey, *Admiral Halsey's Story,* pp. 240–41.

255 "Captain Plage, officers": William F. Halsey speech to officers and men of USS *Tabberer,* Dec. 29, 1944.

255 "blush from hairline . . . salute or shake": Adamson and Kosco, *Halsey's Typhoons,* pp. 143–44.

255 "flabbergasted": Henry L. Plage letter, Dec. 29, 1944.

256 "Chief, pack your . . . in charge of . . . hated to lose . . . good for him": Phillips interview.

256–57 "What type . . . that hurt . . . our type fighting . . . Destroyer escort . . . sedan with a . . . looking very stern . . . Plage, did . . . Yes, sir . . . put his head . . . one up . . . My battleship buddy": Henry Plage, second reunion, USS *Tabberer,* Aug. 30, 1986.

256 "outstanding leadership . . . ": Walden L. Ainsworth biography, Naval Historical Center.

257 "best ship in . . . in correct position . . . the alertness of . . . a part of": Henry L. Plage letter, Jan. 13, 1944.

257–58 "outstanding when compared . . . For extremely meritorious": Navy Unit Commendation, Naval Historical Center.

Chapter Nineteen   *pages 259–302*

259 "received a direct . . . lay unexploded": Log of *New Jersey,* Dec. 24, 1944.

259 "very concerned": "The Reminiscences of Vice Admiral Truman J. Hedding," Naval Historical Center.

259 "had expected": Reynolds, *The Fast Carriers,* p. 283; *New York Times,* Dec. 16, 1944.

260 "typhoon-delayed": Thomas, *Sea of Thunder,* p. 345.

260 "the dangerous semicircle . . . seamanship . . . practically tore": "The Reminiscences of Vice Admiral Truman J. Hedding."

260   "to know the": Thomas B. Buell, *Master of Sea Power*, p. 492.

260   "the greatest loss": Potter, *Nimitz*, p. 349.

261   "Genial John . . . dour personality . . . capable enforcer . . . not be browbeaten":
      Buell, *Master of Sea Power*, pp. 80–81.

261   "or as soon . . . inquiring into all . . . as a result . . . full statement . . . any
      offenses . . . specifically recommend": C. W. Nimitz letter, Dec. 25, 1944.

261   "more inquisitorial . . . serious affairs . . . cripple or wreck": Melton, *Sea Cobra*,
      pp. 205.

262   "reacted to storm . . . in the thick . . . casual observers": Calhoun, *Typhoon*,
      p. 123.

262–63 "sit with closed . . . fleet witnesses . . . in view of . . . defendant . . . an interest
      in . . . ": Record of the Court of Inquiry, 1944–45, p. 1.

263   "pressing home numerous": *Washington Post*, Jan. 3, 1990.

263–67 Testimony of Robert B. Carney, Record of the Court of Inquiry, 1944–45,
      pp. 2–6.

267*n*. "MacArthur was counting": Taussig, *A Warrior for Freedom*, p. 114.

267–69 Testimony of George Kosco, Record of the Court of Inquiry, 1944–45, pp. 11–18.

270   Testimony of Stuart Ingersoll, Record of the Court of Inquiry, 1944–45,
      pp. 18–23.

270   Testimony of George H. DeBaun, Record of the Court of Inquiry, 1944–45,
      pp. 23–26.

270–71 Testimony of Jasper Acuff, Record of the Court of Inquiry, 1944–45, pp. 36–41.

271   Testimony of William T. Kenny, Record of the Court of Inquiry, 1944–45,
      pp. 47–51.

271–72 Testimony of Michael Kernodle, Record of the Court of Inquiry, 1944–45,
      pp. 54–57.

272–75 Testimony of Preston Mercer, Record of the Court of Inquiry, 1944–45,
      pp. 57–63.

273*n*. "most critical": Calhoun, *Typhoon*, p. 189.

275–76 Testimony of Frederick Sherman, Record of the Court of Inquiry, 1944–45,
      pp. 68–69.

275*n*. "held the limelight": Korth diary, Dec. 30, 1944.

276–77 Testimony of Gerald Bogan, Record of the Court of Inquiry, 1944–45,
      pp. 70–72.

277   Testimony of John S. McCain, Record of the Court of Inquiry, 1944–45,
      pp. 72–73.

278–80 "classified as secret": Record of the Court of Inquiry, 1944–45, p. 6.

278   Testimony of William F. Halsey, Record of the Court of Inquiry, 1944–45, pp. 74–78.

280   "*My God* . . . southern gentleman . . . Glad you": Watkins interview.

281 "narrative statement": Record of the Court of Inquiry, 1944–45, p. 94.

281 "the question and": James Marks letter, Jan. 5, 1944.

281–82 "he might have . . . I lost my . . . call Arthur": Virginia Marks interview.

282 "records, papers": Narrative Statement of James Marks, p. 5.

282–83 Narrative Statement of James Marks, pp. 2–4.

283–84 "true statement . . . It is . . . Have you . . . I do not . . . Have you . . . about 30 minutes . . . At all times . . . cutting in on . . . we took a . . . all empty . . . our fueling . . . a storm warning": Record of the Court of Inquiry, 1944–45, pp. 95–96.

285 "lowly enlisted": DeRyckere interview.

285 "get home as . . . incompetent . . . not assessed the . . . at that particular . . . trying to stay . . . a little": Watkins interview.

285 "I would have": W. F. Halsey letter, June 25, 1954.

286–88 Testimony of James Marks, Record of the Court of Inquiry, 1944–45, pp. 96–101.

288 Testimony of George H. Sharp, Record of the Court of Inquiry, 1944–45, pp. 101–3.

288 Testimony of Roy Lester, Record of the Court of Inquiry, 1944–45, pp. 103–5.

288 Testimony of Ray G. Morgan, Record of the Court of Inquiry, 1944–45, p. 105.

289–90 Testimony of George Kosco, Record of the Court of Inquiry, 1944–45, pp. 107–8.

290–91 Narrative Statement of Joseph McCrane, pp. 1–6.

290n. "never told a": Drury and Clavin, *Halsey's Typhoon,* p. 288.

290n. "There's only": Fenn interview.

291 "lay to the . . . The only thing": Record of the Court of Inquiry, 1944–45, pp. 110–11.

291 Testimony of James T. Story, Record of the Court of Inquiry, 1944–45, pp. 117–18.

292 "an interested party": Record of the Court of Inquiry, 1944–45, p. 119.

292 "shivering figures . . . our hopes dashed . . . What took": Krauchunas, "USS *Spence:* The Typhoon and the Senior Survivor."

292 "dry bunk and": Miller interview.

293 "missing while": Chief of Naval Personnel telegram, Jan. 3, 1945.

293 "there is no . . . lost his life": Chief of Naval Personnel telegram, Feb. 9, 1945.

293 "pebbles striking": Richard Strand interview.

293–94 Narrative Statement of Al Krauchunas, p. 103.

294 "how many men": *Greenfield* (Mass.) *Recorder,* Oct. 1983.

294 Testimony of Edward Traceski, Record of the Court of Inquiry, 1944–45, pp. 121–24.

294 Testimony of Jim Felty, Record of the Court of Inquiry, 1944–45, pp. 127–28.

295 Testimony of George Johnson, Record of the Court of Inquiry, 1944–45, pp. 125–26.

295–96 Testimony of Wilbur Lockhart, Record of the Court of Inquiry, 1944–45, pp. 132–35.

296n. "We don't . . . six or seven . . . Are you . . . Did you . . . Well, I was": Bryson interview.

296n. "All those years": Bryson, "Typhoon Forecasting, 1944, or, The Making of a Cynic," *Bulletin of the American Meteorological Society,* Oct. 2000.

297–99 Facts and Opinions, Record of the Court of Inquiry, 1944–45, pp. 146–68.

299 "column of ships": Watkins interview.

300 "figuring it would . . . made my peace . . . a ship . . . men walking . . . one of our . . . another low point . . . mad at God . . . *God doesn't torture* . . . hard right turn . . . completely worn . . . You've done all . . . half a body": Rust interview.

300–301 "Oh, yeah . . . Guess I won't . . . You guys want": Watkins interview.

301 "a commendable desire . . . the mistakes made": Record of the Court of Inquiry, 1944–45, appendix.

301 "Lessons of Damage . . . steps must be . . . a ship's safety": C. W. Nimitz letter, Feb. 13, 1945.

302 "stable . . . no major alterations . . . again rolled": Calhoun, *Typhoon,* p. 197–8.

302 "unfit for other . . . the poor stability": Report of Material Inspection and Survey of *Aylwin,* Oct. 17, 1945.

Postscript

pages 303–305

303–304 "Japanese origin . . . digested all . . . would not . . . best course . . . maintain the . . . You are . . . in shallow . . . assuming that . . . safely . . . less than 100 . . . surpassed in . . . much too late . . . with almost . . . fared well": Adamson and Kosco, *Halsey's Typhoons,* pp. 172–76, 187, 189.

304 "deserved a general": Buell, *Master of Sea Power,* p. 492.

304–305 "in the face . . . spirit and letter . . . serious consideration . . . more experienced . . . disapproved . . . skill and determination . . . services of great": Record of the Court of Inquiry, 1945.

305 "inept in acting . . . no stomach for . . . ruined the": Buell, *Master of Sea Power,* p. 492.

305 "on the point . . . national hero": Potter, *Nimitz,* p. 377.

Dramatis Personae

pages 307–312

307 "I think I . . . She always . . . According to Mom . . . ": Mahood-Cochran interview.

307–308 "Halsey disaster . . . The colonel . . . could have . . . sudden and undetectable . . . I think Halsey": Bryson interview.

308 "a wonderful ship": Arleigh Burke, "Spirit of the *Spence*," March 1983.

309 "kicked out of . . . Unfortunately, he became": Consolvo Jr., interview.

309 "We were just . . . We felt overwhelmed": DeRyckere interview.

309 "tragic experience": Halsey, *Admiral Halsey's Story*, p. 253.

310 "only imagine . . . tossed our enormous . . . most of our": Halsey, *Admiral Halsey's Story, p 239.*

310 "big wingding": Bernadette Kosco interview.

310–11 "The greatest . . . When a mother . . . finding it hard": Al Krauchunas letter, March 7, 1945.

311 "the great love . . . right up until . . . only 77 days": Portia Albee interview.

312 "not a good . . . He should have": Virginia Marks interview.

312 "any other . . . could have saved . . . one hundred percent . . . survivor's guilt . . . spared from dying": Rust interview.

Bibliography

Adamson, Hans Christian, and George Francis Kosco. *Halsey's Typhoons*. New York: Crown Publishers, 1967.

Baldwin, Hanson W. *Sea Fights and Shipwrecks*. New York: Hanover House, 1956.

Bogan, Gerald F. *Reminiscences of Vice Admiral Gerald F. Bogan*. Annapolis, Md.: U.S. Naval Institute, 1986.

Bowditch, Nathaniel. *American Practical Navigator* (No. 9). Washington, D.C.: 1943.

Buell, Thomas B. *Master of Sea Power*. Boston: Little, Brown and Co., 1980.

Calhoun, C. Raymond. *Typhoon: The Other Enemy*. Annapolis, Md.: United States Naval Institute, 1981.

Candelaria, Gary Joseph. "Tin Cans: The USS *Monaghan* and the Destroyer War in the Pacific." Master's thesis, School of Liberal Arts, University of Oklahoma, 1993.

Cant, Gilbert. *America's Navy in World War II*. New York: John Day Co., 1943.

Clark, Joseph J., and Dwight H. Barnes. *Sea Power and Its Meaning*. New York: Franklin Watts, 1966.

Clark, Joseph J., with Clark G. Reynolds. *Carrier Admiral*. New York: David McKay, 1967.

Conrad, Joseph. *Typhoon*. New York: Signet, 1963.

Davis, Martin. *Traditions and Tales of the Navy*. Montana: Pictorial Histories Publishing, 2000.

Dawes, Robert A., Jr., *The Dragon's Breath*. Annapolis, Md.: Naval Institute Press, 1996.

Drury, Bob, and Tom Clavin. *Halsey's Typhoon*. New York: Atlantic Monthly Press, 2007.

DuCharme, George W. "Recollections of 7 December 1941." San Diego, 2005.

Findley, Dean E. *The Life of a Tin Can Sailor*. Colorado: Aviation Forum Co., 2001.

Friedman, Norman. *U.S. Destroyers, an Illustrated Design History*. Annapolis: United States Naval Institute, 1982.

Halsey, William F., and J. Bryan III. *Admiral Halsey's Story*. New York: Whittlesey House, 1947.

Hammel, Eric M., and John E. Lane. *Bloody Tarawa.* New York: Zenith Press, 2006.

Holland, W. J., Jr., *The Navy.* Washington, D.C.: Naval Historical Foundation, 2000.

Hornfischer, James D. *The Last Stand of the Tin Can Sailors.* New York: Bantam, 2004.

Hoyt, Edwin P. *The Typhoon That Stopped a War.* New York: Van Ress Press, 1968.

Jernigan, E. J. *Tin Can Man.* Arlington, Va.: Vandamere Press, 1993.

Jones, Ken. *Destroyer Squadron 23.* Annapolis, Md.: Naval Institute Press, 1997.

Jones, Ken, and Hubert Kelley, Jr. *Admiral Arleigh (31-Knot) Burke.* New York: Bantam Books, 1985.

King, Ernest J., and Walter Muir Whitehill. *Fleet Admiral King.* New York: W. W. Norton, 1952.

Knight, Austin M. *Modern Seamanship.* 10th ed. New York: D. Van Nostrand Co., 1941.

Kotsch, William J., and Richard Henderson. *Heavy Weather Guide.* Annapolis, Md.: Naval Institute Press, 1984.

Krauchunas, Alphonso. "USS *Spence*: The Typhoon and the Senior Survivor." Battle Creek, Mi., 1992.

Lemont, Harrison E. *Never Alone Until Admiral Halsey Left.* Pittsburgh: Dorrance Publishing, 2003.

Lord, Walter. *Day of Infamy.* New York: Holt, Rinehart and Winston, 1957.

McCain, John, and Mark Salter. *Faith of My Fathers.* New York: Random House, 1999.

Melton, Buckner F., Jr. *Sea Cobra.* New York: Lyons Press, 2007.

Morison, Samuel Eliot. *History of United States Naval Operations in World War II,* Volume 3: *The Rising Sun.* Edison, N.J.: Castle Books, 2001.

————. *History of United States Naval Operations in World War II,* Volume 5: *The Struggle for Guadalcanal.* Edison, N.J.: Castle Books, 2001.

————. *History of United States Naval Operations in World War II,* Volume 6: *Breaking the Bismarcks Barrier.* Edison, N.J.: Castle Books, 2001.

————. *History of United States Naval Operations in World War II,* Volume 7: *Aleutians, Gilberts and Marshalls.* Edison, N.J.: Castle Books, 2001.

————. *History of United States Naval Operations in World War II,* Volume 12: *Leyte.* Edison, N.J.: Castle Books, 2001.

Owin, Theodore F. *Typhoon Cobra.* Lake Havasu, Ariz.: Theodore F. Owin, 1995.

Parkin, Robert Sinclair. *Blood on the Sea.* New York: Da Capo Press, 2001.

Potter, E. B. *Bull Halsey.* Annapolis, Md.: Naval Institute Press, 1985.

————. *Admiral Arleigh Burke.* Annapolis, Md.: Naval Institute Press, 1990.

————. *Nimitz.* Annapolis, Md.: Naval Institute Press, 1976.

Prange, Gordon W., with Donald M. Goldstein and Katherine V. Dillon. *December 7, 1941.* New York: Warner Books, 1988.

Reader's Digest. *Great World Atlas.* New York: Reader's Digest Association, 1997.

Reynolds, Clark G. *The Fast Carriers: The Forging of an Air Navy.* Annapolis, Md.: Naval Institute Press, 1992.

Reynolds, Clark G. *The Fighting Lady*. Montana: Pictorial Histories Publishing, 1986.

Roscoe, Theodore. *United States Destroyer Operations in World War II*. Annapolis, Md.: United States Naval Institute, 1953.

Ryan, Cornelius. *The Longest Day*. New York: Popular Library, 1959.

Schom, Alan. *The Eagle and the Rising Sun*. New York: W. W. Norton, 2004.

Shafter, Richard A. *Destroyers in Action*. New York: Cornell Maritime Press, 1945.

Solberg, Carl. *Decision and Dissent*. Annapolis, Md.: Naval Institute Press, 1995.

Taussig, Betty Carney. *A Warrior for Freedom*. Manhattan, Kan.: Sunflower University Press, 1995.

Thomas, Evan. *Sea of Thunder*. New York: Simon and Schuster, 2006.

United States Naval Academy Alumni Association. *Register of Alumni, 1845–1986*. Annapolis, Md.: USNA Alumni Association, 1986.

White, David Fairbank. *Bitter Ocean*. New York: Simon and Schuster, 2006.

Woodward, C. Vann. *The Battle for Leyte Gulf*. New York: Ballantine Books, 1947.

Wouk, Herman. *The Caine Mutiny*. New York: Doubleday & Co., 1951.

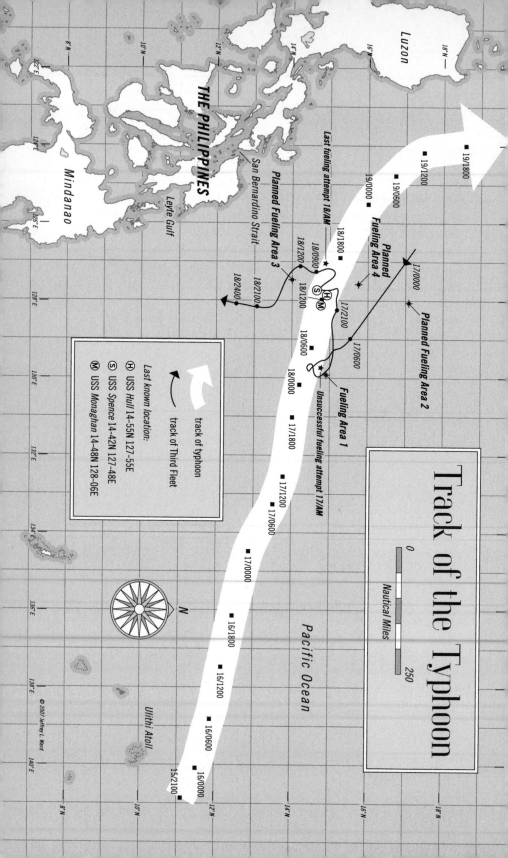

Track of the Typhoon

THE PHILIPPINES

Luzon

Mindanao

Leyte Gulf

San Bernardino Strait

Pacific Ocean

Ulithi Atoll

Nautical Miles

0   250

Last fueling attempt 18/AM

Planned Fueling Area 3

Planned Fueling Area 4

Planned Fueling Area 2

Fueling Area 1

Unsuccessful fueling attempt 17/AM

18/0900
18/1200
18/2400
18/2100
18/1200
18/0600
18/0000

17/1800
17/1200
17/0600
17/0000

16/1800
16/1200
16/0600
16/0000
15/2100

17/2100
17/0600
17/0000

19/1800
19/1200
19/0600
19/0000
18/1800

Legend:
track of typhoon
track of Third Fleet

Last known location:
Ⓗ USS *Hull* 14-55N 127-55E
Ⓢ USS *Spence* 14-42N 127-48E
Ⓜ USS *Monaghan* 14-48N 128-06E

© 2007 Jeffrey L. Ward

N

Appendix:
Crew Muster Rolls for Lost Destroyers

USS Hull (DD-350)

Lt. Cmdr. James A. Marks
Lt. Greil I. Gerstley *
Lt. (MC) Hira C. Baker *
Lt. (j.g.) Frank L. Snodgrass *
Lt. (j.g.) Arthur L. Fabrick
Lt. (j.g.) Kenneth W. Kappus *
Lt. (j.g.) George H. Sharp
Lt. (j.g.) Myron S. Wall
Lt. (j.g.) H. Norman Byrant *
Lt. (j.g.) Lloyd G. Rust, Jr.
Lt. (j.g.) Donald C. Watkins
Lt. (j.g.) Edwin B. Brooks, Jr.
Lt. (j.g.) G. C. Nelson *
Lt (j.g.) Felix G. Smart *
Ensign W. G. Johnson *
Ensign Alfred A. Anido *
Ensign D. L. Korinko *
Ensign O. E. Schuerman *
Abreu, Manuel, Jr. *
Ackerman, John F. *
Adams, Malcolm *

Alexander, Scott C. *
Anderson, Harold E. *
Anderson, James W. *
Ashley, Felix *
Banes, Donald R. *
Banfill, Daniel P. *
Beck, Robert C.
Belanger, Francis
Belden, James I. *
Bokn, Peter *
Boldman, William
Bollin, William H.
Burleson, Earl G. *
Carlstrom, Robert D. *
Coleman, Homer W.
Colville, Thomas J., Jr.
Cosgrove, Gleason J. *
Cothran, Charles R. *
Cotten, J. O. *
Crossman, Reuben T.
Dean, Billy Bob *
Derthick, Gordon K.
DeRyckere, Archie G.

* Killed in Pacific typhoon of December 18, 1944

De Vaney, Willis F., Jr. *
Douhan, Patrick H.
Drummond, Kenneth L.
Eichen, Douglas J. *
Eisele, Edwin F. *
Eisinbach, Arthur E. *
Farrell, Edward V. *
Fenderson, Robert N. *
Fernandez, Angel *
Fisnefska, George L., Jr. *
Fowler, Robert B. *
Foster, Henry L. *
Franchak, Michael
Gabler, Darrell M. *
Gaulke, Donald J. *
Geddert, George F. *
Gilberg, John A. *
Gillette, William F. *
Gipson, Roy B. *
Gower, Waldo H. *
Granby, Haywood *
Green, Orville L. *
Gruner, Henry *
Gundel, Herbert E. *
Guthrie, Ray M. *
Guy, George A.
Hahn, Lecil A. *
Haislip, Claude J., Jr. *
Hall, Henry S.
Halladay, Junior Ray *
Hart, William F. *
Haydon, Martin R. *
Henkel, Edward *
Henry, George W. *
Hicks, Lyman F., Jr. *
Hicks, Robert H. *
Hill, Lyle G. *
Hixon, Merritt B. *
Hoghaug, Gilman *

Horton, John T. *
Houghton, Don C., Jr. *
Howland, Walter K. *
Hoy, Marshall L. *
Hughes, Samuel N. *
Ingraham, Theodore M. *
Jambor, Joseph J. *
Juson, Ralph D. *
Karnopp, Wallace E. *
Kay, David M.
Kelly, Chester R.
Kelly, Keith W. *
Kelly, William L. *
Kendrovics, A. *
Kennedy, William H., Jr. *
Kinderman, Robert R.
King, James P. *
Knadler, G. F. *
Korponai, Z. A. *
Kraus, George W. *
Kreidler, John E. *
Kunz, Edward W. *
Lane, Alton B. *
La Gow, Lawson S. *
Leabo, Donald C. *
Leombrone, Joseph *
Lester, Roy
Lewis, George J. *
Lindquist, August H.
Loney, Dewey W. *
Mabius, John, Jr. *
Macchio, Joseph *
Mackenzie, Roderick J.
Makris, Manalows
Martin, Francis B.
Martin, Lawrence
Martin, Paul A., Jr. *
May, Kent J.
Mc Donough, Walter T. *

Mc Gee, T. M. *
Mc Gill, Charles J. *
Mc Glaughn, W. M. *
Mc Intyre, J. F. *
Mc Invale, R. R. *
Mc Keehan, L. *
Melancon, Charles F. *
Miller, Ernest H. *
Millsaps, Sylvan *
Mink, Martin S. *
Mohr, Otto E., Jr. *
Moon, Roland R. *
Moreno, John *
Morgan, Roy G.
Mullins, Lawrence W. *
Mullins, Lester C. *
Murie, John H. *
Nagurney, Nicklas
Newsom, Earl T. *
Newsom, Geral
Niss, Arnold W. *
Novak, Rudolph R.
Oakley, Robert D. *
Ostlee, Mareus L. *
Palmer, Roger A.
Papcke, Frank A. *
Parker, Robert E. *
Parrott, Robert T., Jr. *
Peek, John T. *
Peterson, John M. *
Pererson, Verl D.
Pfeifer, John W. *
Phillippi, Elden A. *
Phillips, Fred F. *
Piccin, Ernest R. *
Pickett, Frank M., Jr. *
Pierce, Clyde W. *
Pintvers, John J.
Pippins, Wilburn E. *

Poe, Clinton L. *
Porter, Charles W.
Powell, Vernon D.
Pratt, Glen L. *
Preisser, Thomas E. *
Preston, Alton R. *
Price, Charles E. *
Price, Edward J.
Price, Ernie E.
Pruitt, Paul G. *
Rattaro, John J. *
Rotter, Ira F. *
Ruby, Clarence E., Jr. *
Sackmaster, Harold E. *
Scarletta, Frank *
Scherer, Kenwood R. *
Scherzinger, Roy, Jr. *
Schmiderer, Joseph A., Jr. *
Schrader, Harold D. *
Schultz, John Ray
Simons, Martin E. *
Smit, Cornelius G.
Smith, Perry A., Jr. *
Smith, Johnny H. *
Smith, William W., Jr. *
Solano, Howard J. *
Spencer, Olin G. *
Spohn, Thomas L., Jr.
Stacy, Allen J. *
Stealey, Thomas A., Jr.
Stennett, James *
Stephens, James H. *
Stercula, Walter V. *
Stevens, Paul Q. *
Stilwell, James M.
Stipetich, George L. *
Stoddard, Kenneth C. *
Szente, George A. *
Tabor, Hollis G. *

Talmadge, James S.

Taylor, Albert J.

Teiffel, Wesley S. *

Tenholder, Alvin J. *

Thornton, Kermitt G. *

Tong, Raymond H.

Torkildson, Keith B. *

Trapp, Edward G. *

Travis, Austin H. *

Trujillo, Orlando B.

Updegraff, George J. *

Valverde, John

Vaughn, Archie L. *

Vernon, William H. *

Vulcano, Samuel J. *

Wall, Carl J. *

Ward, Elbert M. *

Weathers, John L., Jr.

Webb, Carl T.

Webb, Willard *

Webster, Leroy S.

Weimers, Gerald L. *

Weiss, Victor J. *

Whitney, Robert A. *

Wilkerson, Glenn H.

Williams, Calvin E. *

Williams, Ernest L.

Wilson, Robert *

Winton, Albert J.

Wolfe, Gideon J., Jr. *

Wolff, Thomas A. *

Wolkins, Marshall W. *

Wright, James E. *

Yarbrough, Harvey C. *

Yonts, Charles W. *

Youmans, T. K., Jr.

Young, Joseph A., Jr.

Zawne, Robert C. *

Zentner, W. H., Jr.

Zielinski, Leonard A. *

USS *Monaghan* (DD-354)

Lt. Cmdr. F. Bruce Garrett, Jr. *

Lt. F. J. Elliott, Jr. *

Lt. Donald H. Van Iderstine *

Lt. R. C. Mills *

Lt. E. W. Siegesmund *

Lt. S. F. Stark *

Lt. (j.g.) J. W. Prather *

Lt. (j.g.) R. C. Burden *

Lt. (j.g.) R. F. Pedersen *

Lt. (j.g.) C. H. Blittersdorf *

Lt. (j.g.) L. A. Carlson *

Lt. (j.g.) Robert E. Nolop *

Lt. (j.g.) R. K. Hargrave *

Lt. (j.g.) (MC) R. C. Gavin *

Ensign H. E. Davidson *

Ensign J. Dubpernell *

Ensign E. M. Cochran *

Ensign H. H. Weber *

Adler, Victor L. *

Alessi, Joseph R. *

Allen, Fred A. *

Amis, Hugh B. *

Anspach, Roy R. *

Anthony, James C. *

Attaway, James R. *

Aubrey, John T. *

Austin, William J. *

Bailey, Clarence A. *

Baker, Willard R. *

Barber, Horace L. *

Bard, Richard F. *

Barszczewski, Edward J. *

Bassett, William M. *

Beach, Gurney A. *

Benfatti, John J. *

Berger, George W. *

Binner, Clyde R. *

Borouff, Jerome J. *

Brunkow, Ferdinand *
Britton, Thomas J. *
Bryant, Leonard R. *
Burgess, Fred J. *
Burtch, Charles D. *
Burnett, Raymond O. *
Busch, Martin L. *
Butler, Ralph J. *
Cain, Frank A. *
Campbell, Bruce *
Camunez, James C. *
Carbone, Richard E. *
Carpenter, Doil T.
Carriker, Robert R. *
Casson, Orval K. *
Caswell, Floyd A., Jr. *
Chapin, Gail W. *
Cieszynski, Stanley J. *
Cimino, Ceasar D. *
Clark, George M. *
Cochran, James H. *
Coleman, William D. *
Cooper, Carl L. *
Conrad, Carl E. *
Corley, Howard W. *
Cote, Leonard G. *
Cotton, Almer A. *
Cradduck, Raymond A. *
Crater, Ray E. *
Croft, Edward J. *
Culp, Donald A. *
Darden, Robert J.
Debord, John G. *
Dedmore, Gale A. *
Delis, John S. *
Devault, Francois L. J. *
Doyle, George W. *
Driscoll, Charles J. *
Dumas, Frank S. *
Eggen, Wendell J. *

Ellis, John H., Jr. *
Entrican, Carroll V. *
Etheridge, Lindsey E. *
Evans, Paul R. *
Fantozzi, Frederick C. *
Faught, Laurence M. *
Felton, Russel B. *
Fenn, Evan *
Ferrero, Robert E. *
Ferrier, James J. *
Finch, Lester J. *
Fisher, Roland D. *
Foster, Victor W. *
Fox, Loyd S. *
Gastmann, Harm *
Geier, Donald C. *
Genest, Dayton *
Givens, Robert P. *
Godfrey, Harvey H. *
Goetz, Otto J. *
Gostyla, Bruno *
Green, James B. *
Grubham, Fred W. *
Guio, Joseph, Jr.
Haight, Benjamin E. *
Hally, William *
Hankins, Marvin R. *
Harris, Harry L. *
Heitner, Henry *
Heflin, Dewey L., Jr. *
Higginbotham, Sherman *
Hill, Horace C. *
Hirsch, Leonard H. *
Holland, Will Ben *
Holt, Jack L. *
Hommel, John W. *
Hudgins, Wendell R. *
Iler, George, Jr. *
Ingoe, James E. *
Jeffery, Charles W. *

Johnson, William E. *
Johnson, William L. *
Jones, Alfred H. *
Jones, Forest E. *
Jordan, Harold L. *
Karns, Forest E. *
Karpinski, Milford J. *
Kell, Ralph O. *
Kimmel, Donald *
Kowalk, Bernard A. *
Kramer, William F.
Kurikjan, Edward A. *
Lamb, Arthur *
Larson, Joseph J. *
Larson, Simon P. *
Lee, James S. *
Lenig, Walter L., Jr. *
Lewellen, Roy L. *
Litton, John A. *
Lindsey, David F. *
Lindsey, Ben H., Jr. *
Lomax, Clenn C. *
Lopez, Jose E. *
Lowe, Jefferson T. *
Lyste, Chester M. *
Maldonado, Alex J. *
Malveau, Chester *
Mann, Boyd *
Manochi, William M. *
Martin, Hubert M. *
Mattern, Robert A. *
Matzener, John F. *
Maxwell, Earl T. *
Mayer, Kenneth G. *
McConnell, Harley M. *
McCormick, William *
McCrane, Joseph C.
McFalls, Charles J. *
McGough, Joseph N. *
McIntosh, R. C. P. *

McNally, Joseph L. *
Meisen, Francis P. *
Messa, Louis *
Mickelson, Max L. *
Mills, Houston E. *
Moore, Carlus B. *
Moore, John C., Jr. *
Moran, Voy G. *
Morrison, Melroy E. *
Nichols, Robert J. *
Nutting, Roland B. *
Ogborn, John "J" *
Osterby, Albin E.
Panas, Peter *
Peterson, Wayne L. *
Peerenboom, Donald G. *
Perry, Joe E. *
Phister, Harold C. *
Platt, Bryce L. *
Ragland, Ralph K. *
Ravnell, Stacks *
Richmond, Ganis J. *
Ricker, Doyle L. *
Bingham, Eugene D. *
River, Francis N. *
Rogers, Harry A. *
Ross, Charles L. *
Sables, John A., Jr. *
Sanford, Clyde H. *
Sanford, James L., Jr. *
Schubert, Elmer H. *
Schuler, Carl F., Jr. *
Shaner, Leonard P. *
Shalkowski, Louis P. *
Sherrill, Harvey "J" *
Shoemate, Roy C. *
Sigafoos, Jack R. *
Simmons, William W. *
Smith, Dewey L. *
Smrkolj, Harry *

Spence, Louis L. *
Stark, Robert M.
Stapleton, James *
Stevenson, James H. *
Stewart, Harold B. *
Stewart, Robert F. *
Stewart, Robert L. *
Still, Frederic P. *
Stimolo, Sam, Jr. *
Stone, Edward
Stoner, Howard A. *
Stopher, Robert H. *
Story, James T.
Storz, Andrew "J" *
Stowe, Marvin E. *
Strahan, Everett N. *
Stravinksi, John J. *
Strength, Luther N. *
Strickland, Henry P., Jr. *
Strickland, J. M. *
Stricklin, Arthur G., Jr. *
Stutes, Lee R. J. *
Sullivan, Earl P. *
Sutherin, Donald G. *
Swartwood, James N. *
Thomas, Cyril J. E. *
Tondreau, Joseph A. *
Tripp, Alonzo W. *
Trostel, William J. *
Tschimperle, Matthew J. *
Tyler, Fred D. *
Tyler, George "L" *
Varazo, Siver C. *
Varvil, Kenneth E. *
Villanueva, Gustavo B. *
Wallace, Ronald J. *
Warren, Coleman Y. *
Warren, Robert F. *
Walters, William, Jr. *
Weaver, William D. *

Werbelow, Louis C. *
White, Clifton J. *
Willenborg, Ervin E. *
Willhite, Lawrence W. *
Willis, Edward *
Wilson, Willie *
Witcher, Leroy *
Working, James W. *
Wright, Evert *
Wright, Paul *
Wright, Ralph J. *

USS Spence (DD-512)

Lt. Cmdr. James P. Andrea *
Lt. Frank Van Dyke Andrews *
Lt. (j.g.) John J. Bellion *
Lt. (j.g.) John Whalen *
Lt. (j.g.) Lester V. Robinson *
Lt. (j.g.) Vincent McClelland *
Lt. (j.g.) Lawrence D. Sundin *
Lt (j.g.) George P. Isham *
Lt (j.g.) David H. Collins *
Lt (j.g.) Paul R. Harnish *
Lt (j.g.) John C. Gaffney *
Lt (j.g.) Alphonso S. Krauchunas
Ensign Charles P. Bickel *
Ensign Donald H. Smith *
Ensign William I. F. Sellers *
Ensign George W. Poer *
Ensign Rexal H. Kissinger *
Ensign Robert W. Brightman *
Ensign Herbert Luebbe *
Ensign William R. Ligon *
Adams, Vern O. *
Adams, William J. *
Akers, James R. *
Alarie, Earl J. *
Aldrich, Raymond E. *

Alexander, George J. *
Alexander, Thornton R. *
Allen, Edgar A. *
Alley, George W. *
Anastoff, Carl *
Anderson, James *
Arvold, Arthur A. *
Ashcraft, William D. *
Ashworth, Ernest, Jr. *
Asmus, Charles B. *
Aulenbach, Edward J. *
Armbrust, Almon L. *
Aube, Harold N. *
Ayers, Harley R.
Baeder, Francis Elmer *
Bailey, Walter R., Jr. *
Bair, Allen G. *
Bair, Lewis J. *
Balliett, Floyd
Baren, Lionel *
Barr, Robert N. *
Barras, Sidney Theo *
Beachy, Jonas E. *
Bean, Charles R. *
Bearrows, Charles G. *
Beaty, Douglas C. *
Beaumont, Cletus E. *
Beeman, Nyle J.
Billingsley, Ernest E., Jr. *
Blackburn, Oliver *
Blanton, Warren T. *
Bloom, Axel W., Jr. *
Bloomquist, Harold E. *
Bodaq, Michael *
Boone, Joseph P. *
Bowman, Richard G. *
Bradshaw, Raymond, Jr. *
Brandl, Harry G. *
Britzman, Glen G. *
Broch, Peter *

Brochu, Edward A. *
Bronis, Henry *
Brown, Paul T. *
Browning, Alva L. *
Bruchert, Albert J., Jr. *
Bryant, Harold L. *
Buck, Walton G. *
Buckley, William J. *
Buczek, Edward C. *
Burian, Carlton R.
Busby, Wallace H. *
Butts, Harry L. *
Byrd, Joseph E. *
Cahill, Christopher D. *
Callier, Lawrence
Campton, Gilbert G. *
Carls, Jack W. *
Carrigan, Harlan K. *
Cayer, Albert F. *
Chalfant, George R. *
Champagne, William A. *
Chastain, James W. *
Christiansen, Wesley J. *
Christensen, George L. *
Clark, Donald C. *
Clark, Stillman *
Cloutier, Marcel G. *
Cobb, Robert M. *
Coleman, Manuel W. *
Connolly, John E. *
Cook, William W., Jr. *
Cooper, Frederick *
Cope, Robert E. *
Copeland, Roosevelt *
Coppens, William A. *
Corder, Frank B. *
Craver, Charles R. *
Crawford, Newton C. *
Criteser, Gordon L. *
Crooks, Fred T. *

Crump, Walter R. *
Culler, Alvin J. *
Dalrymple, Arthur G., Jr. *
David, Jerome A. *
Davis, Calvin M. *
Davis, Edward N. *
Deeters, Henry J.
De Vaughan, James F. *
Dewan, Benjamin T., Jr. *
Di Stanislao, Frank *
Dommeleers, Daniel V. *
Duda, Alfred F. *
Edwards, Simpson *
Elburn, Charles F. *
Ellison, Lowell E. *
Ellwanger, Fred O. *
Elwell, Roland D., Jr. *
Esler, Gilbert J. *
Faust, Kyle H. *
Felty, James M.
Ferguson, George J. *
Filannino, Frank C. *
Finch, David L. *
Finch, Hugh A. *
Finn, Martin F. *
Ford, Joel L. *
Forry, Dudley C. *
Frost, Frank B. *
Frost, Myron D. *
Gary, Leroy J. *
Gehle, Willard F. *
Gehr, John T. *
Gibbons, William H. *
Giffin, Freeman F. *
Gill, Francis J. *
Goodwin, Edward W. *
Gordon, Jack *
Governile, Frank P. *
Grabowski, Joseph C. *
Graddy, George W. *

Gray, John M. *
Gray, William H. *
Greenwald, Henry W., Jr. *
Grillo, Roy R. *
Grounds, Charles L.
Haak, Menno M. *
Hakenson, Warren I. *
Hampton, David D. *
Hansen, John H. *
Hariskevich, Rudolph *
Harris, Earl R. *
Healy, Lorin C. *
Heater, James P. *
Heaverlo, Clarence J. *
Heckart, Billy Dean *
Hensley, George W. *
Hill, Harold H. *
Holland, Clarence E. *
Horkey, Franklin W. *
Hosford, Leonard D. *
Huffman, Walter L. *
Huppert, Carlton J. *
Jackson, Lawrence N. *
Jankowski, Stanley E. *
Johnson, Clarence E. *
Johnson, George A. *
Johnson, George W.
Jones, Bennie Joe *
Jorden, Harry E., Jr. *
Karl, John W. *
Kaufman, Harold D. *
Kaufman, John C. *
Keith, William E.
Kelin, Vernon *
Kelley, Thomas R., Jr. *
Kelly, Brady L. *
Kelly, Carlyle X. *
Kelly, John J. *
Kendall, Leonard F. *
Kestle, James R. *

Kittle, James R. *
Kleckley, William W. *
Kosters, Henry A. *
Lancaster, Roger O. L.
Lara, Eddie M. *
Lasorsa, Louis J. *
Leschinski, Thaddeus J. *
Lesher, Robert L. *
Lewis, John F. *
Lindseth, Theodore S. *
Loughery, Edward A. *
Lundgren, Arthur L. *
Lundren, Lloyd O. *
Luther, Francis E. *
Lybrand, Robert L. *
Marconi, Carl C. *
Mahaffee, Lawrence M. *
Mahan, Henry F. *
Manchisi, Peter P. *
Marget, Glenn E. *
Martin, Andrew J. *
Marvel, Robert E. *
Massa, John D. *
Mathes, Frank E. *
McAllister, Lyndon M.
McCrary, Hubert E. *
McEachern, Glen E. *
McFaddin, William C. *
McKinstry, Albert LeRoy
McQuerry, George P. *
Mergenthaler, Elmer J. *
Merritt, Roy L. *
Miley, Frank B. *
Miller, Cecil B. *
Miller, Edward A., Jr.
Monnig, Harold A. *
Moore, David *
Moore, John *
Murphy, Hugh *
Myers, Frederick P. *

Naquin, Roy J. *
Neal, Ralph D. *
Neely, Clifford *
Netolicky, George S. *
Nevins, William B. *
Nose, Allen J. *
Oesau, Andrew J. *
Orasi, Harold W. *
Owens, Muriel G. *
Paleski, Edward M.
Pauer, Joseph A. *
Payton, Russell *
Penpraese, George H. *
Pera, Aldo *
Polhemus, John D. *
Powell, Curtis T. *
Powell, Luther K. *
Powers, Dorman M. *
Purvis, Joe E. *
Reams, Charles H. *
Redmond, James *
Richards, Donald G. *
Ridge, Edwin G. *
Ringuiso, Francesco C. *
Roberts, Harvey D. *
Rosen, Samuel *
Rosley, Albert, Jr.
Rossi, Joseph A. *
Rutter, James V. *
Ryan, Harry J. *
Sarli, Michael *
Saxon, John J. *
Scalese, James P.
Schmidt, Frank A. *
Scheider, Benjamin M. *
Schnell, George F. *
Schoen, Andrew S. *
Schult, Howard A. *
Schwartz, David E. *
Schwarz, Paul F. *

Sehnert, Maurice D.
Selders, Everett E. *
Sepanski, Alexander L. *
Shepherd, Marion *
Sherer, Edward G. *
Shropshier, Louis H. *
Shupek, John *
Slagle, Benjamin E. *
Slaughter, Layton W. *
Small, Norman E. *
Smith, Charles A. *
Smith, Carl G., Jr. *
Smith, Donald L. *
Smith, Jerome H. *
Smith, Robert E. *
Snyder, Karl H. *
Spotts, Virgil D. *
Spradling, Frenchy *
Springbett, William G. *
Staffen, Howard F. *
Stalder, Duane J. *
Stein, Warren J. *
Stepp, Edgar E. *
Stevens, Max J. *
Stevenson, Richard L. *
Stewart, David L., Jr. *
Stewart, Robert W. *
Strand, Robert L. *
Straszynski, Philip J. *
Stromberg, Vernon A. *
Stroup, Bernard C. *
Sullivan, Francis M. *
Sullivan, Robert A. *
Sundbeck, Clarence V. *
Svouros, George A. *
Swerzbin, Ralph J. *

Tagg, Henry O. *
Tapovic, Peter *
Tate, Ernest J. *
Taylor, Robert J. *
Thompson, Frank A. *
Thompson, Thomas N. *
Thorpe, Errol M. *
Tighe, Rody J. *
Traceski, Edward F.
Tryon, Perry R. *
Turner, Claude *
Vinald, Willard H. *
Vining, James O. *
Votta, Gerald V. *
Wallace, John J. *
Waters, Earl V. *
Weaver, Carl R. *
Weiser, Edwin M. *
Weissenborn, Wayne W. *
Whited, Cleo P. *
Widmeyer, Richard C. *
Wildenhain, Leo J. *
Wilhelm, William E. *
Williams, James B. *
Williams, Joel J. *
Williamson, Burnis F. *
Wilson, Frank E. *
Wohlleb, Charles F.
Woitkovich, John L. *
Wool, Frank F., Jr. *
Wright, Robert A. *
Wright, William P. *
Yaksich, George F., Jr. *
Young, Frank F.
Zaszdil, Ramon J.
Zinamon, Morris L. *

Index